INTERNATIONAL TABLE OF ATOMIC WEIGHTS (1973)

Based on relative atomic mass of $^{12}C = 12$. Includes 19

The following values apply to elements as they exist in m[illegible] artificial elements. When used with the due regard to footno[illegible] digit, or ±3 when followed by an asterisk (*). Value in parentheses is the mass number of the isotope of longest half-life.

	Symbol	Atomic number	Atomic weight
Actinium	Ac	89	(227)
Aluminum	Al	13	26.98154[a]
Americium	Am	95	(243)
Antimony	Sb	51	121.75*
Argon	Ar	18	39.948[b,c,d,g]
Arsenic	As	33	74.9216[a]
Astatine	At	85	(210)
Barium	Ba	56	137.33
Berkelium	Bk	97	(247)
Beryllium	Be	4	9.01218[a]
Bismuth	Bi	83	208.9804[a]
Boron	B	5	10.81[c,d,e]
Bromine	Br	35	79.904[c]
Cadmium	Cd	48	112.41
Calcium	Ca	20	40.08[g]
Californium	Cf	98	(251)
Carbon	C	6	12.011[b,d]
Cerium	Ce	58	140.12
Cesium	Cs	55	132.9054[a]
Chlorine	Cl	17	35.453[c]
Chromium	Cr	24	51.996[c]
Cobalt	Co	27	58.9332[a]
Copper	Cu	29	63.546*[c,d]
Curium	Cm	96	(247)
Dysprosium	Dy	66	162.50*
Einsteinium	Es	99	(254)
Erbium	Er	68	167.26*
Europium	Eu	63	151.96
Fermium	Fm	100	(257)
Fluorine	F	9	18.998403[a]
Francium	Fr	87	(223)
Gadolinium	Gd	64	157.25*
Gallium	Ga	31	69.72
Germanium	Ge	32	72.58*
Gold	Au	79	196.9665[a]
Hafnium	Hf	72	178.49*
Helium	He	2	4.00260[b,c]
Holmium	Ho	67	164.9304[a]
Hydrogen	H	1	1.0079[b,d]
Indium	In	49	114.82
Iodine	I	53	126.9045[a]
Iridium	Ir	77	192.22*
Iron	Fe	26	55.847*
Krypton	Kr	36	83.80[e]
Lanthanum	La	57	138.9055*[b]
Lawrencium	Lr	103	(260)
Lead	Pb	82	207.2[d,g]
Lithium	Li	3	6.941*[c,d,e,g]
Lutetium	Lu	71	174.967*
Magnesium	Mg	12	24.305[c,g]
Manganese	Mn	25	54.9380[a]
Mendelevium	Md	101	(258)
Mercury	Hg	80	200.59*
Molybdenum	Mo	42	95.94
Neodymium	Nd	60	144.24*
Neon	Ne	10	20.179*[c,e]
Neptunium	Np	93	237.0482[f]
Nickel	Ni	28	58.70
Niobium	Nb	41	92.9064[a]
Nitrogen	N	7	14.0067[b,c]
Nobelium	No	102	(255)
Osmium	Os	76	190.2[g]
Oxygen	O	8	15.9994[b,c,d]
Palladium	Pd	46	106.4
Phosphorus	P	15	30.97376[a]
Platinum	Pt	78	195.09*
Plutonium	Pu	94	(244)
Polonium	Po	84	(209)
Potassium	K	19	39.0983*
Praseodymium	Pr	59	140.9077[a]
Promethium	Pm	61	(145)
Protactinium	Pa	91	231.0359[a,f]
Radium	Ra	88	226.0254[f,g]
Radon	Rn	86	(222)
Rhenium	Re	75	186.207[c]
Rhodium	Rh	45	102.9055[a]
Rubidium	Rb	37	85.4678*[c]
Ruthenium	Ru	44	101.07*
Samarium	Sm	62	150.4
Scandium	Sc	21	44.9559[a]
Selenium	Se	34	78.96*
Silicon	Si	14	28.0855*[d]
Silver	Ag	47	107.868[c]
Sodium	Na	11	22.9877[a]
Strontium	Sr	38	87.62[g]
Sulfur	S	16	32.06[d]
Tantalum	Ta	73	180.9479*[b]
Technetium	Tc	43	(97)
Tellurium	Te	52	127.60*
Terbium	Tb	65	158.9254[a]
Thallium	Tl	81	204.37*
Thorium	Th	90	232.0381[f,g]
Thulium	Tm	69	168.9342[a]
Tin	Sn	50	118.69*
Titanium	Ti	22	47.90*
Tungsten	W	74	183.85*
Uranium	U	92	238.029[b,c,e,g]
Vanadium	V	23	50.9415[b,c]
Wolfram	W	74	183.85*
Xenon	Xe	54	131.30[e]
Ytterbium	Yb	70	173.04*
Yttrium	Y	39	88.9059[a]
Zinc	Zn	30	65.38
Zirconium	Zr	40	91.22

[a] Elements with only one stable nuclide.

[b] Element with one predominant isotope (about 99 to 100% abundance).

[c] Element for which the atomic weight is based on calibrated measurements.

[d] Element for which known variation in isotopic abundance in terrestrial samples limits the precision of the atomic weight given.

[e] Element for which users are cautioned against the possibility of large variations in atomic weight due to inadvertent or undisclosed artificial isotopic separation in commercially available materials.

[f] Most commonly available long-lived isotope.

[g] In some geological specimens this element has a highly anomalous isotopic composition, corresponding to an atomic weight significantly different from that given.

$$\frac{\text{Exp } 3}{\text{Exp } 1} = \frac{4.0 \times 10^{-4}}{4.0 \times 10^{-4}} = \frac{k[.20][.10][.30]}{k[.20][.10][.1]}$$

$Z = 0$ $\qquad 1 = 3^x$

~~$\frac{3}{2}$ $\frac{4.0 \times 10^{-4}}{P}$~~

$\frac{4}{3}$ $\quad \frac{2}{1} = \frac{1.2 \times 10^{-3}}{4.0 \times 10^{-4}} = \frac{k[.20][.30]}{k[.20][.10]}$

$y = 1$ $\qquad 3 = 3^y$

$\frac{4}{2}$ $\quad \frac{3.6 \times 10^{-3}}{1.2 \times 10^{-3}} = \frac{.60}{.20}$

$3 = 3^x$

$Rate = k[A][B]$

Problem Solving in GENERAL CHEMISTRY

SECOND EDITION

Kenneth W. Whitten
Kenneth D. Gailey
University of Georgia

Keyed to
General Chemistry, second edition
and
General Chemistry with Qualitative Analysis, second edition
by Whitten and Gailey

and

Principles of Chemistry
by Davis, Gailey & Whitten

Saunders Golden Sunburst Series
SAUNDERS COLLEGE PUBLISHING
Philadelphia New York Chicago
San Francisco Montreal Toronto
London Sydney Tokyo Mexico City
Rio de Janeiro Madrid

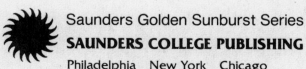

Address orders to:
383 Madison Avenue
New York, NY 10017

Address all editorial correspondence to:
West Washington Square
Philadelphia, PA 19105

PROBLEM SOLVING IN GENERAL CHEMISTRY ISBN 0-03-063576-4

567 073 6543

CBS COLLEGE PUBLISHING
Saunders College Publishing
Holt, Rinehart and Winston
The Dryden Press

FOREWORD

PROBLEM-SOLVING IN GENERAL CHEMISTRY contains nineteen chapters that correspond to the first nineteen chapters in our texts, GENERAL CHEMISTRY and GENERAL CHEMISTRY WITH QUALITATIVE ANALYSIS, 2nd Eds. Although it was designed to be used with these texts, it may be used with other texts since the coverage corresponds to the "common core" of most general chemistry courses for science majors.

In particular, this book can be used to accompany PRINCIPLES OF CHEMISTRY by Davis, Gailey, and Whitten. The first page of each chapter contains references to the related sections in PRINCIPLES. For example, the following notation is found on page 14, the first page of Chapter 2: (D, G, W: Sec. 1-12 through 1-15; Chapter 2). This indicates that Chapter 2 in PROBLEM SOLVING is related to material found in Sections 12 through 15 of Chapter 1 and all of Chapter 2 in PRINCIPLES. Additionally, pages iv and v contain a complete listing of such correlations for all chapters for easy reference.

Each chapter contains (1) a brief discussion of the appropriate topics, (2) several illustrative examples, (3) a set of exercises, and (4) the answers to all these exercises.

Throughout this book we make references to tables, figures, and appendices in the text. ALL SUCH REFERENCES REFER TO TABLES, FIGURES, AND APPENDICES IN OUR TEXTS, GENERAL CHEMISTRY AND GENERAL CHEMISTRY WITH QUALITATIVE ANALYSIS. All such items also can be found easily within the PRINCIPLES sections referred to on the opening page of each chapter.

Both authors have checked every example and every answer, but experience indicates that some errors creep through even after the most careful proofreading. We shall be happy to learn of any errors you detect and to receive suggestions for improvements.

CORRELATION OF CHAPTERS WITH RELATED SECTIONS AND CHAPTERS IN PRINCIPLES OF CHEMISTRY BY DAVIS, GAILEY, AND WHITTEN

PROBLEM SOLVING IN GENERAL CHEMISTRY	correlates with	PRINCIPLES OF CHEMISTRY
Chapter 1		D, G, W: Sec. 1-1 through 1-11; Sec. 12-12
Chapter 2		D, G, W: Sec. 1-12 through 1-15; Chap. 2
Chapter 3		D, G, W: Chap. 4 and 5
Chapter 4		D, G, W: Sec. 6-1 through 6-10 and 6-12; Sec. 1-16 and 1-17
Chapter 5		D, G, W: Sec. 6-11 through 6-14; Sec. 7-1 through 7-10; Sec. 8-1 through 8-8
Chapter 6		D, G, W: Sec. 8-9 through 8-14
Chapter 7		D, G, W: Sec. 9-4 through 9-7
Chapter 8		D, G, W: Chap. 11
Chapter 9		D, G, W: Chap. 12
Chapter 10		D, G, W: Chap. 13
Chapter 11		D, G, W: Sec. 9-4.2, .3, .4 and 9-6.1, .2; Sec. 17-1 through 17-4 and Sec. 17-10
Chapter 12		D, G, W: Sec. 9-7.1 and Sec. 10-5
Chapter 13		D, G, W: Chap. 3; Sec. 7-11; Sec. 12-11.3; Chap. 14
Chapter 14		D, G, W: Chap. 15

Chapter 15	D, G, W: Chap. 16
Chapter 16	D, G, W: Sec. 17-1; Sec. 17-5 through 17-8; Sec. 18-1 through 18-5.1
Chapter 17	D, G, W: Sec. 17-9 and Sec. 18-5.2
Chapter 18	D, G, W: Sec. 18-6 through 18-11
Chapter 19	D, G, W: Chap. 19

CONTENTS

Chapter One

THE FOUNDATIONS OF CHEMISTRY

D,G,W: Sec. 1-1 through 1-11; Sec. 12-12

1-1 Units of Measurement

Most measurements in the sciences are reported in units of the metric system or the International System (SI), which was adopted by the National Bureau of Standards in 1964. However, widespread acceptance of a new system of units occurs very slowly, and, as a result, it is necessary to learn to perform conversions among various kinds of units. The seven fundamental base units of the SI system, from which other units are derived, are given in Table 1-1.

Table 1-1

Physical Property	Name of Unit	Symbol
Length	Meter	m
Mass	Kilogram	kg
Time	Second	s
Electric current	Ampere	A
Thermodynamic temperature	Kelvin	K
Luminous intensity	Candela	cd
Quantity of substance	Mole	mol

Both the metric and SI systems utilize a series of prefixes to indicate fractions and multiples of ten. These prefixes have the same meaning regardless of the unit they precede. Several of them are given, with examples, in Table 1-2.

Table 1-2

Some Metric Prefixes

Prefix	Abbreviation	Meaning	Example
Mega-	M	10^6	1 megagram (Mg) = 1×10^6 g
Kilo-	k	10^3	1 kilometer (km) = 1×10^3 m
Deci-	d	10^{-1}	1 decigram (dg) = 1×10^{-1} g
Centi-	c	10^{-2}	1 centimeter (cm) = 1×10^{-2} m
Milli-	m	10^{-3}	1 millisecond (ms) = 1×10^{-3} s
Micro-	µ	10^{-6}	1 micrometer (µm) = 1×10^{-6} m
Nano-	n	10^{-9}	1 nanogram (ng) = 1×10^{-9} g
Pico-	p	10^{-12}	1 picosecond (ps) = 1×10^{-12} s

Appendix C in the text gives conversion factors relating several English units to metric and SI units. The disadvantage of the English system, relative to metric and SI systems, is that there is no obvious relationship between similar units. Learning the units and prefixes in Tables 1-1 and 1-2 will be helpful. At least one equality each relating English and metric units of volume, length, and mass or weight, should also be learned.

	Metric		English
Volume:	1.00 liter (L)	=	1.06 qt
	(or 0.946 L	=	1.00 qt)
Length:	2.54 cm	=	1.00 in (exactly)
Mass or Weight:	454 g	=	1.00 lb

1-2 Significant Figures

Significant figures are numbers believed to be correct by a competent person who makes a measurement. The result of a measurement should always be reported to the maximum number of significant figures allowed by the measurement and no more. All digits known to be correct plus the first estimated digit constitute the proper number of significant figures.

Exact numbers are numbers that are known to be absolutely correct. For example, you count the number of coins in your pocket. If there are five there is no doubt that there are exactly five coins and not 4.9 or 5.1. Such numbers, called exact numbers, are understood to have an infinite number of significant figures.

However, the results of most measurements yield inexact numbers, numbers whose accuracy is limited by the accuracy of the measuring device. For example, the smallest units marked on most triple-beam balances are 0.1

gram units. Therefore, the first estimated digit in a mass or weight obtained from such a balance would be in the hundredths of grams column. The mass of a small block of wood weighed on a triple-beam balance could be reported as 39.42 grams. This number has four significant figures and it indicates that the person who weighed the block is certain that its weight is between 39.4 and 39.5 grams. To the best of his/her ability to estimate, the weight is 39.42 grams. The digit 2 is considered a significant figure because it represents the best estimate of the person who weighed the block of wood.

Analytical balances are much more accurate and expensive than triple-beam balances. Most have smallest divisions corresponding to thousandths of grams and, therefore, weigh to the nearest ten-thousandth of a gram, i.e., to the nearest 0.0001 gram. If the same block of wood were weighed on an analytical balance its mass might be reported as 39.4167 grams, which has six significant figures --- five exactly known digits and one estimated digit, 7.

Accuracy refers to how closely a measured value agrees with the correct value. Precision refers to how closely the results of individual measurements agree with each other. Because there is always some estimation involved in measurements, several measurements are usually performed and the results are then averaged to minimize error.

The following are simple conventions regarding significant figures.

1. Zeroes used only to place the decimal are not significant. Consider the following examples.

   a. 0.00682 — three significant figures
      (0.00 = place holders; 68 = known exactly; 2 = estimated)

   b. 1.072 — four significant figures

   c. 13.0600 — six significant figures

   d. 30.4 — three significant figures

Three simple statements cover all cases involving zeroes.

Leading zeroes (those that preceed the first nonzero digit) are never significant.

Imbedded zeroes (those between nonzero digits) are always significant.

Trailing zeroes (those that follow the last nonzero digit) are significant only if the decimal point is specified (shown).

2. In addition and subtraction the last digit retained in the sum or difference is determined by the first doubtful digit.

3. In multiplication and division an answer contains no more significant figures than the least accurately known number used in the operation.

The following examples illustrate the use of significant figures.

Example 1-1: What is the total mass of three samples of silver that weigh 2.3162 g, 0.212 g, and 14.10 g respectively?

$$\begin{array}{rl} & 2.316\underline{2}\ \text{g} \\ & 0.21\underline{2}\ \text{g} \\ & 14.1\underline{0}\ \text{g} \\ \hline \text{(calculator result)} \rightarrow & 16.6\underline{2}\underline{8}\underline{2}\ \text{g} = \underline{16.63\ \text{g}} \leftarrow \text{(correct answer)} \end{array}$$

(sum of exact 1, exact 1 and inexact (but significant) zero; first uncertain digit of the sum is the last significant digit.)

Example 1-2: What is the volume of a carton with a basal area (area of the base) of 162 cm^2 and a height of 24 cm? The volume is the basal area, a, times the height, h.

$$\text{Volume} = a \times h = 162\ cm^2 \times 24\ cm = \underline{\underline{3.9 \times 10^3\ cm^3}} \quad \text{(3888 is calculator result)}$$

Electronic calculations "assume" all numbers to be exact, i.e., all digits not given are taken to be exactly known zeroes. This, of course, is not the case and we must apply the rules of significant figures to report correctly the results of calculations done by electronic calculators. The following longhand multiplication of the same two numbers shows why only two significant figures are justified in this calculation. All estimated digits, and all digits obtained from them, are underlined.

$$\begin{array}{r} 16\underline{2}\ cm^2 \\ 2\underline{4}\ cm \\ \hline 6\underline{4}\underline{8} \\ 32\underline{4} \\ \hline 3\underline{8}\underline{8}\underline{8}\ cm^3 \end{array} = \underline{\underline{3.9 \times 10^3\ cm^3}}$$

(product of an inexact 2 and an exact 2 -- must be inexact.)

first uncertain digit - The sum of an inexactly known 6 and an exactly known 2 must be inexact.

Example 1-3: Perform the following operations:

a. 6.120×18.1

b. $46.106 \div 14$

c. $210.4 \div 111$

d. $64.308 + 72.1776$

e. $142.4 \div 384$

f. $816.24 - 1.00 \times 10^3$

g. $438.6 \div 0.0102$

The correct answers are underlined and calculator results are given in parentheses.

a. $6.120 \times 18.1 = \underline{\underline{111}}$
(110.772)

b. $46.106 \div 14 = \underline{\underline{3.3}}$
(3.2932857)

c. $210.4 \div 111 = \underline{\underline{1.90}}$
(1.8954955)

d. $64.308 + 72.1776 = \underline{\underline{136.486}}$
(136.4856)

e. $142.4 \div 384 = \underline{\underline{0.371}}$
(0.3708333)

f. $816.24 - 1.00 \times 10^3 = \underline{\underline{-1.8 \times 10^2}}$
(-1.8376×10^2)

g. $438.6 \div 0.0102 = \underline{\underline{4.30 \times 10^4}}$
(43000)

1-3 Unit Factors and Dimensional Analysis

Units are an integral part of most calculations and should not be neglected. In fact one of the easiest and most general methods of solving problems, dimensional analysis or the unit factor method, places great reliance on units. The method involves converting a given unit into the desired unit by multiplication of the given number and unit by a factor, or factors, that relate one unit to another. The factors (such as 2.54 cm/in), which are constructed from conversion equalities (such as 2.54 cm = 1 inch), are called unit factors because both the numerator and the denominator of any such factor represent equal amounts of the same thing. Therefore, multiplying by a unit factor is really "multiplying by one".

Consider the equality 2.54 cm = 1 inch. Division of both sides by 1 inch yields a unit factor. (Since 1 inch is exactly 2.54 cm, it is not necessary to write 1.00 inch.)

$$\frac{2.54 \text{ cm}}{1 \text{ in}} = \frac{1 \text{ in}}{1 \text{ in}} \quad \text{or} \quad \frac{2.54 \text{ cm}}{1 \text{ in}} = 1$$

Division of both sides of the original equality by 2.54 cm gives another unit factor, the reciprocal of the first:

$$\frac{1\ \text{in}}{2.54\ \text{cm}} = \frac{2.54\ \text{cm}}{2.54\ \text{cm}} \quad \text{or} \quad \frac{1\ \text{in}}{2.54\ \text{cm}} = 1$$

Two unit factors can be constructed from any equality. The use of unit factors is illustrated in the next few examples. Most of the factors are constructed from the conversion equalities given in Appendix C in the text.

Example 1-4: How many centimeters are there in 15.2 inches?

We must convert a given number of inches to the corresponding number of centimeters. We begin the problem as follows:

$$\underline{?}\ \text{cm} = 15.2\ \text{in}$$

We then multiply the length, 15.2 inches, by the unit factor that converts inches to centimeters. Therefore the appropriate unit factor has inches in the denominator and centimeters in the numerator so that after multiplication the "inches cancel" and centimeters remain as the desired unit.

$$\underline{?}\ \text{cm} = 15.2\ \text{in} \times \frac{2.54\ \text{cm}}{1\ \text{in}} = \underline{\underline{38.6\ \text{cm}}}$$

(38.608)

Example 1-5: How many inches are there in 174 centimeters?

We are given a number of centimeters and want to know the corresponding number of inches.

$$\underline{?}\ \text{in} = 174\ \text{cm}$$

We use the reciprocal of the factor used in Example 1-4.

$$\underline{?}\ \text{in} = 174\ \text{cm} \times \frac{1\ \text{in}}{2.54\ \text{cm}} = \underline{\underline{68.5\ \text{in}}}$$

(68.503937)

Example 1-6: How many square millimeters, mm^2, are there in 7.6 square inches?

We know that 2.54 cm = 1 inch and therefore, by squaring both sides of this equation we see that, $(2.54\ \text{cm})^2 = (1\ \text{inch})^2$. We also know that 10.0 mm = 1.00 cm and therefore $(10.0\ \text{mm})^2 = (1.00\ \text{cm})^2$.

$$\underline{?}\ \text{mm}^2 = 7.6\ \text{in}^2 \times \left(\frac{2.54\ \text{cm}}{1\ \text{in}}\right)^2 \times \left(\frac{10\ \text{mm}}{1\ \text{cm}}\right)^2 = \underline{\underline{4.9 \times 10^3\ \text{mm}^2}}$$

(4903.216)

These factors contain two exact numbers.

Example 1-7: How many gallons are there in a volume of 6.82×10^6 cubic millimeters?

$$\underline{?}\,\text{gal} = 6.82 \times 10^6\ \text{mm}^3 \times \left(\frac{1\ \text{cm}}{10\ \text{mm}}\right)^3 \times \frac{1\ \text{L}}{1000\ \text{cm}^3} \times \frac{1.06\ \text{qt}}{1\ \text{L}} \times \frac{1\ \text{gal}}{4\ \text{qt}} = \underline{\underline{1.81\ \text{gal}}}$$

(1.8073)

(These factors contain only exact numbers)

1-4 Density and Specific Gravity

The density of a substance or object is its mass per unit volume.

$$\text{Density} = \frac{\text{mass}}{\text{volume}} \qquad \text{or} \qquad D = \frac{M}{V}$$

Densities of substances vary with temperatures. They are often expressed in g/cm^3 or g/mL for solids and liquids and in g/L for gases. Density is an intensive property, i.e., it does not depend on the size of the sample as do extensive properties, such as mass and volume.

Example 1-8: What is the density of a solid object having a mass of 4.687 g if it displaces a volume of 1.84 mL of water in a graduated cylinder?

$$D = \frac{M}{V} = \frac{4.687\ \text{g}}{1.84\ \text{mL}} = \underline{\underline{2.55\ \text{g/mL}}}$$

(2.5472826)

Since 1 mL is 1 cm^3, the density is also $\underline{\underline{2.55\ g/cm^3}}$.

It is not necessary to memorize the formula D = M/V if dimensional analysis is used.

$$\frac{?\ \text{g}}{\text{mL}} = \frac{4.687\ \text{g}}{1.84\ \text{mL}} = \underline{\underline{2.55\ \text{g/mL}}}$$

Example 1-9: What volume of a substance has a mass of 12.42 grams if its density is 4.16 g/cm^3?

We convert the mass of the substance to the volume it occupies:

$$\underline{?}\ \text{cm}^3 = 12.42\ \text{g} \times \frac{1\ \text{cm}^3}{4.16\ \text{g}} = \underline{\underline{2.99\ \text{cm}^3}}$$

(2.9855769)

Volume and mass were known to 3 sig. figs. to report density to 3 sig. figures. The numerator is understood to contain 3 sig. figs.

Specific gravity is the density of a substance relative to the density of a standard. For liquids and solids the standard is water at 4°C, which has a density of 1.00 g/cm^3 (or 1.00 g/mL). Thus, in the metric system, specific gravity is numerically equal to density (without units).

Example 1-10: What is the specific gravity of sulfur, which has a density of 2.07 g/cm^3 ?

$$\text{Specific gravity} = \frac{\text{Density of sulfur}}{\text{Density of water}} = \frac{2.07\ \text{g/cm}^3}{1.00\ \text{g/cm}^3} = \underline{\underline{2.07}}$$

(2.07)

1-5 Heat and Temperature Scales

Heat is a form of energy and therefore it can be expressed in any energy units (see Appendix C). The SI unit of energy is the joule (1 joule = 1 kg·m^2/s^2 or 1 newton-meter); another commonly used unit is the calorie (1 calorie = 4.184 joules). Temperature is a measure of heat intensity. A large object at 65.0°C possesses more heat than a smaller object of the same material at 65.0°C. The three most commonly used temperature scales are the Fahrenheit (°F), Celsius (°C) and absolute or Kelvin (K) scales. The reference points for the Celsius scale are chosen as the freezing and boiling points of water at one atmosphere pressure (average pressure at sea level at 0°C). The scales are compared in Figure 1-1.

Figure 1-1

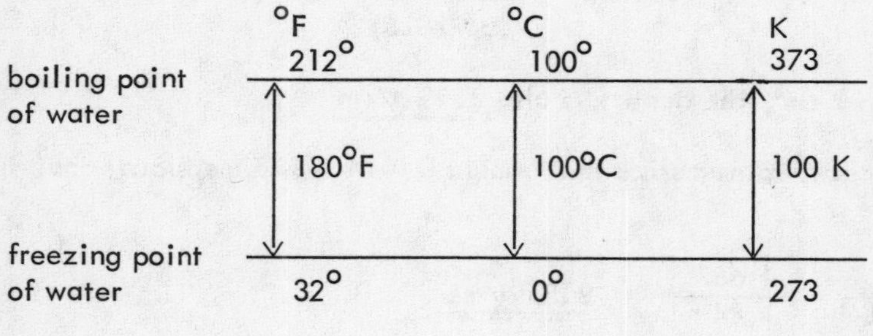

As the figure shows the Celsius degree is 180/100 or 9/5 as large as the Fahrenheit degree and there is a 32° difference between the scales at the freezing point of water. From this information the following relation can be derived.

$$°F = 9/5°C + 32° \qquad \text{or} \qquad °C = 5/9\,(°F - 32°)$$

The absolute temperature scale is based on the observed behavior of matter. A temperature of 0 K is, theoretically, the lowest temperature attainable, the temperature at which all molecular motion ceases. It corresponds to -273.15°C. The size of the Kevin degree is the same as that of the Celsius degree and, therefore, the relationship between K and $^\circ C$ is simple. (No degree sign is used to indicate temperatures on the Kelvin scale.)

$$K = {}^\circ C + 273 \qquad \text{or} \qquad {}^\circ C = K - 273$$

Example 1-11: The melting point of lead is 328°C. What is its melting point in Fahrenheit degrees? in K?

$$^\circ F = 9/5\ {}^\circ C + 32^\circ$$

$$= 9/5\,(328^\circ) + 32^\circ = 590^\circ + 32^\circ = \underline{\underline{622^\circ F}}$$

$$K = {}^\circ C + 273 = 328^\circ + 273 = \underline{\underline{601\ K}}$$

Example 1-12: Oxygen is a gas at room temperature and atmospheric pressure. But when cooled it will liquify at -297°F which is, of course, its boiling point at atmospheric pressure. What is its boiling point in °C?

$$^\circ C = 5/9\ ({}^\circ F - 32^\circ) = 5/9\,(-297^\circ - 32^\circ) = \underline{\underline{-183^\circ C}}$$

1-6 Heat Transfer and the Measurement of Heat

The specific heat of a substance is the amount of heat required to raise the temperature of one gram of a substance 1°C with no change in state. The amount of heat transferred when a substance is heated (or cooled) is given by the simple relationship in which Δt represents the temperature change.

$$\text{amount of heat} = (\text{mass}) \times (\text{sp. heat}) \times (\Delta t)$$

Example 1-13: The specific heat of lead is 0.159 J/g°C. How much heat is required to raise the temperature of one pound of lead from 20.0°C to 35.0°C?

One pound is 454 grams, and the temperature change is 15.0°C.

$$\underline{?}\ \text{cal} = 454\ \text{g Pb} \times \frac{0.159\ \text{J}}{\text{g Pb}^\circ\text{C}} \times 15.0^\circ\text{C} = \underline{\underline{1.08 \times 10^3\ \text{J}}}$$

Example 1-14: If 500 grams of lead at 200.0°C are placed in 100 grams of water at 20.0°C in an insulated container, what will be the temperature when the lead and water are at the same temperature? The specific heat of water is 4.18 J/g°C.

Let: $t\ ^\circ C$ = final temperature

Heat gain by water = Heat loss by lead

$$(100\ g\ H_2O)(4.18\ J/g\ H_2O\ ^\circ C)(t - 20.0)^\circ C =$$

$$(500\ g\ Pb)(0.159\ J/g\ Pb\ ^\circ C)(200.0 - t)^\circ C$$

$$418\ t - 8360 = 15900 - 79.5\ t$$

$$498\ t = 24260$$

$$\underline{\underline{t = 48.7^\circ C}}$$

EXERCISES

Any of the appendices in the text may be consulted.

1. Express the following numbers in scientific notation to the parenthetically indicated number of significant figures.

   a. 324,000 (2) b. 609,764,210 (3) c. 1.00876 (3)

   d. 0.004387 (4) e. 0.674×10^{-6} (3) f. 60.3430 (4)

   g. 30074 (5)

2. How many significant figures do the following numbers contain? Assume that these numbers represent measurements.

   a. 62,000 b. 60.1042 c. 0.08040 d. 182,107.2

   e. 6.41×10^{-5} f. 26,300 g. 1.0064 h. 0.0064

3. Perform the indicated operations and express the results to the proper number of significant figures. Assume that these numbers represent measurements.

   a. $487 \div 32$

   b. $26.204 + 7.16$

   c. $4.640 \times 8.23 \times 10^{-4}$

   d. $66.314 + 14.72 - 0.116$

   e. $6.48 \div 1.184 \times 10^{-3}$

   f. $1082 + 0.681$

   g. $16.382 \times 4.64 \times 10^{4}$

   h. $14.3 \div 4.11$

   i. $0.6814 - 1.416 - 0.8332$

   j. $4.17 \times 10^{-4} \times 3.6 \times 10^{3}$

4. How many grams are contained in 6.00 pounds?

5. Convert 48.4 tons to kilograms. (Short ton, unless other is specified.)

6. What is the area of a surface that is 34.211 millimeters long and 112.4 centimeters wide?

7. The volume of a cube is 542 mm^3. How long is each side in centimeters?

8. Convert 1.74×10^{5} calories to (a) joules, (b) kilocalories, and (c) ergs.

9. How many milliliters are contained in 5.14 quarts?

10. How many pints are contained in 57,200 cubic centimeters of liquid?

11. Express 5.43×10^{4} liters in (a) cubic meters, (b) cubic yards and (c) gallons.

12. Express 4.936×10^5 grams in tons.

13. How many milligrams are contained in 1.00 pound?

14. How many ounces are contained in 600 grams?

15. Express 2.50 miles in (a) kilometers, (b) meters, and (c) millimeters.

16. What is the volume in liters of a cylinder that has a radius of 4.13 cm and a height of 41.64 decimeters?

17. The density of carbon tetrachloride is 1.595 g/mL.
    (a) What is the mass of 50.0 mL of carbon tetrachloride?
    (b) How many milliliters of carbon tetrachloride have a mass of 350 grams?

18. A piece of metal that has a mass of 16.862 grams is placed into 24.36 mL of water in a graduated cylinder. The volume of the metal plus the water is 26.94 mL. What is the specific gravity of the metal?

19. The "lift" of a balloon is the difference between the weight of the gas in the balloon and the weight of an equal volume of air. How many pounds, including its own weight, can be lifted by a spherical balloon, filled with helium, that has a diameter of 150 feet? The density of air is 1.205 g/L and that of helium is 0.178 g/L.

20. The specific gravity of gold is 19.32. What is the mass in grams of a cube of gold that is 0.25 mm on an edge (a very tiny cube)?

21. If platinum costs $1793 per ounce what would be the cost (to the nearest dollar) of a platinum crucible and cover having a mass of 11.6841 grams?

22. The boiling point of liquid helium (also the liquefaction point of gaseous helium) is $-269^\circ C$. What is this temperature on the Kelvin temperature scale?

23. Mercury freezes at $-39^\circ C$. What is its freezing point in Fahrenheit degrees?

24. The temperature at Gila Bend, Arizona was $121^\circ F$ one day. What is this temperature on the Celsius and Kelvin scales?

25. The specific heat of liquid mercury is $0.138\ J/g^\circ C$.
    a. How much heat is required to raise the temperature of 15.0 grams of mercury from $20.0^\circ C$ to $100.0^\circ C$.
    b. Repeat the calculation in (a) for water, rather than mercury.
    c. What is the ratio of your answer for (b) to your answer for (a)?
    d. What is the ratio of the specific heat of water (liquid) to the specific heat of mercury (liquid)?
    e. Do your answers for (c) and (d) agree exactly? Why?

26. If 100 grams of water at $20.0^\circ C$ is mixed with 200 grams of water at $90.0^\circ C$ in an insulated container, what will be the temperature of the mixture?

27. If 100 grams of mercury at 20.0°C is placed in 200 grams of water at 90 0°C in an insulated container, what will be the temperature when both mercury and water reach the same temperature?

ANSWERS TO EXERCISES

1. (a) 3.2×10^5 (b) 6.10×10^8 (c) 1.01 (d) 4.387×10^{-3} (e) 6.74×10^{-7} (f) 6.034×10^1 (g) 3.0074×10^4 2. (a) 5 (b) 6 (c) 4 (d) 7 (e) 3 (f) 5 (g) 5 (h) 2 3. Calculator results are given in parentheses following the correct answers. (a) 15 (15.21875) (b) 33.36 (33.364) (c) 3.82×10^{-3} (3.81872×10^{-3}) (d) 80.92 (80.918) (e) 5.47×10^3 (5.472973×10^3) (f) 1083 (1082.681) (g) 7.60×10^5 (7.601248×10^5) (h) 3.48 (3.4793187) (i) -1.568 (-1.5678) (j) 1.5 (1.5012) 4. 2.72×10^3 g 5. 4.39×10^4 kg 6. 384.5 cm^2 7. 0.815 cm 8. (a) 7.28×10^5 joules (b) 1.74×10^2 kcal (c) 7.28×10^{12} ergs 9. 4.85×10^3 mL 10. 121 pints 11. (a) 54.3 m^3 (b) 71.0 yd^3 (c) 1.44×10^4 gal 12. 0.5436 ton 13. 4.54×10^5 mg 14. 21.1 oz 15. (a) 4.02 km (b) 4.02×10^3 m (c) 4.02×10^6 mm 16. 2.23×10^4 cm^3 17. (a) 79.8 g (b) 219 mL 18. 6.54 19. 1.13×10^5 lb 20. 3.01×10^{-4} g 21. $738 22. 4 K 23. -38°C 24. 49°C, 322 K 25. (a) 166 J (b) 5.02×10^3 J (c) 30.2:1 (d) 30.3:1 (e) No, but agreement is good; difference is due to round-off error. 26. 66.7°C 27. 88.9°C

Chapter Two

STOICHIOMETRY

D, G, W: Sec. 1-12 through 1-15; Chap. 2

Stoichiometry describes the quantitative relationships among elements and compounds, and among substances as they undergo chemical changes. Reference can be made to the stoichiometry of a compound, that is to the composition of a compound, or to the stoichiometry of a reaction, i.e., the quantitative relationships between different species involved in a chemical reaction.

2-1 Symbols and Formulas

The symbols for the elements consist of either a capital letter or a capital letter followed by a lower case letter. For example, C represents carbon and Fe is iron. Formulas for chemical substances indicate the relative proportions of elements that make up the substances. Elements may exist as monatomic, diatomic, or polyatomic molecules such as Ne, H_2, and S_8.

Since compounds contain two or more different elements in combination, their formulas indicate the elements present and their proportions. For example, sulfur trioxide molecules, SO_3, consist of sulfur and oxygen atoms in a 1:3 ratio. Each formula unit of calcium phosphate, $Ca_3(PO_4)_2$, consists of three calcium atoms, two phosphorus atoms, and eight oxygen atoms.

2-2 Atomic Weights and The Mole

The relative masses of atoms are expressed in terms of atomic mass units on an atomic weight scale in which an arbitrary value of exactly 12 atomic mass units is assigned to a carbon atom containing six protons, six neutrons, and six electrons (carbon-12). Protons, neutrons, and electrons are subatomic particles that we shall discuss in Chapter 3. An average hydrogen atom has an atomic weight of 1.0079 atomic mass units and is about 1/12 as heavy as a ^{12}C atom. An average magnesium atom has an atomic weight of 24.305 atomic mass units and is about twice as heavy as a ^{12}C atom.

While the atomic mass unit (amu) is an arbitrary but convenient unit of mass for individual atoms, its use as a unit of mass (weight) in the laboratory is impractical. It is much more convenient to measure mass (weight) in grams.

The number of atoms of an element in the mass numerically equal to its atomic weight in grams is always the same, 6.022×10^{23}. There are

6.022×10^{23} atoms of hydrogen in 1.0079 grams of hydrogen, 6.022×10^{23} magnesium atoms in 24.305 grams of magnesium, and 6.022×10^{23} lead atoms in 207.2 grams of lead.

This number, 6.022×10^{23}, is also known as Avogadro's number or the mole. One mole of atoms of any sort is 6.022×10^{23} atoms. One mole of molecules of any sort is 6.022×10^{23} molecules, and one mole of Na^+ ions is 6.022×10^{23} Na^+ ions. One mole of carrots is 6.022×10^{23} carrots. This is analogous to the fact that one dozen "things" is always twelve "things", regardless of what the "things" are. The abbreviation of mole is mol. We frequently round the mole to 6.02×10^{23}.

Example 2-1: Calculate the number of moles of phosphorus atoms in 127.6 g of phosphorus.

One mole of phosphorus atoms is 30.97 g of phosphorus.

$$? \text{ mol P atoms} = 127.6 \text{ g P} \times \frac{1 \text{ mol P atoms}}{30.97 \text{ g P}} = \underline{\underline{4.120 \text{ mol P atoms}}}$$

Example 2-2: Calculate the number of chlorine atoms in 96.3 g of chlorine, Cl_2.

One mole of chlorine atoms is 35.45 g of chlorine.

$$? \text{ Cl atoms} = 96.3 \text{ g } Cl_2 \times \frac{1 \text{ mol } Cl_2 \text{ molecules}}{70.9 \text{ g } Cl_2} \times \frac{2 \text{ mol Cl atoms}}{1 \text{ mol } Cl_2 \text{ molecules}}$$

$$\times \frac{6.02 \times 10^{23} \text{ Cl atoms}}{1 \text{ mol Cl atoms}} = \underline{\underline{1.64 \times 10^{24} \text{ Cl atoms}}}$$

Example 2-3: Calculate the ratio of the number of calcium atoms in 1.00 gram of calcium to the number of mercury atoms in 1.00 gram of mercury.

First we calculate the number of atoms in 1.00 gram of each element.

$$? \text{ Ca atoms} = 1.00 \text{ g Ca} \times \frac{6.02 \times 10^{23} \text{ Ca atoms}}{40.08 \text{ g Ca}} = \underline{1.50 \times 10^{22} \text{ Ca atoms}}$$

$$? \text{ Hg atoms} = 1.00 \text{ g Hg} \times \frac{6.02 \times 10^{23} \text{ Hg atoms}}{200.6 \text{ g Hg}} = \underline{\underline{3.00 \times 10^{21} \text{ Hg atoms}}}$$

$$\text{Ratio} = \frac{1.50 \times 10^{22} \text{ Ca atoms}}{3.00 \times 10^{21} \text{ Hg atoms}} = \underline{\underline{5 \text{ Ca atoms/Hg atom}}}$$

This calculation tells us that Hg atoms are five times as heavy as Ca atoms.

Example 2-4: What is the atomic weight of an element if 5.0 grams of it contain 4.2×10^{22} atoms?

$$\frac{?\ g}{\text{mol atoms}} = \frac{5.0\ g}{4.2 \times 10^{22}\ \text{atoms}} \times \frac{6.02 \times 10^{23}\ \text{atoms}}{1\ \text{mol atoms}} = \underline{\underline{72\ g/\text{mol atoms}}}$$

Atomic weight = 72 amu

Example 2-5: How many grams of gold contain ten trillion (1.00×10^{13}) atoms of gold?

$$?\ g\ Au = 1.00 \times 10^{13}\ Au\ \text{atoms} \times \frac{197.0\ g\ Au}{6.02 \times 10^{23}\ Au\ \text{atoms}}$$

$$= \underline{\underline{3.27 \times 10^{-9}\ g\ Au}}$$

or 0.00000000327 g Au (too small to be weighed)

2-3 Formula Weights and Molecular Weights

The mass of one formula unit of a substance in amu is its formula weight. It is simply the sum of the masses of its individual atoms. For non-ionic substances the term molecular weight is often used instead. The mass of one mole of molecules in grams is numerically equal to the molecular (formula) weight in amu.

Example 2-6: What is the molecular weight of thionyl chloride, $SOCl_2$?

kind of atoms	moles of atoms per mole of molecules	x	mass of 1 mole of atoms	=	contribution
S	1	x	32.06 grams	=	32.06 grams
O	1	x	16.00 grams	=	16.00 grams
Cl	2	x	35.45 grams	=	70.90 grams
			mass of one mole of molecules	=	118.96 grams
			molecular weight	=	118.96 amu

Example 2-7: Determine the formula weight and mass of one mole of ammonium sulfate $(NH_4)_2SO_4$, an ionic compound that consists of NH_4^+ and SO_4^{2-} ions in a 2:1 ratio.

kind of atoms	number of atoms per formula unit	x	atomic weight	=	contribution
N	2	x	14.01 amu	=	28.02 amu
H	8	x	1.01 amu	=	8.08 amu
S	1	x	32.06 amu	=	32.06 amu
O	4	x	16.00 amu	=	64.00 amu
			formula weight	=	132.16 amu
			1 mole	=	132.16 grams

The following tables should help put the above concepts into perspective.

Table 2-1

One Mole of Some Common Elements

Element	Mass of 1 mole of atoms	Contains
sodium	22.99 g Na	6.022×10^{23} Na atoms
nickel	58.70 g Ni	6.022×10^{23} Ni atoms
chlorine*	35.45 g Cl_2	6.022×10^{23} Cl atoms (3.011×10^{23} Cl_2 molecules)

Table 2-2

One Mole of Some Common Covalent Substances

Substance	Mass of 1 mole of molecules	Contains
fluorine*, F_2	38.00 g F_2	6.022×10^{23} F_2 molecules (1.204×10^{24} F atoms)
glucose, $C_6H_{12}O_6$	180.2 g $C_6H_{12}O_6$	6.022×10^{23} $C_6H_{12}O_6$ molecules $6(6.022 \times 10^{23})$ C atoms $12(6.022 \times 10^{23})$ H atoms $6(6.022 \times 10^{23})$ O atoms

*Note that elemental fluorine and chlorine are diatomic molecules and fit into two categories.

Table 2-3

One Mole of Some Common Ionic Compounds

Compound	Mass of 1 mole of formula units	Contains
magnesium bromide, $MgBr_2$	184.1 g $MgBr_2$	6.022×10^{23} Mg^{2+} ions $2(6.022 \times 10^{23})$ Br^- ions
aluminum carbonate, $Al_2(CO_3)_3$	234.0 g $Al_2(CO_3)_3$	$2(6.022 \times 10^{23})$ Al^{3+} ions $3(6.022 \times 10^{23})$ CO_3^{2-} ions

2-4 Percent Composition

If the formula for a compound is known its composition may be expressed in terms of the percent by weight of its constituent elements. Ammonia, NH_3, consists of three hydrogen atoms and one nitrogen atom per molecule. The percent of the total mass of ammonia molecules due to hydrogen is the mass of three hydrogen atoms divided by the mass of an ammonia molecule, 17.03 amu, multiplied by 100%.

$$\%H = \frac{3 \times H}{NH_3} \times 100\% = \frac{3 \times 1.008 \text{ amu}}{17.03 \text{ amu}} \times 100\% = \underline{\underline{17.76\% \text{ H}}}$$

The percentage of nitrogen is calculated similarly.

$$\%N = \frac{1 \times N}{NH_3} = \frac{1 \times 14.01 \text{ amu}}{17.03 \text{ amu}} \times 100\% = \underline{\underline{82.27\% \text{ N}}}$$

All percentages for any compound should add to 100%, although very slight apparent deviations may arise due to round-off errors in calculations as this example demonstrates.

Example 2-8: Calculate the percent composition of potassium chromate, K_2CrO_4.

First let's calculate the formula weight of potassium chromate.

K	2 x 39.10 amu	=	78.20 amu
Cr	1 x 52.00 amu	=	52.00 amu
O	4 x 16.00 amu	=	64.00 amu
	Formula Weight	=	194.20 amu

We then determine the percentages of the constituent elements.

$$\%K = \frac{2 \times 39.10 \text{ amu}}{194.20 \text{ amu}} \times 100\% = 40.27\% \text{ K}$$

$$\%Cr = \frac{52.00 \text{ amu}}{194.20 \text{ amu}} \times 100\% = 26.78\% \text{ Cr}$$

$$\%O = \frac{4 \times 16.00 \text{ amu}}{194.20 \text{ amu}} \times 100\% = \underline{32.96\% \text{ O}}$$

$$100.01\%$$

2-5 Determination of Simplest Formulas from Percent Composition

If the elemental percentages of a compound are known, the procedure used in the above problems may be reversed and the simplest or empirical formula of a compound may be determined. The empirical or simplest formula of a compound indicates the simplest whole number ratio of atoms in the compound. If the formula (molecular) weight is also known, the true formula, the actual number of atoms or ions per molecule, may be determined.

Let us illustrate the difference between simplest and true formulas. The simplest formula of the hydrocarbon cyclohexane is CH_2. This indicates that the compound is made up of carbon and hydrogen in a 1:2 atom ratio. However, the molecular weight of cyclohexane is 84 amu. The true (molecular) formula must be an integral multiple of the simplest formula, i.e., $(CH_2)_n$. Since one atom of carbon plus two atoms of hydrogen weigh 14 amu, and 84/14 = 6, then there must be six of these combinations per cyclohexane molecule, $(CH_2)_6$. The true molecular formula for cyclohexane is C_6H_{12}. On the other hand, for many compounds such as H_2O, H_2SO_4, and nitrobenzene, $C_6H_5NO_2$, the simplest and true formulas are identical.

Example 2-9: Hydrogen peroxide is 5.93% hydrogen and 94.07% oxygen by weight. What is its empirical formula?

The ratio of the number of atoms per structural unit is equal to the ratio of the number of moles of atoms in a sample, and the latter ratio is calculated. For convenience, we decl with 100.00 g of compound. Thus, there are 5.93 grams of H and 94.07 grams of O in 100.00 grams of hydrogen peroxide.

$$\underline{?} \text{ of mol H atoms} = 5.94 \text{ g H atoms} \times \frac{1 \text{ mol H atoms}}{1.008 \text{ g H atoms}} = 5.89 \text{ mol H atoms}$$

$$\underline{?} \text{ of mol O atoms} = 94.07 \text{ g O atoms} \times \frac{1 \text{ mol O atoms}}{16.00 \text{ g O atoms}} = 5.88 \text{ mol O atoms}$$

The ratio of mol H atoms : mol O atoms is $\frac{5.89}{5.88}$ or 1:1 within round-off accuracy. So the simplest formula for hydrogen peroxide is $\underline{\underline{HO}}$.

Example 2-10: The molecular weight of hydrogen peroxide is 34 amu. What is its true molecular formula? See Example 2-9.

The true formula is a multiple of the simplest formula, HO.

$$\text{True formula} = (HO)_n$$

$$\begin{aligned} 1\,H &= 1.0 \text{ amu} \\ 1\,O &= 16.0 \text{ amu} \\ \hline 1\,HO &= 17.0 \text{ amu} \\ (17.0 \text{ amu})(n) &= 34.0 \text{ amu} \\ n &= 2 \end{aligned}$$

$$\text{True formula} = (HO)_n = (HO)_2 = \underline{\underline{H_2O_2}}$$

Example 2-11: Calcium permanganate is 14.41% calcium, 39.53% manganese, and 46.05% oxygen by mass. Determine its simplest formula.

The number of moles of atoms of each element in 100 g of compound is calculated first.

$$\underline{?}\text{ mol Ca atoms} = 14.41 \text{ g Ca} \times \frac{1 \text{ mol Ca atoms}}{40.08 \text{ g Ca}} = 0.360 \text{ mol Ca atoms}$$

$$\underline{?}\text{ mol Mn atoms} = 39.53 \text{ g Mn} \times \frac{1 \text{ mol Mn atoms}}{54.94 \text{ g Mn}} = 0.720 \text{ mol Mn atoms}$$

$$\underline{?}\text{ mol O atoms} = 46.05 \text{ g O} \times \frac{1 \text{ mol O atoms}}{16.00 \text{ g O}} = 2.88 \text{ mol O atoms}$$

We are interested in establishing the simplest whole number ratio of the elements. If we divide each of the three numbers of moles of atoms by the smallest of them, at least one of the numbers in the resulting ratio will be unity.

$$\frac{0.360}{0.360} = 1.00\ \text{Ca}$$

$$\frac{0.720}{0.360} = 2.00\ \text{Mn} \longrightarrow CaMn_2O_8 \quad \text{or} \quad \underline{\underline{Ca(MnO_4)_2}}$$

$$\frac{2.88}{0.360} = 8.00\ \text{C}$$

2-6 Chemical Equations

Chemical equations provide a shorthand description of chemical reactions. The reactants are shown on the left and the products on the right. According to the Law of Conservation of Matter, there is no gain or loss of mass during any ordinary chemical reaction. Thus, we must have the same number of atoms of each element on the left and right sides of an equation. The Law of Definite Proportions (Constant Composition) requires that the formula for each substance always be written in the same way since different samples of the same compound always have the same composition. Consider the reaction of sodium chlorite with gaseous chlorine to produce sodium chloride and gaseous chlorine dioxide. The unbalanced equation is:

$$NaClO_2 + Cl_2 \longrightarrow NaCl + ClO_2$$

Since there must be the same number of each kind of atom on both sides, coefficients of 2 must precede $NaClO_2$, $NaCl$, and ClO_2.

$2\ NaClO_2$	+	Cl_2	$\longrightarrow$	$2\ NaCl$	+	$2\ ClO_2$
2 formula units*		1 molecule		2 formula units*		2 molecules

The equation is now balanced.

The coefficients of the species always show the ratios in which substances react and are produced, but not necessarily the ratios of amounts of reactants that are mixed together. They indicate the relative numbers of molecules, formula units, atoms, and/or ions involved in a reaction, and the relative numbers of moles of each.

$2\ NaClO_2$	+	Cl_2	$\longrightarrow$	$2\ NaCl$	+	$2\ ClO_2$
$2(6.02 \times 10^{23})$		6.02×10^{23}		$2(6.02 \times 10^{23})$		$2(6.02 \times 10^{23})$
formula units		molecules		formula units		molecules
or		or		or		or
2 moles		1 mole		2 moles		2 moles

*Since $NaClO_2$ and $NaCl$ are ionic compounds, we refer to formula units (not molecules) of $NaClO_2$ and $NaCl$.

Example 2-12: Balance the following equation.

$$Al + F_2 \longrightarrow AlF_3$$

On the left side there are 2 F atoms and on the right side there are 3 F atoms. The least common multiple is 6 and we must have 6 F on each side. This requires 2 AlF_3 on the right and therefore 2 Al and 3 F_2 on the left.

$$2\ Al + 3\ F_2 \longrightarrow 2\ AlF_3$$

Example 2-13: Balance the following equation for the complete combustion of the hydrocarbon propane, C_3H_8.

$$C_3H_8 + O_2 \longrightarrow CO_2 + H_2O$$

Note that the equation, as written, involves an even number (2) of oxygen atoms on the left, while on the right oxygen atoms occur in groups of 2 in CO_2 and 1 in H_2O. Clearly there must be an even number of H_2O molecules on the right side of the balanced equation. Each molecule of C_3H_8 contains 8 H atoms and therefore we choose 1 and 4 as the coefficients of C_3H_8 and H_2O, respectively.

$$1\ C_3H_8 + O_2 \longrightarrow CO_2 + 4\ H_2O$$

Now there are 3 carbon atoms on the left and only 1 on the right, and we place a coefficient of three before carbon dioxide to balance the carbon atoms.

$$1\ C_3H_8 + O_2 \longrightarrow 3\ CO_2 + 4\ H_2O$$

The balancing operation is completed by placing a coefficient of 5 in front of O_2 to balance the 10 oxygen atoms on the right.

$$1\ C_3H_8 + 5\ O_2 \longrightarrow 3\ CO_2 + 4\ H_2O$$

Balancing the oxygen atoms was intentionally left until last because the number of oxygen atoms on the left is the sum of the number appearing in 3 CO_2 and 4 H_2O.

After balancing is complete, check the set of coefficients to see if all are divisible by the same integer. For example, assume the following equation results from the balancing procedure.

$$12\ FeO + 2\ O_2 \longrightarrow 4\ Fe_3O_4$$

Every coefficient is evenly divisible by two and the equation is properly expressed with the smallest possible whole number coefficients, i.e.,

$$6\ FeO + O_2 \longrightarrow 2\ Fe_3O_4$$

2-7 Interpretation of Chemical Equations

The reaction of gaseous ammonia with red-hot copper(II) oxide is described by the balanced equation given below.

$$2\ NH_3 + 3\ CuO \xrightarrow{\Delta} 3\ H_2O + N_2 + 3\ Cu$$

As we have seen, the reaction ratio on the molecular level is the same as on the mole level, i.e.,

$2\ NH_3$ +	$3\ CuO$ $\xrightarrow{\Delta}$	$3\ H_2O$ +	N_2 +	$3\ Cu$
2 molecules	3 formula units	3 molecules	1 molecule	3 atoms
		or		
2 mol	3 mol	3 mol	1 mol	3 mol
2(17.03 g)	3(79.55 g)	3(18.02 g)	28.01 g	3(63.55 g)
		or		
34.06 g	238.65 g	54.06 g	28.01 g	190.65 g
		or		
34.06 mass units	238.65 mass units	54.06 g mass units	28.01 mass units	190.65 g mass units

Although grams are commonly used as mass units, any unit of mass or weight, can be used. This illustrates the quantitative precision and versatility of chemical equations. Remember, however, that the coefficients in no way limit the amounts of reactants that may be mixed for reaction purposes, but they do define the ratios in which the substances actually react and are produced. A large number of unit factors can be constructed from the information given above. Uses of some of these unit factors are illustrated in the following examples which are based on the balanced equation given above.

Example 2-14: How many molecules of NH_3 are required to react with 4.14×10^{22} formula units (fu) of CuO?

$$? \ NH_3 \text{ molecules} = 4.14 \times 10^{22} \text{ CuO fu} \times \frac{2 \ NH_3 \text{ molecules}}{3 \text{ CuO fu}} =$$

$$\underline{\underline{2.76 \times 10^{22} \ NH_3 \text{ molecules}}}$$

Example 2-15: How many formula units of CuO are required to react completely with 2.76×10^{22} molecules of NH_3?

$$? \text{ CuO fu} = 2.76 \times 10^{22} \ NH_3 \text{ molecules} \times \frac{3 \text{ CuO fu}}{2 \ NH_3 \text{ molecules}} =$$

$$\underline{\underline{4.14 \times 10^{22} \text{ CuO fu}}}$$

Note that the unit factors, which represent the chemical equivalence between NH_3 and CuO, are inverted in Examples 2-14 and 2-15.

Example 2-16: How many copper atoms can be produced from the reaction of an excess of CuO with 2.76×10^{22} NH_3 molecules?

$$? \text{ Cu atoms} = 2.76 \times 10^{22} \ NH_3 \text{ molecules} \times \frac{3 \text{ Cu atoms}}{2 \ NH_3 \text{ molecules}} =$$

$$\underline{\underline{4.14 \times 10^{22} \text{ Cu atoms}}}$$

(You might ask yourself why the answers to Examples 2-15 and 2-16 are numerically equal. Refer back to the balanced equation and the answer is obvious.)

Example 2-17: How many grams of N_2 can be produced from the reaction of 4.14×10^{22} formula units of CuO with an excess of NH_3?

$$? \text{ g } N_2 = 4.14 \times 10^{22} \text{ CuO fu} \times \frac{1 \text{ mol CuO}}{6.02 \times 10^{23} \text{ CuO fu}} \times \frac{1 \text{ mol } N_2}{3 \text{ mol CuO}}$$

$$\times \frac{28.01 \text{ g } N_2}{1 \text{ mol } N_2} = \underline{\underline{0.642 \text{ g } N_2}}$$

Example 2-18: How many moles of N_2 can be produced from the reaction of 4.14×10^{22} formula units of CuO with an excess of NH_3?

$$? \text{ mol } N_2 = 4.14 \times 10^{22} \text{ CuO fu} \times \frac{1 \text{ mol CuO}}{6.02 \times 10^{23} \text{ CuO fu}} \times \frac{1 \text{ mol } N_2}{3 \text{ mol CuO}}$$

$$= \underline{\underline{0.0229 \text{ mol } N_2}}$$

Example 2-19: How many moles of N_2 can be prepared by the reaction of 5.24 grams of CuO with an excess of ammonia?

$$? \text{ mol } N_2 = 5.24 \text{ g CuO} \times \frac{1 \text{ mol CuO}}{79.55 \text{ g CuO}} \times \frac{1 \text{ mol } N_2}{3 \text{ mol CuO}} = \underline{\underline{0.0220 \text{ mol } N_2}}$$

Example 2-20: How many grams of water can be produced from the reaction of 6.67 grams of NH_3 with an excess of CuO?

$$\underline{?} \text{ g } H_2O = 6.67 \text{ g } NH_3 \times \frac{1 \text{ mol } NH_3}{17.03 \text{ g } NH_3} \times \frac{3 \text{ mol } H_2O}{2 \text{ mol } NH_3} \times \frac{18.02 \text{ g } H_2O}{1 \text{ mol } H_2O} = \underline{\underline{10.6 \text{ g } H_2O}}$$

Alternate solution:

$$\underline{?} \text{ g } H_2O = 6.67 \text{ g } NH_3 \times \frac{3(18.02) \text{ g } H_2O}{2(17.03) \text{ g } NH_3} = \underline{\underline{10.6 \text{ g } H_2O}}$$

Example 2-21: How many pounds of copper can be prepared by the reaction of 8.44 pounds of copper(II) oxide with an excess of ammonia?

$$\underline{?} \text{ lb Cu} = 8.44 \text{ lb CuO} \times \frac{3(63.55) \text{ lb Cu}}{3(79.55) \text{ lb CuO}} = \underline{\underline{6.74 \text{ lb Cu}}}$$

2-8 Limiting Reagent

In the previous examples one of the reactants was in excess and the other (the limiting reagent) reacted completely. Or, in other cases, the minimum amount of one reactant necessary for complete reaction with the other was calculated. In cases when nonstoichiometric quantities of reactants are mixed it is often necessary to determine which reactant is the limiting reagent, i.e., the one that will be used up completely, and therefore the reactant that limits the extent to which the reaction can occur.

Example 2-22: If 5.40 grams of sodium sulfide and 16.0 grams of bismuth nitrate are dissolved in separate beakers of water which are then poured together, what is the maximum mass of bismuth sulfide, an insoluble compound, that could precipitate?

Let us interpret the balanced equation for this reaction:

$3\ Na_2S$	+	$2\ Bi(NO_3)_3$	$\longrightarrow$	$6\ NaNO_3$	+	$Bi_2S_3\ (s)$
3 mol		2 mol		6 mol		1 mol
3(78.04 g)		2(395.0 g)		6(85.00 g)		514.1 g

To identify the limiting reagent one of two calculations may be carried out. We may determine how much $Bi(NO_3)_3$ is necessary to react completely with 5.40 grams of Na_2S.

$$\underline{?}\ g\ Bi(NO_3)_2 = 5.40\ g\ Na_2S \times \frac{1\ mol\ Na_2S}{78.04\ g\ Na_2S} \times \frac{2\ mol\ Bi(NO_3)_3}{3\ mol\ Na_2S}$$

$$\times \frac{395.0\ g\ Bi(NO_3)_3}{1\ mol\ \ Bi(NO_3)_3} = 18.2\ g\ Bi(NO_3)_3$$

Since only 16.0 grams of $Be(NO_3)_3$ are available it is impossible for all 5.40 grams of Na_2S to react. Bismuth nitrate is the limiting reagent and no more bismuth sulfide can be produced after the 16.0 grams of $Bi(NO_3)_3$ have reacted.

Alternatively, we may determine how much sodium sulfide is necessary to react with 16.0 grams of bismuth nitrate.

$$\underline{?}\ g\ Na_2S = 16.0\ g\ Bi(NO_3)_3 \times \frac{1\ mol\ Bi(NO_3)_3}{395.0\ g\ Bi(NO_3)_3} \times \frac{3\ mol\ Na_2S}{2\ mol\ Bi(NO_3)_3}$$

$$\times \frac{78.04\ g\ Na_2S}{1\ mol\ \ Na_2S} = 4.74\ g\ Na_2S$$

This calculation only reinforces the first one. It tells us that 16.0 g of $Bi(NO_3)_3$ react with only 4.74 grams of Na_2S. Since there are 5.40 grams of Na_2S, some of it remains unreacted, and we conclude that $Bi(NO_3)_3$ is the limiting reagent. The calculation of the amount of bismuth sulfide precipitated must be based on the amount of bismuth nitrate available.

$$\underline{?}\ g\ Bi_2S_3 = 16.0\ g\ Bi(NO_3)_3 \times \frac{1\ mol\ Bi(NO_3)_3}{395.0\ g\ Bi(NO_3)_3} \times \frac{1\ mol\ Bi_2S_3}{2\ mol\ Bi(NO_3)_3}$$

$$\times \frac{514.1\ g\ Bi_2S_3}{1\ mol\ Bi_2S_3} = \underline{\underline{10.4\ g\ Bi_2S_3}}$$

2-9 Theoretical Yield and Percentage Yield

In previous examples we have assumed that all of at least one of the reactants is completely converted to products. However, in many instances some of each of the reactants remains unreacted. In other instances one or more competing side reactions may use up some of the reactants, thus lowering the amount of desired product obtained from given amounts of reactants.

The theoretical yield of a product in a reaction is defined as the amount of product which would be obtained if 100% of at least one reactant were converted to products in the desired reaction and if 100% recovery of the desired pure product were achieved. The actual yield is the amount of desired pure product actually obtained. The percentage yield is defined in terms of theoretical and actual yields.

$$\% \text{ yield} = \frac{\text{actual yield}}{\text{theoretical yield}} \times 100\%$$

Example 2-23: What are the theoretical and percent yields of elemental sulfur in the following reaction if 4.05 grams of sulfur are obtained from the reaction of 3.14 grams of H_2S with 6.00 grams of SO_2?

$$2\,H_2S \quad + \quad SO_2 \quad \longrightarrow \quad 2\,H_2O \quad + \quad 3\,S$$

We determine the limiting reagent first by calculating the mass of H_2S that can react with 6.00 g SO_2.

$2\,H_2S$	+	SO_2	$\longrightarrow$	$2\,H_2O$	+	$3\,S$
2 mol		1 mol		2 mol		3 mol
2(34.1 g)		64.1 g		2(18.0 g)		3(32.1 g)

$$\underline{?}\text{ g } H_2S = 6.00\text{ g } SO_2 \times \frac{1\text{ mol } SO_2}{64.1\text{ g } SO_2} \times \frac{2\text{ mol } H_2S}{1\text{ mol } SO_2} \times \frac{34.1\text{ g } H_2S}{1\text{ mol } H_2S} = 6.38\text{ g } H_2S$$

This calculation shows that 6.38 g of H_2S are necessary to react with 6.00 g SO_2, while only 3.14 g of H_2S are present. Therefore H_2S is the limiting reagent, and we calculate the theoretical yield of sulfur assuming all the H_2S reacts.

$$\underline{?}\text{ g S} = 3.14\text{ g } H_2S \times \frac{1\text{ mol } H_2S}{34.1\text{ g } H_2S} \times \frac{3\text{ mol S}}{2\text{ mol } H_2S} \times \frac{32.1\text{ g S}}{1\text{ mol S}} = \underline{\underline{4.43\text{ g S}}}$$

(theoretical)

$$\% \text{ yield} = \frac{\text{actual yield}}{\text{theoretical yield}} \times 100\% = \frac{4.05\text{ g S}}{4.43\text{ g S}} \times 100\% = \underline{\underline{91.4\% \text{ yield}}}$$

2-10 Percentage Purity

Very few substances can be obtained in exactly 100% purity. If commercially purchased anhydrous $CoCl_2$ is 99.9% pure, this minimum purity is stated on the label of the container. The remaining 0.1% could be made up of impurities such as iron, manganese, nickel, copper, zinc, bromine, iodine, and water.

Example 2-24: How much calcium carbonate and magnesium carbonate are contained in 521 grams of a mixture that is 88.2% $CaCO_3$ and 11.8% $MgCO_3$?

$$\underline{?}\text{ g } CaCO_3 = 521 \text{ g mixture} \times \frac{88.2 \text{ g } CaCO_3}{100 \text{ g mixture}} = \underline{\underline{460 \text{ g } CaCO_3}}$$

$$\underline{?}\text{ g } MgCO_3 = 521 \text{ g mixture} \times \frac{11.8 \text{ g } MgCO_3}{100 \text{ g mixture}} = \underline{\underline{61 \text{ g } MgCO_3}}$$

Or, the amount of $MgCO_3$ could be obtained by difference since the sample is known to contain only $CaCO_3$ and $MgCO_3$.

$$\underline{?}\text{ g } MgCO_3 = 521 \text{ g} - 460 \text{ g} = \underline{\underline{61 \text{ g } MgCO_3}}$$

Example 2-25: How many kilograms of a mixture of the composition described in Example 2-24 contain 3.50 kilograms of $CaCO_3$?

$$\underline{?}\text{ kg mixture} = 3.50 \text{ kg } CaCO_3 \times \frac{100 \text{ kg mixture}}{88.2 \text{ kg } CaCO_3} = \underline{\underline{3.97 \text{ kg mixture}}}$$

2-11 Simultaneous Reactions

Often two or more substances are difficult to separate because they have similar chemical properties. However, their chemical similarity can be used as a basis for determining the percent purity or percent composition of the sample, as the next example illustrates.

Example 2-26: A 12.61 gram sample of anhydrous solid consists of only $MgSO_3$ and $CaSO_3$. Both undergo thermal decomposition reactions to liberate gaseous SO_2 at high temperatures. Analysis indicates that complete thermal decomposition of the sample leaves 5.41 grams of solid residue. What weights of $MgSO_3$ and $CaSO_3$ were present in the sample?

$$CaSO_3\,(s) \xrightarrow{\Delta} CaO\,(s) + SO_2\,(g)$$

$$MgSO_3\,(s) \xrightarrow{\Delta} MgO\,(s) + SO_2\,(g)$$

Let us interpret the balanced equations.

$CaSO_3$	$\xrightarrow{\Delta}$	CaO	+	SO_2
1 mol		1 mol		1 mol
120.2 g		56.1 g		64.1 g

$$MgSO_3 \xrightarrow{\Delta} MgO + SO_2$$

$MgSO_3$	MgO	SO_2
1 mol	1 mol	1 mol
104.4 g	40.3 g	64.1 g

We do not know how much of the 12.61 grams of original sample is $CaSO_3$ and how much is $MgSO_3$. Let us arbitrarily define the amount of one of them, say $CaSO_3$, to be x g. The amount of the other is (12.61 - x)g.

Let: $x = g\ CaSO_3$,

$(12.61 - x) = g\ MgSO_3$

Since the decomposition is complete, no $CaSO_3$ or $MgSO_3$ remains and the 5.41 grams of solid residue is a mixture of CaO and MgO. The total mass of gaseous SO_2 liberated is the difference between the weight of the original sample and the residue.

$$\underline{?}\,g\ SO_2 = (12.61\ g - 5.41)g = 7.20\ g\ SO_2$$

Because SO_2 is produced in both reactions we are able to relate the total mass of SO_2 produced to the amounts produced by the individual reactions.

$$\text{total g } SO_2 = \text{g } SO_2 \text{ from } CaSO_3 + \text{g } SO_2 \text{ from } MgSO_3$$

$$7.20\ g\ SO_2 = \underbrace{(x\ g\ CaSO_3)\left(\frac{1\ mol\ CaSO_3}{120.2\ g\ CaSO_3}\right)\left(\frac{1\ mol\ SO_2}{1\ mol\ CaSO_3}\right)\left(\frac{64.1\ g\ SO_2}{1\ mol\ SO_2}\right)}_{g\ SO_2\ \text{from}\ CaSO_3}$$

$$+ \underbrace{[(12.61 - x)g\ MgSO_3]\left(\frac{1\ mol\ MgSO_3}{104.4\ g\ MgSO_3}\right)\left(\frac{1\ mol\ SO_2}{1\ mol\ MgSO_3}\right)\left(\frac{64.1\ g\ SO_2}{1\ mol\ SO_2}\right)}_{g\ SO_2\ \text{from}\ MgSO_3}$$

Clearing fractions gives:

$$7.20 = 0.533x + 7.74 - 0.614x$$

Solving for x gives the mass of $CaSO_3$ in the sample.

$$0.081\ x = 0.54$$

$$x = \underline{\underline{6.67\ g\ CaSO_3}}$$

$$\underline{?}\,g\ MgSO_3 = 12.61\ g - 6.67\ g = \underline{\underline{5.94\ g\ MgSO_3}}$$

SOLUTIONS

Solutions are homogeneous mixtures consisting of a solvent (dissolving medium) and one or more solutes (dissolved species). The term concentration refers to the amount of solute in a given amount of solvent or solution. Two common methods of expressing concentrations are percent by mass of solute and molarity ($\underline{M}$).

2-12 Percent by Mass

The percent by mass of a solute in solution is 100% times the mass of solute divided by the mass of solution in which it is contained.

$$\text{weight \%} = \frac{\text{g solute}}{\text{g solution}} \times 100\%$$

For example, 100 grams of a 15.0% solution of NaCl contains 15.0 grams of NaCl and 85.0 grams of H_2O.* From this information alone six factors can be constructed:

$$\frac{15.0 \text{ g NaCl}}{100 \text{ g solution}} \quad \text{and its reciprocal}$$

$$\frac{85.0 \text{ g } H_2O}{100 \text{ g solution}} \quad \text{and its reciprocal}$$

$$\frac{15.0 \text{ g NaCl}}{85.0 \text{ g } H_2O} \quad \text{and its reciprocal}$$

In each case the numerator and the denominator represent equal amounts of solution. A few examples illustrate the use of these factors.

Example 2-27: How many grams of NaCl are contained in 125 grams of a 15.0%** aqueous solution of NaCl?

$$\underline{?} \text{ g NaCl} = 125 \text{ g solution} \times \frac{15.0 \text{ g NaCl}}{100 \text{ g solution}} = \underline{\underline{18.8 \text{ g NaCl}}}$$

---

*Unless we specify otherwise, solutions may be assumed to be aqueous solutions.

**Unless we specifiy otherwise, percent refers to percent by mass.

Example 2-28: How many grams of water are contained in 125 grams of a 15.0% aqueous solution of NaCl?

$$\underline{?}\text{ g } H_2O = 125\text{ g solution} \times \frac{85.0\text{ g } H_2O}{100\text{ g solution}} = \underline{\underline{106\text{ g } H_2O}}$$

Example 2-29: What mass of a 15.0% solution of NaCl contains 75.0 grams of NaCl?

$$\underline{?}\text{ g solution} = 75.0\text{ g NaCl} \times \frac{100\text{ g solution}}{15.0\text{ g NaCl}} = \underline{\underline{500\text{ g solution}}}$$

Amounts of solutions are usually measured by volume in graduated cylinders or in burets rather than by weight. The next two examples illustrate how the density of a solution can be of use in measuring volumes, rather than masses, of solutions of known percent by mass.

Example 2-30: The density of a 15.0% solution of NaCl is 1.108 g/mL. How many milliliters contain 30.0 grams of NaCl?

The density tells us that one milliliter of solution weighs 1.108 grams. This information can be used to construct two unit factors, $\frac{1.108\text{ g solution}}{\text{mL solution}}$ and $\frac{1\text{ mL solution}}{1.108\text{ g solution}}$. The latter is used here.

$$\underline{?}\text{ mL solution} = 30.0\text{ g NaCl} \times \frac{100\text{ g solution}}{15.0\text{ g NaCl}} \times \frac{1\text{ mL solution}}{1.108\text{ g solution}} = \underline{\underline{181\text{ mL solution}}}$$

Example 2-31: How many grams of NaCl are contained in 300 mL of a 15.0% solution of NaCl which has a density of 1.108 g/mL?

$$\underline{?}\text{ g NaCl} = 300\text{ mL solution} \times \frac{1.108\text{ g solution}}{\text{mL solution}} \times \frac{15.0\text{ g NaCl}}{100\text{ g solution}} = \underline{\underline{49.9\text{ g NaCl}}}$$

2-13 Molarity (Molar Concentration)

The molarity ($\underline{M}$) of a solute is the number of moles of solute per liter of solution.

$$\underline{M} = \frac{\text{moles solute}}{\text{liter solution}} = \frac{\text{millimoles solute}}{\text{milliliter solution}}$$

This is the method most often used for expressing concentrations of solutions in chemistry laboratories.

Example 2-32: How many grams of $Ca(OH)_2$ must be dissolved in water to give 250 mL of 0.0200 molar $Ca(OH)_2$ solution?

The formula weight of $Ca(OH)_2$ is 74.1 so 74.1 g $Ca(OH)_2$/1 mole $Ca(OH)_2$ and its reciprocal are unit factors. We also know that since the solution is 0.0200 $\underline{M}$ $Ca(OH)_2$, other unit factors are 0.0200 mol $Ca(OH)_2$/liter solution and its reciprocal.

$$\underline{?}\ g\ Ca(OH)_2 = 0.250\ L\ solution \times \frac{0.0200\ mol\ Ca(OH)_2}{L\ solution} \times \frac{74.1\ g\ Ca(OH)_2}{mol\ Ca(OH)_2}$$

$$= \underline{\underline{0.371\ g\ Ca(OH)_2}}$$

Example 2-33: How many moles and how many grams of $Ca(OH)_2$ are contained in 500 mL of 0.0200 $\underline{M}$ $Ca(OH)_2$?

$$\underline{?}\ mol\ Ca(OH)_2 = 500\ mL\ solution \times \frac{0.0200\ mol\ Ca(OH)_2}{1000\ mL} =$$

$$\underline{\underline{0.0100\ mol\ Ca(OH)_2}}$$

$$\underline{?}\ g\ Ca(OH)_2 = 0.0100\ mol\ Ca(OH)_2 \times \frac{74.1\ g\ Ca(OH)_2}{mol\ Ca(OH)_2} =$$

$$\underline{\underline{0.741\ g\ Ca(OH)_2}}$$

Example 2-34 illustrates how we can calculate the molarity of a solution if we know its weight percent and density.

Example 2-34: What is the molarity of a 20.0% solution of sodium nitrate, $NaNO_3$? Its density is 1.143 g/mL.

We use a series of unit factors to convert from grams of $NaNO_3$/gram of solution to moles of $NaNO_3$/liter of solution.

$$\frac{?\ \text{mol NaNO}_3}{\text{L solution}} = \frac{20.0\ \text{g NaNO}_3}{100\ \text{g solution}} \times \frac{1.143\ \text{g solution}}{\text{mL solution}} \times \frac{1000\ \text{mL}}{\text{L}} \times \frac{1\ \text{mol NaNO}_3}{85.0\ \text{g NaNO}_3}$$

$$= \frac{2.69\ \text{mol NaNO}_3}{\text{L solution}} = 2.69\ \underline{\text{M}}\ \text{NaNO}_3$$

Thus a 20.0% solution of $NaNO_3$ is also 2.69 M $NaNO_3$.

2-14 Dilution and Volumes of Solutions Required for Reactions

Dilution is the process of making a solution less concentrated by adding some of the solution to more solvent or mixing two solutions that contain the same solute. In all dilutions, the number of moles of solute remains the same. The number of moles (or millimoles) of solute in a given volume of a solution of known concentration can always be determined by multiplying the volume times the molarity.

$$\text{liters} \times \frac{\text{moles solute}}{\text{liter}} = \text{moles solute}$$

$$\text{Volume} \times \text{Molarity} = \text{amount of solute}$$

$$\text{milliliters} \times \frac{\text{millimoles solute}}{\text{milliliter}} = \text{millimoles solute}$$

After dilution, the new molarity is simply the number of moles of solute present divided by the total volume of solution. This is illustrated in Example 2-35.

Example 2-35: What is the molarity of the solution resulting from the addition of 600 mL of 2.00 M K_2SO_4 solution to 300 mL of water?

The number of moles of K_2SO_4 present is:

$$?\ \text{mol K}_2\text{SO}_4 = 600\ \text{mL} \times \frac{2.00\ \text{mol K}_2\text{SO}_4}{1000\ \text{mL}} = 1.20\ \text{mol K}_2\text{SO}_4$$

Thus, 1.20 moles of K_2SO_4 are contained in 300 mL + 600 mL = 900 mL or 0.900 L of solution after dilution and the molarity is:

$$\frac{?\ \text{mol K}_2\text{SO}_4}{\text{L solution}} = \frac{1.20\ \text{mol K}_2\text{SO}_4}{0.900\ \text{L solution}} = 1.33\ \underline{\text{M}}\ \text{K}_2\text{SO}_4$$

Since the number of moles of solute does not change and since liters x (moles/liter) = moles, the following relationship applies in the dilution process. The subscripts "1" and "2" refer to the solution before and after dilution.

$$L_1 \times \underline{M}_1 = L_2 \times \underline{M}_2 \qquad \text{or} \qquad mL_1 \times \underline{M}_1 = mL_2 \times \underline{M}_2$$

Example 2-36: What is the molarity of the solution prepared by mixing 750 mL of water and 350 mL of 3.00 $\underline{M}$ HCl?

$$mL_1 \times \underline{M}_1 = mL_2 \times \underline{M}_2$$

We must calculate the final molarity, $\underline{M}_2$, and since we already know the other terms, we solve for $\underline{M}_2$ and substitute in the other values.

$$\underline{M}_2 = \frac{mL_1 \times \underline{M}_1}{mL_2} = \frac{350\ mL \times 3.00\ \underline{M}\ HCl}{1100\ mL} = 0.955\ \underline{M}\ HCl$$

If two solutions containing the same solute in different concentrations are mixed, we must calculate the total number of moles of solute present, and then divide by the total volume of solution (after mixing) to determine the new molarity. Example 2-37 shows such a calculation.

Example 2-37: What is the molarity of the solution produced by mixing 1200 mL of 6.00 $\underline{M}$ NH_4Cl with 200 mL of 1.00 $\underline{M}$ NH_4Cl?

We first calculate the total number of moles of NH_4Cl in each solution before mixing.

$$\underline{?}\ mol\ NH_4Cl = 1200\ mL \times \frac{6.00\ mol\ NH_4Cl}{1000\ mL} = 7.20\ mol\ NH_4Cl$$

$$\underline{?}\ mol\ NH_4Cl = 200\ mL \times \frac{1.00\ mol\ NH_4Cl}{1000\ mL} = 0.200\ mol\ NH_4Cl$$

The final solution contains 7.20 moles + 0.200 moles = 7.40 moles NH_4Cl. The total volume is 1200 mL + 200 mL = 1400 mL or 1.40 L, and so the new molarity is:

$$\frac{?\ mol\ NH_4Cl}{L} = \frac{7.40\ mol\ NH_4Cl}{1.40\ L} = 5.29\ \underline{M}\ NH_4Cl$$

Example 2-38: (a) What volume of 0.420 M potassium chromate, K_2CrO_4, solution is required to react with an aqueous solution containing 3.50 grams of silver nitrate, $AgNO_3$, according to the following equation? (b) What mass of silver chromate precipitates, assuming the reaction goes to completion?

$$2\ AgNO_3\ (aq) + K_2CrO_4\ (aq) \longrightarrow Ag_2CrO_4\ (s) + 2\ KNO_3\ (aq)$$

(a) First, we interpret the balanced equation.

$2\ AgNO_3$ (aq)	+	K_2CrO_4 (aq)	$\longrightarrow$	Ag_2CrO_4 (s)	+	$2\ KNO_3$ (aq)
2 mol		1 mol		1 mol		2 mol
2(169.9 g)				331.8 g		

Now we calculate the number of moles of $AgNO_3$ in 3.50 g $AgNO_3$.

$$?\ \text{mol } AgNO_3 = 3.50\ \text{g } AgNO_3 \times \frac{1\ \text{mol } AgNO_3}{169.9\ \text{g } AgNO_3} = 0.0206\ AgNO_3$$

We use the information obtained from the balanced equation to convert from moles of $AgNO_3$ to moles of K_2CrO_4.

$$?\ \text{mol } K_2CrO_4 = 0.0206\ \text{mol } AgNO_3 \times \frac{1\ \text{mol } K_2CrO_4}{2\ \text{mol } AgNO_3}$$

$$= 0.0103\ \text{mol } K_2CrO_4$$

Finally, we use a unit factor constructed from the molar concentration of the K_2CrO_4 solution to determine the volume of the solution that contains 0.0103 moles of K_2CrO_4.

$$?\ \text{L } K_2CrO_4\ \text{sol'n} = 0.0103\ \text{mol } K_2CrO_4 \times \frac{1\ \text{L } K_2CrO_4\ \text{sol'n}}{0.420\ \text{mol } K_2CrO_4}$$

$$= \underline{\underline{0.0245\ \text{L}}} = \underline{\underline{24.5\ \text{mL } K_2CrO_4\ \text{sol'n}}}$$

If we combine all these steps into one expression we have

$$?\ \text{L } K_2CrO_4\ \text{sol'n} = 3.50\ AgNO_3 \times \frac{1\ \text{mol } AgNO_3}{169.9\ \text{g } AgNO_3} \times \frac{1\ \text{mol } K_2CrO_4}{2\ \text{mol } AgNO_3}$$

$$\times \frac{1\ \text{L } K_2CrO_4\ \text{sol'n}}{0.420\ \text{mol } K_2CrO_4} = \underline{\underline{0.0245\ \text{L } K_2CrO_4\ \text{sol'n}}}$$

(b) We calculate the mass of Ag_2CrO_4 that precipitates from the complete reaction of 3.50 grams of $AgNO_3$ by using the information obtained from the balanced equation for the reaction.

$$? \text{ g } Ag_2CrO_4 = 3.50 \text{ g } AgNO_3 \times \frac{1 \text{ mol } AgNO_3}{169.9 \text{ g } AgNO_3} \times \frac{1 \text{ mol } Ag_2CrO_4}{2 \text{ mol } AgNO_3}$$

$$\times \frac{331.8 \text{ g } Ag_2CrO_4}{1 \text{ mol } Ag_2CrO_4}$$

$$= \underline{\underline{3.42\ Ag_2CrO_4}}$$

Exercises

1. Calculate:

   (a) the number of phosphorus atoms in 3.00 moles of P_4.
   (b) the number of grams in 3.00 moles of P_4.
   (c) the number of moles of P_4 in 10.0 grams of P_4.
   (d) the number of atoms of phosphorus in 20.0 grams of phosphorus.

2. Calculate:

   (a) the number of moles of magnesium in 26.4 grams of magnesium.
   (b) the number of grams in 2.86 moles of sodium atoms.
   (c) the number of aluminum atoms in 32.0 grams of aluminum.
   (d) the number of moles of calcium atoms in 8.42×10^{24} calcium atoms.
   (e) the number of atomic mass units in 1.00 gram of silicon.

3. Calculate the molecular weight of:

   (a) methane, CH_4
   (b) ethane, C_2H_6
   (c) phosphoric acid, H_3PO_4
   (d) sulfuric acid, H_2SO_4
   (e) boron trifluoride, BF_3
   (f) benzene, C_6H_6

4. Calculate the formula weight of:

   (a) rubidium chloride, RbCl
   (b) ammonium bromide, NH_4Br
   (c) sodium carbonate, Na_2CO_3
   (d) aluminum sulfate, $Al_2(SO_4)_3$
   (e) nickel sulfate hexahydrate, $NiSO_4 \cdot 6H_2O$
   (f) parathion (an insecticide), $C_{10}H_{14}SNO_5P$?

5. Calculate:

   (a) the number of moles of nitric acid, HNO_3, in 10.0 grams of HNO_3.
   (b) the number of moles of calcium chloride, $CaCl_2$, in 10.0 grams of $CaCl_2$.
   (c) the number of grams of sodium phosphate, Na_3PO_4, in 13.6 moles of Na_3PO_4.
   (d) the number of nitrogen dioxide, NO_2, molecules in 10.0 moles of NO_2.
   (e) the number of sulfur trioxide, SO_3, molecules in 15.0 grams of SO_3.

(f) the number of chlorine atoms in 22.0 grams of phosphorus trichloride, PCl_3.

6. Calculate:

(a) the number of bromide ions, Br^-, in 6.00 moles of sodium bromide, NaBr, an ionic compound.
(b) the number of sulfate ions, SO_4^{2-}, in 8.13 grams of potassium sulfate, K_2SO_4, an ionic compound.
(c) the number of grams of boron trichloride, BCl_3, in 5.12 moles of BCl_3.
(d) the number of moles of ammonium ions, NH_4^+, in 1.50 grams of ammonium sulfate, $(NH_4)_2SO_4$, an ionic compound.
(e) the number of moles of NH_4^+ ions in 1.50 moles of $(NH_4)_2SO_4$.
(f) the number of moles of lithium carbonate, Li_2CO_3, in 23.0 grams of Li_2CO_3, an ionic compound.

7. Calculate the percentage composition of the following compounds.

(a) KCl
(b) Na_2O
(c) $HClO_4$
(d) $K_2Cr_2O_7$
(e) $Cr(OH)_3$
(f) $Ni(CH_3COO)_2$
(g) $NiSO_4 \cdot 6H_2O$
(h) $C_6H_5NO_2$

8. A compound was found to be 27.27% carbon and 72.73% oxygen by mass. What is the simplest formula for the compound?

9. A compound is 16.53% silicon and 83.47% chlorine by mass. What is its simplest formula?

10. A 2.1640 gram sample of a solid compound was found to contain 0.7033 gram of iron, 0.6548 gram of chromium and 0.8059 gram of oxygen. What is the simplest formula for the compound?

11. How many grams of potassium, manganese, and oxygen are contained in 6.216 grams of potassium permanganate, $KMnO_4$?

12. A compound is 85.62% carbon and 14.38% hydrogen by mass. Its molecular weight is 56.064. What is its molecular (true) formula?

13. A 3.5471 gram sample of a compound contains 0.1153 gram of hydrogen, 1.6018 grams of nitrogen and 1.8300 grams of oxygen. One mole has a mass of 62.032 grams. What is the true formula of the compound?

14. A compound is 12.72% magnesium, 37.08% chlorine, and 50.20% oxygen by mass. Its formula weight is 191.2 g/mol. What is its true formula?

15. Supply coefficients to balance the following equations.

(a) $Mg + Br_2 \longrightarrow MgBr_2$

(b) $Cr + Cl_2 \longrightarrow CrCl_3$

(c) $NH_3 + H_2SO_4 \longrightarrow (NH_4)_2SO_4$

(d) $Mg(OH)_2 + H_3AsO_4 \longrightarrow Mg_3(AsO_4)_2 + H_2O$

(e) $Ag_2N_2O_2 + HCl \longrightarrow AgCl + H_2N_2O_2$

(f) $NH_4NO_3 \longrightarrow N_2O + H_2O$

(g) $P_4 + NO \longrightarrow P_4O_6 + N_2$

(h) $PCl_3 + H_2O \longrightarrow H_3PO_3 + HCl$

(i) $C_4H_{10} + O_2 \longrightarrow CO_2 + H_2O$

(j) $C_4H_{10}O + O_2 \longrightarrow CO_2 + H_2O$

(k) $CO + H_2 \longrightarrow CH_3OH$

(l) $CO_2 + H_2 \longrightarrow CH_3OH + H_2O$

(m) $NaBF_4 + H_2O \longrightarrow H_3BO_3 + NaF + HF$

(n) $BF_3 + LiAlH_4 \longrightarrow B_2H_6 + LiF + AlF_3$

(o) $SiO_2 + Na_2CO_3 \longrightarrow Na_2SiO_3 + CO_2$

(p) $SiF_4 + H_2O \longrightarrow H_4SiO_4 + H_2SiF_6$

(q) $Cu(OH)_2 + NH_3 \longrightarrow Cu(NH_3)_4(OH)_2$

(r) $Sb_2S_3 + Fe \longrightarrow Sb + FeS$

(s) $SO_2 + O_2 + H_2O \longrightarrow H_2SO_4$

(t) $Zn + HNO_3 \longrightarrow Zn(NO_3)_2 + NH_4NO_3 + H_2O$

(u) $Cu + HNO_3 \longrightarrow Cu(NO_3)_2 + NO_2 + H_2O$

(v) $Cu + HNO_3 \longrightarrow Cu(NO_3)_2 + NO + H_2O$

16. Oxygen can be generated in the laboratory by thermal decomposition of potassium chlorate, $KClO_3$.

$$2\ KClO_3\ (s) \xrightarrow{\Delta} 2\ KCl\ (s) + 3\ O_2\ (g)$$

(a) How many grams of $KClO_3$ must be decomposed to produce 3.00 moles of O_2?

(b) How many grams of oxygen, O_2, can be prepared by complete decomposition of 2.56 grams of $KClO_3$?

(c) How many moles of $KClO_3$ must be decomposed to prepare 10.0 grams of O_2?

(d) How many grams of KCl are produced during a decomposition that also produces 10.0 grams of O_2?

17. Hydrogen iodide, HI, can be produced by the reaction of sodium iodide, NaI, with phosphoric acid, H_3PO_4.

$$3\ NaI\ (s) + H_3PO_4\ (\ell) \xrightarrow{\Delta} 3\ HI\ (g) + Na_3PO_4\ (s)$$

(a) How many moles of HI can be produced by the reaction of 10.0 grams of NaI with excess H_3PO_4?
(b) How many grams of HI can be prepared by the reaction of 2.00 moles of NaI with an excess of H_3PO_4?
(c) How many grams of H_3PO_4 are required to react completely with 6.40 grams of NaI?
(d) How many grams of Na_3PO_4 are produced in a reaction that also produces 3.66 grams of HI?

18. What is the maximum mass of gaseous chlorine dioxide, ClO_2, that could be prepared from the reaction of 56.0 grams of Cl_2 with 63.4 grams of sodium chlorite, $NaClO_2$?

$$2\ NaClO_2 + Cl_2 \longrightarrow 2\ NaCl + ClO_2$$

19. Refer to question 18. What is the maximum mass of ClO_2 that could be produced from 37.2 grams of Cl_2 and 106 grams of $NaClO_2$?

20. If 3.00 grams of dinitrogen trioxide decompose completely to nitrogen oxide and nitrogen dioxide:

(a) what will be the ratio of grams of NO to grams of NO_2 produced?
(b) what would be the ratio if 4.00 grams of N_2O_3 decomposed?
(c) what would be the ratio if 70.0% yields of both NO and NO_2 were obtained in the decomposition of 3.00 grams of N_2O_3?

$$N_2O_3 \longrightarrow NO + NO_2$$

21. If 12.3 grams of carbon tetrachloride, CCl_4, are produced from the reaction of 18.0 grams of carbon disulfide with 22.0 grams of Cl_2, what percentage yield of CCl_4 is obtained?

$$CS_2 + 3\ Cl_2 \longrightarrow CCl_4 + S_2Cl_2$$

22. If 3.68 grams of CCl_4 are produced from the reaction of 6.30 grams of CS_2 with 22.0 grams of Cl_2, what percentage yield of CCl_4 is obtained? Refer to question 21.

23. A 15.0 gram sample of calcium sulfite, $CaSO_3$, undergoes complete thermal decomposition. How many grams of sulfur dioxide, SO_2, are produced?

$$CaSO_3 \xrightarrow{\Delta} CaO + SO_2$$

24. An impure 15.0 gram sample of $CaSO_3$ undergoes thermal decomposition to produce 5.60 grams of SO_2. Assuming the impurities do not produce SO_2 when heated, what percentage of the original sample is due to $CaSO_3$? Refer to question 23.

25. A 3.46 gram sample consisting of only strontium bromide, $SrBr_2$, and barium bromide, $BaBr_2$, was treated with excess silver nitrate, $AgNO_3$, in aqueous solution. A light yellow precipitate of silver bromide, AgBr, weighing 5.13 grams was formed. Calculate the masses and percentages of $SrBr_2$ and $BaBr_2$ in the original sample.

$$SrBr_2 + 2\,AgNO_3 \longrightarrow 2\,AgBr\,(s) + Sr(NO_3)_2$$

$$BaBr_2 + 2\,AgNO_3 \longrightarrow 2\,AgBr\,(s) + Ba(NO_3)_2$$

26. A 1.83 gram mixture containing only calcium sulfite and magnesium sulfite was heated to complete decomposition. The residue, a mixture of calcium oxide and magnesium oxide, weighed 0.82 gram. What were the masses and percentages of $CaSO_3$ and $MgSO_3$ in the original sample?

$$CaSO_3\,(s) \xrightarrow{\Delta} CaO\,(s) + SO_2\,(g)$$

$$MgSO_3\,(s) \xrightarrow{\Delta} MgO\,(s) + SO_2\,(g)$$

27. If the poisonous gas phosphine, PH_3, is produced by the reaction of 5.42 grams of calcium phosphide, Ca_3P_2, with excess water, what is the maximum mass of phosphoric acid, H_3PO_4, that could be prepared by reaction of the phosphine produced with excess oxygen, O_2?

$$Ca_3P_2\,(s) + 6\,H_2O\,(\ell) \longrightarrow 2\,PH_3\,(g) + 3\,Ca(OH)_2\,(aq)$$

$$PH_3\,(g) + 2\,O_2\,(g) \longrightarrow H_3PO_4\,(\ell)$$

28. Consider the reactions of question 27. If the first reaction produces an 87.4% yield of phosphine, PH_3, and the second gives a 74.2% yield, what mass of H_3PO_4 will be produced?

29. Consider the reactions of question 27. If 3.44 grams of phosphoric acid, H_3PO_4, are actually obtained from 5.42 grams of calcium phosphide, Ca_3P_2, what is the overall percentage yield?

30. An aqueous solution is 18.0 percent calcium bromide, $CaBr_2$, by mass.

(a) How many grams of $CaBr_2$ are contained in 150 grams of solution?
(b) How many grams of water are contained in 125 grams of solution?
(c) How many grams of solution contain 75.0 grams of $CaBr_2$?
(d) How many grams of solution contain 100 grams of water?

31. A 24.0 percent aqueous solution of potassium iodide, KI, has a density of 1.208 g/mL.

(a) How many grams of KI are contained in 75.0 mL of solution?
(b) How many grams of water are contained in 50.0 grams of the solution?
(c) How many grams of solution contain 35.0 grams of KI?
(d) How many milliliters of the solution contain 25.0 grams of KI?
(e) How many grams of KI are contained in 50.0 mL of solution?

32. What is the molarity of each solution described below?

(a) 25.0 grams of NaOH in 100 mL of solution
(b) 13.0 grams of $K_2Cr_2O_7$ in 250 mL of solution
(c) $CuSO_4$ solution prepared by dissolving 60.0 g of $CuSO_4 \cdot 5H_2O$ in enough water to give 800 mL of solution
(d) 38.0 grams of H_2SO_4 in 1500 mL of solution
(e) 46.4 grams of $Ca(NO_3)_2$ in 672 mL of solution

33. How many moles of solute are contained in:

(a) 250 mL of 2.00 $\underline{M}$ $CaCl_2$?
(b) 50.0 mL of 0.0500 $\underline{M}$ $Ba(OH)_2$?
(c) 1500 mL of 0.150 $\underline{M}$ $(NH_4)_2SO_4$?
(d) 175 mL of 0.200 $\underline{M}$ HCl?

34. What is the molarity of 15.00% aqueous sodium chloride, NaCl, solution? Its density is 1.111 g/mL.

35. What is the percent by mass of NH_4Cl in 1.53 $\underline{M}$ NH_4Cl whose specific gravity is 1.025?

36. How many grams of solute are contained in:

(a) 150 mL of 2.00 $\underline{M}$ NaCl?
(b) 80.0 mL of 1.33 $\underline{M}$ HBr?
(c) 100 mL of 0.750 $\underline{M}$ $NiCl_2$?
(d) 1250 mL of 0.500 $\underline{M}$ $Pb(NO_3)_2$?
(e) 300 mL of 0.100 $\underline{M}$ NH_4Br?

37. What is the molarity of the solution resulting from the addition of:

(a) 100 mL of water to 250 mL of 1.00 M NaCl?
(b) 50.0 mL of water to 500 mL of 0.750 M NH_4NO_3?
(c) 150 mL of 0.150 M HCl to 150 mL of 0.300 M HCl?
(d) 300 mL of 12.0 M HNO_3 to 100 mL of 3.00 M HNO_3?

38. What volume of 0.106 M KOH solution is just sufficient to react completely with 3.16 grams of copper(II) sulfate pentahydrate, $CuSO_4 \cdot 5H_2O$, to form insoluble copper(II) hydroxide? Solid $CuSO_4 \cdot 5H_2O$ dissolves as the KOH solution is added.

$$CuSO_4\ (aq) + 2\ KOH\ (aq) \longrightarrow Cu(OH)_2\ (s) + K_2SO_4\ (aq)$$

39. What volume of 0.600 M H_2SO_4 is required to react with an excess of barium chloride, $BaCl_2$, to form 5.00 grams of insoluble barium sulfate?

$$H_2SO_4\ (aq) + BaCl_2\ (aq) \longrightarrow BaSO_4\ (s) + 2\ HCl\ (aq)$$

ANSWERS TO EXERCISES

1. (a) 7.22×10^{24} P atoms (b) 372 g P_4 (c) 0.0806 mol P_4 (d) 3.88×10^{23} P atoms 2. (a) 1.09 mol Mg (b) 65.8 g Na (c) 7.13×10^{23} Al atoms (d) 14.0 mol Ca (e) 6.02×10^{23} amu Si 3. (a) 16.043 amu (b) 30.070 amu (c) 97.994 amu (d) 98.07 amu (e) 67.81 amu (f) 78.114 amu 4. (a) 120.92 amu (b) 97.943 amu (c) 105.988 amu (d) 342.14 amu (e) 262.85 amu (f) 291.26 amu 5. (a) 0.159 mol HNO_3 (b) 0.0901 mol $CaCl_2$ (c) 2.23×10^3 g Na_3PO_4 (d) 6.02×10^{24} NO_2 molecules (e) 1.13×10^{23} SO_3 molecules (f) 2.89×10^{23} Cl atoms 6. (a) 3.61×10^{24} Br^- ions (b) 2.81×10^{22} SO_4^{2-} ions (c) 599 g BCl_3 (d) 0.0227 mol NH_4^+ ions (e) 3.00 mol NH_4^+ ions (f) 0.311 mol Li_2CO_3 7. (a) 52.45% K, 47.55% Cl (b) 74.19% Na, 25.81% O (c) 1.01% H, 35.29% Cl, 63.71% O (d) 26.58% K, 35.35% Cr, 38.07% O (e) 50.47% Cr, 46.59% O, 2.94% H (f) 33.20% Ni, 27.17% C, 3.43% H, 36.20% O (g) 22.33% Ni, 12.20% S, 60.86% O, 4.61% H (h) 58.53% C, 4.10% H, 11.38% N, 25.99% O 8. CO_2 9. $SiCl_4$ 10. $FeCrO_4$ 11. 1.538 g K, 2.161 g Mn, 2.517 g O 12. C_4H_8 13. $H_2N_2O_2$ 14. $MgCl_2O_6$ or

$Mg(ClO_3)_2$

15. (a) $Mg + Br_2 \longrightarrow MgBr_2$

(b) $2\ Cr + 3\ Cl_2 \longrightarrow 2\ CrCl_3$

(c) $2\ NH_3 + H_2SO_4 \longrightarrow (NH_4)_2SO_4$

(d) $3\ Mg(OH)_2 + 2\ H_3AsO_4 \longrightarrow Mg_3(AsO_4)_2 + 6\ H_2O$

(e) $Ag_2N_2O_2 + 2\ HCl \longrightarrow 2\ AgCl + H_2N_2O_2$

(f) $NH_4NO_3 \longrightarrow N_2O + 2\ H_2O$

(g) $P_4 + 6\ NO \longrightarrow P_4O_6 + 3\ N_2$

(h) $PCl_3 + 3\ H_2O \longrightarrow H_3PO_3 + 3\ HCl$

(i) $2\ C_4H_{10} + 13\ O_2 \longrightarrow 8\ CO_2 + 10\ H_2O$

(j) $C_4H_{10}O + 6\ O_2 \longrightarrow 4\ CO_2 + 5\ H_2O$

(k) $CO + 2\ H_2 \longrightarrow CH_3OH$

(l) $CO_2 + 3\ H_2 \longrightarrow CH_3OH + H_2O$

(m) $NaBF_4 + 3\ H_2O \longrightarrow H_3BO_3 + NaF + 3\ HF$

(n) $4\ BF_3 + 3\ LiAlH_4 \longrightarrow 2\ B_2H_6 + 3\ LiF + 3\ AlF_3$

(o) $SiO_2 + Na_2CO_3 \longrightarrow Na_2SiO_3 + CO_2$

(p) $3\ SiF_4 + 4\ H_2O \longrightarrow H_4SiO_4 + 2\ H_2SiF_6$

(q) $Cu(OH)_2 + 4\ NH_3 \longrightarrow Cu(NH_3)_4(OH)_2$

(r) $Sb_2S_3 + 3\ Fe \longrightarrow 2\ Sb + 3\ FeS$

(s) $2\ SO_2 + O_2 + 2\ H_2O \longrightarrow 2\ H_2SO_4$

(t) $4\ Zn + 10\ HNO_3 \longrightarrow 4\ Zn(NO_3)_2 + NH_4NO_3 + 3\ H_2O$

(u) $Cu + 4\ HNO_3 \longrightarrow Cu(NO_3)_2 + 2\ NO_2 + 2\ H_2O$

(v) $3\ Cu + 8\ HNO_3 \longrightarrow 3\ Cu(NO_3)_2 + 2\ NO + 4\ H_2O$

16. (a) 245 g $KClO_3$ (b) 1.00 g O_2 (c) 0.208 mol $KClO_3$ (d) 15.5 g KCl 17. (a) 0.0667 mol HI (b) 256 g HI (c) 1.39 g H_3PO_4 (d) 1.56 g Na_3PO_4 18. $NaClO_2$ is limiting reagent; 23.6 g ClO_2

19. Cl_2 is limiting reagent; 35.4 g ClO_2

20. (a) 0.652 g NO/g NO_2 (b) same (c) same

21. Cl_2 is limiting reagent; 77.4% yield

22. CS_2 is limiting reagent; 29.0% yield

23. 8.00 g SO_2 24. 70.0% $CaSO_3$ 25. $3.0\bar{0}$ g $SrBr_2$, 0.46 g $BaBr_2$,

87% $SrBr_2$, 13% $BaBr_2$ 26. $1.3\bar{6}$ g $CaSO_3$, 0.47 g $MgSO_3$, 77% $CaSO_3$, 26% $MgSO_3$ [Note: large (additive) round off errors give percentages that total 103%] 27. 5.84 g H_3PO_4 28. 3.79 g H_3PO_4 29. 58.9% yield 30. (a) 27.0 g $CaBr_2$ (b) 103 g H_2O (c) 417 g solution (d) 122 g solution 31. (a) 21.7 g KI (b) 38.0 g H_2O (c) 146 g solution (d) 86.2 mL solution (e) 14.5 g KI 32. (a) 6.25 $\underline{M}$ NaOH (b) 0.177 $\underline{M}$ $K_2Cr_2O_7$ (c) 0.300 $\underline{M}$ $CuSO_4$ (d) 0.258 $\underline{M}$ H_2SO_4 (e) 0.421 $\underline{M}$ $Ca(NO_3)_2$ 33. (a) 0.500 mol $CaCl_2$ (b) 0.00250 mol $Ba(OH)_2$, (c) 0.225 mol $(NH_4)_2SO_4$ (d) 0.0350 mol HCl 34. 2.851 $\underline{M}$ NaCl 35. 7.99% NH_4Cl by wt. 36. (a) 17.6 g NaCl (b) 8.61 g HBr (c) 9.75 g $NiCl_2$ (d) 207 g $Pb(NO_3)_2$ (e) 2.94 g NH_4Br (a) 0.714 $\underline{M}$ NaCl (b) 0.682 $\underline{M}$ NH_4NO_3 (c) 0.225 $\underline{M}$ HCl (d) 9.75 $\underline{M}$ HNO_3 38. 239 mL 39. 35.8 mL

Chapter Three

ATOMIC STRUCTURE

D, G, W: Chap. 4 and 5

The presently accepted picture of the atom is a nuclear atom in which the protons and neutrons are located in a very small, very dense, positively charged nucleus at the center of the atom. Very lightweight electrons are diffusely distributed in space over relatively great distances from the nucleus. The radii of nuclei and atoms are approximately 10^{-4} and 10 nm, respectively. The masses and charges of the three fundamental particles are tabulated in Table 3-1.

Table 3-1. Fundamental Particles

Particle	Mass (amu)	Charge
electron (e^-)	0.00055	-1
proton (p^+)	1.0073	+1
neutron (n^o)	1.0087	0

3-1 Atomic Number

The atomic number of an element is the number of protons in the nucleus of an atom of the element. All atoms of the same element, by definition, have the same atomic number. For example, all atoms with fourteen protons are silicon atoms and all silicon atoms contain fourteen protons. The atomic number is also equal to the number of electrons in a neutral atom.

3-2 Mass Number

The mass number of an atom is the sum of the number of protons and neutrons in its nucleus, and is an integer. The notation $^{59}_{28}Ni$ refers to a nickel atom, which contains 28 protons and 31 neutrons in its nucleus.

3-3 Isotopes

Most elements occur naturally as mixtures of isotopes, that is, as atoms containing the same number of protons but different numbers of neutrons. For example, naturally occuring antimony consists of two isotopes, $^{121}_{51}Sb$ and $^{123}_{51}Sb$.

Recall that one atomic mass unit (amu) is defined as exactly 1/12 the mass of one $^{12}_{6}C$ atom. Therefore, one atom of the carbon-12 isotope has a mass of exactly 12 amu. An ^{16}O atom is 15.9949/12* times as massive as a ^{12}C atom, and so its mass is 15.9949/12 x 12 amu = 15.9949 amu. The $^{121}_{51}Sb$ isotope weighs 120.9038/12 times as much as a $^{12}_{6}C$ atom, while a $^{123}_{51}Sb$ atom is 122.9041/12 times heavier than a ^{12}C atom. Thus the masses of $^{121}_{51}Sb$ and $^{123}_{51}Sb$ isotopes are 120.9038 amu and 122.9041 amu, respectively.

3-4 Atomic Weights

Accurate and precise values of both percent natural abundances and masses of most known isotopes of the elements have been determined. The atomic weight of an element is the weighted average of the masses (weights) of its constituent isotopes. In other words, the atomic weight of an element can be thought of as the mass of an "average" atom of the element. For elements that occur as mixtures of isotopes, no individual "average" atoms exist, but a collection of naturally occuring atoms of an element may be treated as if each were an "average atom."

Example 3-1: Naturally occuring sulfur consists of the following isotopes in the indicated percentages. (Other isotopes exist, but make up only a total of 0.006% of all sulfur.) Calculate its atomic weight.

^{32}S, 31.972 amu, 95.00%, ^{33}S, 32.971 amu, 0.76%

^{34}S, 33.968 amu, 4.22%, ^{35}S, 35.967 amu, 0.014%

We obtain the contribution of each isotope, and then sum these.

^{32}S	31.972 amu	x	0.9500	=	30.37 amu
^{33}S	32.971 amu	x	0.0076	=	0.25 amu
^{34}S	33.968 amu	x	0.0422	=	1.43 amu
^{36}S	35.967 amu	x	0.00014	=	0.005 amu
			Atomic weight	=	32.06 amu

Note that atomic numbers are integers while atomic weights are not.

Example 3-2: The atomic weight of rubidium is 85.4678 amu. Rubidium occurs as a mixture of two isotopes, ^{85}Rb, mass = 84.9117 amu, and ^{87}Rb, mass = 86.9084 amu. What are the natural abundances of ^{85}Rb and ^{87}Rb?

*This 12 is defined to be an exact number.

Recall that when any quantity is expressed in terms of fractions, the sum of the fractions is unity.

Let x = fraction of one isotope, say ^{85}Rb, i.e., x = fraction of ^{85}Rb, then (1-x) = fraction of ^{87}Rb.

$$(\text{fraction}_1)(\text{mass}_1) + (\text{fraction}_2)(\text{mass}_2) = \text{atomic weight}$$

$$(x)(84.9117 \text{ amu}) + (1-x)(86.9084 \text{ amu}) = 85.4678 \text{ amu}$$

$$84.9117x + 86.9084 - 86.9084x = 85.4678$$

$$1.9967x = 1.4406$$

$$x = 0.7215 = \underline{\underline{72.15\% \ ^{85}\text{Rb}}}$$

$$(1.00 - x) = 0.2785 = \underline{\underline{27.85\% \ ^{87}\text{Rb}}}$$

3-5 Mass Defect

Experiments show that the masses of all atoms are less than the sums of the masses of the individual subatomic particles that make up the atoms. This small mass difference (mass apparently "lost" in the formation of an atom) is called the mass defect, i.e.,

$$\text{mass defect} = (\text{calculated mass}) - (\text{actual mass}).$$

Example 3-3: Calculate the mass defect for a ^{47}Ti atom. Its actual mass is 46.9518 amu.

A ^{47}Ti atom contains 22 protons, 22 electrons, and 47 - 22 = 25 neutrons. The "calculated mass" is the sum of the masses of the subatomic particles.

particle	mass		no. of particles		
proton	1.0073 amu	x	22	=	22.1606 amu
neutron	1.0087 amu	x	25	=	25.2175 amu
electron	0.00055 amu	x	22	=	0.0121 amu
	calculated mass			=	47.3902 amu

Mass defect = (calculated mass) - (actual mass) = (47.3902 - 46.9518) amu

$$= \underline{\underline{0.4384 \text{ amu}}}$$

Thus, we see that the actual mass of a ^{47}Ti atom is 0.4384 amu less than the sum of the masses of its component parts.

3-6 Binding Energy

Much of the energy emanating from the sun is produced by the combination of subatomic particles to form ^4_2He atoms at extremely high temperatures. The mass defect of ^4_2He, calculated below, is 0.0305 amu/atom.

protons:	2 x 1.0073 amu	=	2.0146 amu
neutrons:	2 x 1.0087 amu	=	2.0174 amu
electrons:	2 x 0.00055 amu	=	0.0011 amu
	calculated mass	=	4.0331 amu per atom
	-[actual mass	=	4.0026 amu per atom]

mass defect = (4.0331 - 4.0026) amu/atom = 0.0305 amu/atom

So it appears that 0.0305 amu of mass disappears in the formation of one atom of ^4_2He. This represents less than 0.8 percent of the calculated mass of the ^4_2He atom. Actually, this amount of matter is transformed into energy which is released to the environment as the atom is formed. This transformation is described by the Einstein equation, $E = mc^2$, in which E represents energy released, m represents the amount of matter transformed into energy (mass defect), and c represents the velocity of light in a vacuum, 2.998×10^8 m/s. The 0.0305 amu "lost" per atom may be expressed as its energy equivalent using the Einstein equation.

Recall: 6.022×10^{23} amu = 1.000 g* = 1.000×10^{-3} kg and 1 J = 1 kg·m² s⁻²**

$$E = mc^2 = \left(0.0305\text{ amu} \times \frac{1.000 \times 10^{-3}\text{ kg}}{6.022 \times 10^{23}\text{ amu}}\right)(2.998 \times 10^8 \text{ m/s})^2 = 4.55 \times 10^{-12} \text{ J/atom}$$

Thus, 4.55×10^{-12} joule of energy is released in the formation of one ^4_2He atom, and the same amount is required to separate a ^4_2He atom into its constituent particles. This amount of energy is called the binding energy of ^4_2He. <u>Binding energy is the energy equivalent of the mass defect of an atom</u>.

On an absolute scale 4.55×10^{-12} J does not represent very much energy. However, when one mole (4.0026 grams) of ^4_2He atoms is formed from subatomic particles, 2.74×10^{12} joules (or 6.55×10^8 kilocalories) of energy are liberated as the following calculations demonstrate.

$$\frac{?\text{ J}}{\text{mol}} = 4.55 \times 10^{-12}\frac{\text{J}}{\text{atom}} \times \frac{6.022 \times 10^{23}\text{ atoms}}{\text{mol}} = 2.74 \times 10^{12}\text{ J/mol}$$

---

*See Appendix C in the text.
**By definition.

$$\frac{?\ \text{kcal}}{\text{mol}} = 2.74 \times 10^{12}\ \frac{\text{J}}{\text{mol}} \times \frac{1.000\ \text{kcal}}{4.184 \times 10^{3}\ \text{J}} = 6.55 \times 10^{8}\ \text{kcal/mol}$$

Example 3-4: Calculate the binding energy of $^{47}_{22}Ti$ in joules per atom, kilojoules per mole of $^{47}_{22}Ti$ atoms (46.95 grams), and joules per nucleon (nuclear particle).

From Example 3-3, the mass defect of $^{47}_{22}Ti$ is 0.4384 amu per atom.

$$E = mc^2 = (0.4384\ \text{amu} \times \frac{1.000 \times 10^{-3}\ \text{kg}}{6.022 \times 10^{23}\ \text{amu}})(2.998 \times 10^{8}\ \text{m/s})^2$$

$$= \underline{\underline{6.543 \times 10^{-11}\ \text{J per atom}}}$$

$$\frac{?\ \text{kJ}}{\text{mol}} = \frac{6.543 \times 10^{-11}\ \text{J}}{\text{atom}} \times \frac{1\ \text{kJ}}{1 \times 10^{3}\ \text{J}} \times \frac{6.022 \times 10^{23}\ \text{atoms}}{\text{mol}}$$

$$= \underline{\underline{3.940 \times 10^{10}\ \text{kJ/mol}}}$$

$$\frac{?\ \text{erg}}{\text{nucleon}} = \frac{6.543 \times 10^{-11}\ \text{J}}{\text{atom}} \times \frac{1\ \text{atom}}{47\ \text{nucleons}} = \underline{\underline{1.392 \times 10^{-12}\ \text{J/nucleon}}}$$

3-7 Electromagnetic Radiation and Energy

The most effective probe of electronic structure is electromagnetic radiation. Electrons are classified as matter since they have mass and occupy space, but they are also wave-like and interact with electromagnetic radiation.

Each wave of electromagnetic radiation is characterized by a definite wavelength (λ) and frequency (ν). The wavelengths of electromagnetic radiation cover a wide range from as little as 10^{-12} cm for x-rays up to 10^3 m for radiowaves.

Frequency and wavelength are inversely proportional to each other. The product of the two is equal to the velocity of light in a vacuum, 2.998×10^8 m/s.

$$\nu\lambda = c$$

The energy of a photon of electromagnetic radiation of a given frequency or wavelength is given by the following equation in which h is Planck's constant, 6.626×10^{-34} J·s.

$$E = h\nu = h\left(\frac{c}{\lambda}\right)$$

From these relationships it is apparent that high frequency electromagnetic radiation has low wavelength and high energy and vice-versa as the following examples illustrate.

Example 3-5: Calculate the wavelength and energy of blue-green light of frequency $6.167 \times 10^{14}\ s^{-1}$.

First we calculate the wavelength. $\lambda\nu = c$

$$\lambda = \frac{c}{\nu} = \frac{2.998 \times 10^{8}\ m/s}{6.167 \times 10^{14}\ s^{-1}} = \underline{\underline{4.861 \times 10^{-7}\ m}}$$

Now we calculate the energy of a photon of this light.

$$E = h\nu = (6.626 \times 10^{-34}\ J{\cdot}s)(6.167 \times 10^{14}\ s^{-1})$$

$$= \underline{\underline{4.086 \times 10^{-19}\ J}}$$

Example 3-6: Calculate the frequency and energy of ultraviolet light of wavelength 1.217×10^{-7} m.

We calculate the frequency and then energy of a photon. $\lambda\nu = c$

$$\nu = \frac{c}{\lambda} = \frac{2.998 \times 10^{8}\ m/s}{1.217 \times 10^{-7}\ m} = \underline{\underline{2.463 \times 10^{15}\ s^{-1}}}$$

$$E = h\nu = (6.626 \times 10^{-34}\ J{\cdot}s)(2.463 \times 10^{15}\ s^{-1})$$

$$= \underline{\underline{1.632 \times 10^{-18}\ J}}$$

Comparison of Examples 3-5 and 3-6 shows that the ultraviolet light has shorter wavelength, higher frequency, and higher energy than the blue-green (visible) light.

Radiation is absorbed by an atom only in discrete quanta or packets containing precisely the amount of energy required to promote an electron to a higher energy level. The same amount of energy is emitted by an atom as an electron returns to the lower energy level. These absorptions and emissions of energy provide a basis for studying the characteristics and distributions of electrons in atoms by means of absorption and emission spectroscopy.

Example 3-7: One of the wavelengths of light emitted by excited calcium atoms is 3.93×10^{-7} m. What is the energy separating the two electronic energy levels between which the electron moves in this emission? Calculate the energy in J per atom and in kJ/mol of calcium atoms.

The energy separating the two energy levels involved in the electronic transition is exactly equal to the energy of the light emitted by the excited atoms.

$$E = h\left(\frac{c}{\lambda}\right) = \frac{(6.63 \times 10^{-34}\ \text{J}\cdot\text{s})(3.00 \times 10^{8}\ \text{m/s})}{3.93 \times 10^{-7}\ \text{m}}$$

$$= \underline{\underline{5.06 \times 10^{-19}\ \text{J per atom}}}$$

$$\frac{?\ \text{kJ}}{\text{mol}} = \frac{5.06 \times 10^{-19}\ \text{J}}{\text{atom}} \times \frac{6.02 \times 10^{23}\ \text{atoms}}{\text{mol}} \times \frac{1\ \text{kJ}}{1 \times 10^{3}\ \text{J}}$$

$$= \underline{\underline{305\ \text{kJ/mol}}}$$

This result tells us that 305 kJ of radiant energy is emitted when one mole of excited calcium atoms emit light that has a wavelength of 3.93×10^{-7} m.

3-8 Electronic Structure

The electronic configuration of an element determines its chemical characteristics and reactivity. Chemical reactions and bonding involve shifting or rearranging the outermost electrons in atoms. An understanding and appreciation of periodic trends, chemical reactivity, and bonding is easier if we understand certain concepts associated with the electronic structures of elements.

3-9 Quantum Numbers

The best picture of the electronic structure of atoms is described by quantum mechanics via differential equations. The experimentally supported quantum mechanical treatment takes into consideration the dual particle-like, wave-like characteristics of electrons. A mathematical discussion of quantum mechanics is impossible at the level of this book, so we will summarize the results derived from experimental data and solutions of quantum mechanical equations. These yield atomic orbitals, which are regions in space about the nucleus in which the probability of finding electrons is the greatest.

The orbital residence of an electron may be specified conveniently by a set of four quantum numbers, much like a person's residence may be specified by four components of the address: number, street, city and state. The four types of quantum numbers are described below.

1. Principal quantum number, n – The quantum number n designates the major energy level in which an electron is located. It may take positive integral values beginning with 1.

$$n = 1, 2, 3, 4, \cdots$$

Energy | Increases
$$\underline{n} = 1 \ (K \text{ shell})$$
$$\underline{n} = 2 \ (L \text{ shell})$$
$$\underline{n} = 3 \ (M \text{ shell})$$
$$\underline{n} = 4 \ (N \text{ shell})$$
$$\vdots$$
Increases | Radius

2. Subsidiary quantum number, ℓ – This quantum number specifies a particular sublevel within a major energy level. Electrons in presently-known atoms in their ground states occupy one of four kinds of sublevels, called s, p, d, or f sublevels. The possible values of ℓ depend upon the value of $\underline{n}$ and can range from zero to a maximum of $(\underline{n}-1)$, i.e.,

$$\underline{\ell} = 0, 1, 2, 3, \cdots (\underline{n}-1)$$

The correspondence between $\underline{\ell}$ values and sublevels designated is shown below.

$$\underline{\ell} = 0 \text{ refers to an s sublevel}$$
$$\underline{\ell} = 1 \text{ refers to a p sublevel}$$
$$\underline{\ell} = 2 \text{ refers to a d sublevel}$$
$$\underline{\ell} = 3 \text{ refers to an f sublevel}$$

Example 3-8:   In which sublevel (set of orbitals) is an electron found if its first two quantum numbers are $\underline{n} = 3$, $\underline{\ell} = 1$.

Since $\underline{n} = 3$, the electron is in the third energy level (M shell). Since $\underline{\ell} = 1$ the electron is in a p sublevel, in this case the 3p sublevel.

Example 3-9:   Can an electron occupy a 3f orbital? Why or why not?

No, this would require that $\underline{n} = 3$, and $\underline{\ell} = 3$. Obviously, this would contradict the rule that states that for a given $\underline{n}$, $\underline{\ell}$ must be at least one unit smaller. On a physical basis, this rule indicates that there is no f sublevel until the fourth ($\underline{n} = 4$) energy level.

3. Magnetic quantum number, $m\ell$ – This quantum number designates a specific orbital within a sublevel. Each s sublevel contains one s orbital, each p sublevel three p orbitals, each d sublevel contains five d orbitals,

and each f sublevel contains seven orbitals. The shapes and spatial orientations of the orbitals are described in your text.

The value of $\underline{m}_\ell$ for a given $\underline{\ell}$ may be any of the following integral values.

$$\underline{m}_\ell = -\underline{\ell},\ -(\underline{\ell}-1),\ -(\underline{\ell}-2) \cdots 0 \cdots +(\underline{\ell}-2),\ +(\underline{\ell}-1),\ +\underline{\ell}$$

For a given value of $\underline{\ell}$ there are as many possible values of $\underline{m}_\ell$ as there are orbitals contained within the sublevel designated by $\underline{\ell}$.

$\underline{\ell}$	sublevel	orbitals	$\underline{m}_\ell$	
$\underline{\ell} = 0$	s	s	0	
$\underline{\ell} = 1$	p	p_x	-1	order in which
		p_y	0	electrons fill
		p_z	$+1$	orbitals
$\underline{\ell} = 2$	d*	d_{z^2}	-2	
		$d_{x^2-y^2}$	-1	order in which
		d_{xy}	0	electrons fill
		d_{xz}	$+1$	orbitals
		d_{yz}	$+2$	
$\underline{\ell} = 3$	f		-3	
			-2	
		seven	-1	order in which
		f	0	electrons fill
		orbitals	$+1$	orbitals
			$+2$	
			$+3$	

*It is helpful to remember that the lobes of d_{z^2} and $d_{x^2-y^2}$ orbitals are directed along the axes while lobes of other three d orbitals bisect the axes.

4. Spin quantum number, m_s – Each orbital can hold a maximum of two electrons, if they have opposed spins, that is their magnetic fields must

interact attractively. The spins are designated by the spin quantum numbers m_s. One electron occupying an orbital is assigned $m_s = +1/2$ and another electron in the same orbital is assigned $m_s = -1/2$.

The Pauli Exclusion Principle states that no two electrons in the same atom may have the same four quantum numbers. Such a situation would be the case only if two electrons occupied the same orbital and had the same spin.

3-10 Order of Filling of Atomic Orbitals

Electrons fill available orbitals of lowest energy first. They occupy singly the orbitals of an energetically-equivalent (degenerate) set, such as those of a particular p sublevel, before pairing in any one orbital of the sublevel. The relative energies of the orbitals are shown in Figure 3-24 in the text.

The order of filling is 1s, 2s, 2p, 3s, 3p, 4s, 3d, 4p, 5s, 4d, 5p, 6s, 4f, 5d, 6p, 7s, 5f, 6d, 7p. A useful mnemonic device for remembering the order can be constructed quite easily as shown below.

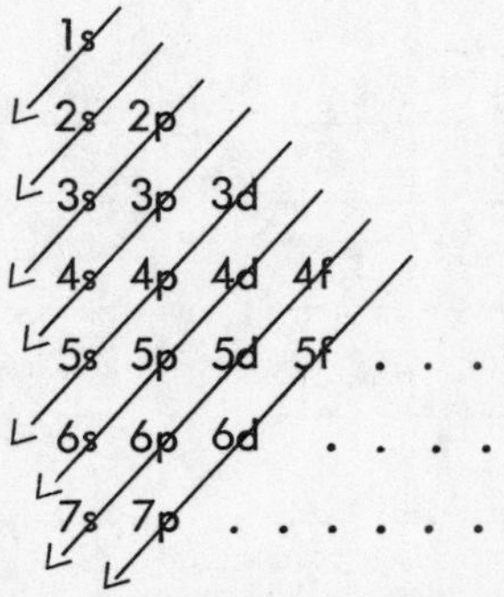

Read arrows from tail to head, top to bottom.

The electronic configuration of any element (or monatomic ion) can be shown by any of three methods as illustrated in the next example.

<u>Example 3-10:</u> Write out the electronic distribution of the sodium atom.

Sodium atoms contain eleven electrons. The lowest energy orbitals are filled first.

Method 1: $_{11}Na$: $1s^2 2s^2 2p^6 3s^1$

The superscripts represent the electronic occupation of atomic orbitals and they must, in this case, add to eleven.

Method 2: $_{11}Na$: $1s^2 2s^2 2p_x{}^2 2p_y{}^2 2p_z{}^2 3s^1$

This method allows designation of the occupany of individual orbitals within a set.

Method 3: $_{11}$Na:

1s	2s	2p			3s
↑↓	↑↓	↑↓	↑↓	↑↓	↑

Each arrow represents an electron. Pairs of arrows pointing in opposite directions represent two electrons in the indicated atomic orbital with paired spins.

We usually focus our attention on the outermost electrons of an atom since these electrons are involved in chemical bonding. In order to avoid writing out all the symbols for each electron the core, electrons are often abbreviated as (noble gas), as shown in the next example.

Example 3-11: What is the electronic configuration for potassium?

$_{19}$K $1s^2 2s^2 2p^6 3s^2 3p^6 4s^1$ or $(Ar)4s^1$ or (Ar) ↑ (4s)

Here, (Ar) represents the electronic configuration of $_{18}$Ar, the preceding noble gas. Since the 4s sublevel has lower energy than the 3d sublevel, it fills first.

Example 3-12: Write out the electronic configurations of elements numbers 21, 24, 28 and 29.

	3d					4s
$_{21}$Sc (Ar)	↑					↑↓
$_{24}$Cr (Ar)	↑	↑	↑	↑	↑	↑
$_{28}$Ni (Ar)	↑↓	↑↓	↑↓	↑	↑	↑↓
$_{29}$Cu (Ar)	↑↓	↑↓	↑↓	↑↓	↑↓	↑

Note: The Cr and Cu configurations involve half-filled and filled sets of 3d 4s orbitals. It is observed that added stability is associated with half-filled and completely filled s and d sublevels.

Example 3-13: Use Method 1 to write out the electronic configurations of all the noble gases (Group 0 elements).

$_{2}$He	$1s^2$		
$_{10}$Ne	$1s^2 2s^2 2p^6$	or	$(He)2s^2 2p^6$
$_{18}$Ar	$1s^2 2s^2 2p^6 3s^2 3p^6$	or	$(Ne)3s^2 3p^6$
$_{36}$Kr	$1s^2 2s^2 2p^6 3s^2 3p^6 3d^{10} 4s^2 4p^6$	or	$(Ar)\ 3d^{10} 4s^2 4p^6$
$_{54}$Xe	$(Kr)5s^2 4d^{10} 5p^6$		
$_{86}$Rn	$(Xe)4f^{14} 5d^{10} 6s^2 6p^6$		

The common characteristic of the noble gases is that all except He have completely filled outer s and p sublevels. This configuration lends exceptional stability to the noble gases and accounts for their relative chemical inertness.

<u>Example 3-14:</u> What are all possible combinations of $\underline{n}$ and $\underline{\ell}$ quantum numbers for the electrons of a nitrogen atom in the ground state?

The electron distribution in nitrogen atoms is: $_7N \quad 1s^2 2s^2 2p^1 2p^1 2p^1$

electrons	n	ℓ	
first + second	1	0	(1s)
third + fourth	2	0	(2s)
last three	2	1	(2p)

<u>Example 3-15:</u> What are the possible values of $\underline{n}$, $\underline{\ell}$, and $\underline{m_\ell}$ for the <u>highest energy</u> electrons in a manganese atom in its ground state?

The electronic configuration of manganese is:

$$_{25}Mn \quad 1s^2 2s^2 2p^6 3s^2 3p^6 3d^1 3d^1 3d^1 3d^1 3d^1 4s^2$$

The highest energy electrons are in 3d orbitals for which $\underline{n} = 3$, and $\underline{\ell} = 2$. There are five electrons in the set of five 3d orbitals and each orbital is singly occupied. The possible values of $\underline{m_\ell}$ are given by:

$$\underline{m_\ell} = -2, -1, 0, +1, +2$$

Thus, there is one $\underline{m_\ell}$ quantum number for each of the 3d orbitals and the five electrons in 3d orbitals in manganese atoms in the ground state have the quantum numbers given below:

n	ℓ	mℓ
3	2	-2
3	2	-1
3	2	0
3	2	+1
3	2	+2

<u>Example 3-16:</u> Write an acceptable set of quantum numbers to describe the electrons in a carbon atom in the ground state.

	1s	2s	2p	2p	2p
The electronic configuration of carbon is: $_6C$	⇅	⇅	↑	↑	___

electrons	n	ℓ	m_ℓ	m_s
1s	1	0	0	+1/2
	1	0	0	-1/2
2s	2	0	0	+1/2
	2	0	0	-1/2
2p	2	1	-1	+1/2
2p	2	1	0	+1/2

Example 3-17: Write an acceptable set of four quantum numbers for the electrons in an oxygen atom in the ground state.

	1s	2s	2p	2p	2p
The electronic configuration of oxygen is: $_8O$	⥮	⥮	⥮	↑	↑

The quantum numbers of the first six electrons are the same as those in carbon atoms. Acceptable quantum numbers of the last two electrons are:

n	ℓ	m_ℓ	m_s	
2	1	+1	+1/2	(2p)
2	1	-1	-1/2	(2p)

Example 3-18: Write an acceptable set of four quantum numbers of the last, i.e., the distinguishing electron of iron which is indicated by a curved arrow.

		3d					4s
$_{26}Fe$	(Ar)	↑↲	↑	↑	↑	↑	⥮

For the "last" electron we could write $n = 3$, $\ell = 2$, $m_\ell = -2$, $m_s = -1/2$, which may be written more simply as 3, 2, -2, -1/2,

EXERCISES

1. Write complete atomic designations ($^{A}_{Z}E$) for the atoms containing the following numbers of subatomic particles.

	protons	electrons	neutrons
(a)	12	12	13
(b)	35	35	45
(c)	45	45	54
(d)	30	30	38
(e)	13	13	14
(f)	15	15	17
(g)	19	19	20
(h)	11	11	12
(i)	31	31	38
(j)	33	33	42
(k)	47	47	60
(l)	53	53	74
(m)	56	56	79
(n)	92	92	143

2. How many electrons, protons, and neutrons are contained in each of the following atoms?

   (a) ^{128}Xe (b) ^{65}Cu (c) ^{28}Si (d) ^{94}Y (e) ^{104}Ru (f) ^{115}Sb (g) ^{138}Ba

3. Naturally occuring silicon consists of the following isotopes: ^{28}Si (27.9769 amu; 92.21%), ^{29}Si (28.9765 amu; 4.70%), and ^{30}Si (29.9738 amu; 3.09%). What is its atomic weight?

4. Naturally occuring zinc consists of the following isotopes: ^{64}Zn (63.9291 amu; 48.89%), ^{66}Zn (65.9260 amu; 27.81%), ^{67}Zn (66.9271 amu; 4.11%), ^{68}Zn (67.9249 amu; 18.57%), and ^{70}Zn (69.9253 amu; 0.62%). Calculate its atomic weight.

5. The atomic weight of gallium is 69.72 amu. Naturally occuring gallium consists of two isotopes, ^{69}Ga (68.926 amu) and ^{71}Ga (70.925 amu). What are the percentage abundances of the two isotopes?

6. Naturally occuring bromine consists of two isotopes, one is ^{79}Br (79.9183 amu; 50.54%). The other is ^{81}Br. The atomic weight of bromine is 79.904 amu. What are the percent abundance and isotopic weight of ^{81}Br?

7. Calculate the mass defect for each of the following isotopes in g/mol: (a) ^{79}Br (b) ^{64}Ni (c) ^{75}As. The actual masses are 78.9183 amu for ^{79}Br, 63.9280 amu for ^{64}Ni, and 74.9216 amu for ^{75}As.

8. Calculate the binding energy in joules per mole, kcal per mole, and kJ/mol for each of the isotopes of problem 7.

9. Convert the following wavelengths to frequencies.
(a) 6.21×10^{-5} cm (b) 5.26×10^{-4} nm (c) 100 Å

10. Convert the following frequencies to wavelengths.
(a) $3.79 \times 10^{15}\ s^{-1}$ (b) $2.41 \times 10^{18}\ s^{-1}$ (c) $5.01 \times 10^{13}\ s^{-1}$

11. Calculate the energy of electromagnetic radiation of frequency:
(a) $2.78 \times 10^{15}\ s^{-1}$ (b) $1.14 \times 10^{17}\ s^{-1}$ (c) $3.17 \times 10^{14}\ s^{-1}$

12. Calculate the energy of light having the following wavelengths.
(a) 5.17×10^{-5} cm (b) 6047 Å (c) 1172 nm

13. Calculate the frequency and wavelength of light having the following energies. (a) 4.38×10^{-19} J/photon (b) 383.6 kJ/mol (c) 8.72 eV/photon

14. Classify each of the kinds of light in problems 9-13 into the proper category listed in the following table.

Wavelength Ranges for Electromagnetic Radiation

Kind	Approximate Wavelength Range (cm)
Cosmic rays	$\sim 10^{-12}$
Gamma rays	$5 \times 10^{-11} - 1 \times 10^{-9}$
X-rays	$1 \times 10^{-9} - 1 \times 10^{-6}$
Ultraviolet	$<4.00 \times 10^{-5}$
Visible:	$4 \times 10^{-5} - 7 \times 10^{-5}$
violet	$4.00 \times 10^{-5} - 4.24 \times 10^{-5}$
blue	$4.24 \times 10^{-5} - 4.91 \times 10^{-5}$
green	$4.91 \times 10^{-5} - 5.75 \times 10^{-5}$
yellow	$5.75 \times 10^{-5} - 5.85 \times 10^{-5}$
orange	$5.85 \times 10^{-5} - 6.47 \times 10^{-5}$
red	$6.47 \times 10^{-5} - 7.00 \times 10^{-5}$
Infrared	$>7.00 \times 10^{-5}$

15. One of the wavelengths of electromagnetic radiation absorbed by boron atoms is 4.94×10^{-7} m. What is this wavelength in angstroms and in nanometers? How large is the energy separation (in J) between the electronic energy levels between which the electron moves during this absorption?

16. The emission spectrum of gold shows a line at 267.6 nanometers. What is the wavelength in centimeters? How much energy is emitted by the excited electron as it falls to a lower energy level (in joules per atom and kJ per mol of atoms).

17. The ionization energy of an element is the energy that must be absorbed by a gaseous atom of the element in order to remove one electron and form a gaseous ion. Lithium is ionized by absorption of light of wavelength 2.31×10^{-7} m.

$$\text{Li (g)} + \text{Ionization Energy} \longrightarrow \text{Li (g)}^{+} + e^{-}$$

Calculate the ionization energy of lithium in J per atom and kJ per mol of lithium atoms.

18. Show the complete distribution of electrons in atomic orbitals for the following atoms by the three methods described in the text and in this book.
(a) Li, (b) F (c) S (d) Ca (e) Ni (f) Cu (g) Ge (h) Pb

19. Show the outermost electron distribution in condensed form for each of the atoms below. Example: As - (Ar) $4s^2\ 4p^3$
(a) H (b) Na (c) S (d) Ca (e) Mn (f) Fe (g) Kr (h) Sr
(i) Ru (j) Te (k) Ba (l) Po

20. Write an acceptable set of the four quantum numbers (n, ℓ, m_ℓ, m_s) for the electrons in the following atoms:
(a) Be (b) B (c) F (d) Ca (e) Sc (f) Fe (g) Zn (h) Se

21. The following six sets of quantum numbers describe electrons of some neutral elements. Some sets are impossible. Which ones? Why?

	n	ℓ	m_ℓ	m_s
(a)	1	0	0	+1/2
(b)	2	1	0	+1/2
(c)	3	2	-2	-1/2
(d)	4	3	0	-1/2
(e)	2	2	-1	+1/2
(f)	4	2	4	+1/2

22. Which elements are described by the following ground state electron configurations. Some of the following configurations are impossible. Which ones? (Note: question asks about atoms, not ions.)

(a) $1s^2\ 2s^2$
(b) $1s^2\ 2s^2\ 2p^3$
(c) $1s^2\ 2s^3\ 2p^4$
(d) $1s^2\ 2s^1\ 3p^5$
(e) $1s^2\ 2s^2\ 2p^6\ 3s^2\ 3p^1\ 3d^2$
(f) $1s^2\ 2s^2\ 2p^6\ 3s^2\ 3p^3$
(g) $1s^2\ 2s^2\ 2p^6\ 3s^2\ 3p^6\ 3d^8$
(h) $1s^2\ 2s^2\ 2p^6\ 3s^2\ 3p^6\ 4s^2\ 4d^{10}\ 5s^1$
(i) $1s^2\ 2s^2\ 2p^6\ 3s^2\ 3p^6\ 4s^1\ 4d^8$
(j) $1s^2\ 2s^2\ 2p^6\ 3s^2\ 3p^6\ 3d^{10}\ 4s^1$

ANSWERS TO EXERCISES

1. (a) $^{25}_{12}Mg$ (b) $^{80}_{35}Br$ (c) $^{99}_{45}Rh$ (d) $^{68}_{30}Zn$ (e) $^{27}_{13}Al$ (f) $^{32}_{15}P$ (g) $^{39}_{19}K$ (h) $^{23}_{11}Na$ (i) $^{69}_{31}Ga$ (j) $^{75}_{33}As$ (k) $^{107}_{47}Ag$ (l) $^{127}_{53}I$ (m) $^{135}_{56}Ba$ (n) $^{235}_{92}U$

2.

	e^-	p^+	n^o
(a)	54	54	74
(b)	29	29	36
(c)	14	14	14
(d)	39	39	55
(e)	44	44	60
(f)	51	51	64
(g)	56	56	82

3. 28.09 amu 4. 65.39 amu 5. 60.3% ^{69}Ga, 39.7% ^{71}Ga
6. 49.46%, 79.88$\bar{9}$ amu 7. (a) 0.7393 g/mol (b) 0.6050 g/mol (c) 0.7029 g/mol 8. (a) 6.645×10^{13} J/mol, 1.588×10^{10} kcal/mol, 6.645×10^{10} kJ/mol, (b) 5.438×10^{13} J/mol, 1.300×10^{10} kcal/mol, 5.438×10^{10} kJ/mol, (c) 6.318×10^{13} J/mol, 1.510×10^{10} kcal/mol, 6.318×10^{10} kJ/mol 9. (a) $4.83 \times 10^{14}\ s^{-1}$ (b) $5.70 \times 10^{20}\ s^{-1}$ (c) $3.00 \times 10^{16}\ s^{-1}$ 10. (a) 7.92×10^{-6} cm (b) 1.24×10^{-8} cm (c) 5.99×10^{-4} cm 11. (a) 1.84×10^{-18} J (b) 7.56×10^{-17} J (c) 2.10×10^{-19} J 12. (a) 3.85×10^{-19} J (b) 3.29×10^{-19} J (c) 1.70×10^{-19} J 13. (a) $\nu = 6.61 \times 10^{14}\ s^{-1}$, $\lambda = 4.54 \times 10^{-5}$ cm (b) $\nu = 9.612 \times 10^{14}\ s^{-1}$, $\lambda = 3.119 \times 10^{-5}$ cm (c) $\nu = 2.10 \times 10^{15}\ s^{-1}$, $\lambda = 1.43 \times 10^{-5}$ cm

14. (9a) visible (orange), (9b) visible (green), (9c) x-ray, (10a) ultraviolet, (10b) x-ray, (10c) infrared, (11a) ultraviolet, (11b) x-ray, (11c) infrared, (12a) visible (green), (12b) visible (orange), (12c) x-ray, (13a) visible (blue), (13b) ultraviolet, (13c) ultraviolet

15. 4.94×10^3 Å, 494 nm, $\Delta E = 4.03 \times 10^{-19}$ J

16. 2.676×10^{-5} cm, $E = 7.423 \times 10^{-19}$ J/atom, 447.3 kJ/mol

17. $E = 8.61 \times 10^{-19}$ J/atom, 519 kJ/mol

18. (a) 1: Li $1s^2\ 2s^1$
2: same
3: Li ↑↓ (1s) ↑ (2s)

(b) 1: F $1s^2\ 2s^2\ 2p^5$
2: F $1s^2\ 2s^2\ 2p^2\ 2p^2\ 2p^1$
3: F ↑↓ (1s) ↑↓ (2s) ↑↓ ↑↓ ↑ (2p)

(c) 1: S $1s^2\ 2s^2\ 2p^6\ 3s^2\ 3p^4$
2: S (Ne) $3s^2\ 3p^2\ 3p^1\ 3p^1$
3: S (Ne) ↑↓ (3s) ↑↓ ↑ ↑ (3p)

(d) 1: Ca $1s^2\ 2s^2\ 2p^6\ 3s^2\ 3p^6\ 4s^2$
2: Ca (Ar) $4s^2$
3: Ca ↑↓ (1s) ↑↓ (2s) ↑↓ ↑↓ ↑↓ (2p) ↑↓ (3s) ↑↓ ↑↓ ↑↓ (3p) ↑↓ (4s)

(e) 1: Ni $1s^2\ 2s^2\ 2p^6\ 3s^2\ 3p^6\ 3d^8\ 4s^2$
2: Ni (Ar) $3d^2\ 3d^2\ 3d^2\ 3d^1\ 3d^1\ 4s^2$
3: Ni (Ar) ↑↓ ↑↓ ↑↓ ↑ ↑ (3d) ↑↓ (4s)

(f) 1: Cu $1s^2\ 2s^2\ 2p^6\ 3s^2\ 3p^6\ 3d^{10}\ 4s^1$
2: Cu (Ar) $3d^2\ 3d^2\ 3d^2\ 3d^2\ 3d^2\ 4s^1$
3: Cu (Ar) ↑↓ ↑↓ ↑↓ ↑↓ ↑↓ (3d) ↓ (4s)

(g) 1: Ge $1s^2\ 2s^2\ 2p^6\ 3s^2\ 3p^6\ 3d^{10}\ 4s^2\ 4p^2$
2: Ge (Ar) $3d^{10}\ 4s^2\ 4p_x^{\ 1}\ 4p_y^{\ 1}$
3: Ge (Ar) ↑↓ ↑↓ ↑↓ ↑↓ ↑↓ (3d) ↑↓ (4s) ↑ ↑ __ (4p)

(h) 1: Pb $1s^2\ 2s^2\ 2p^6\ 3s^2\ 3p^6\ 3d^{10}\ 4s^2\ 4p^6\ 4d^{10}\ 4f^{14}\ 5s^2\ 5p^6\ 5d^{10}\ 6s^2\ 6p^2$
2: Pb (Xe) $4f^{14}\ 5d^{10}\ 6s^2\ 6p^1\ 6p^1$
3: Pb (Xe) ↑↓ ↑↓ ↑↓ ↑↓ ↑↓ ↑↓ ↑↓ (4f) ↑↓ ↑↓ ↑↓ ↑↓ ↑↓ (5d) ↑↓ (6s) ↑ ↑ __ (6p)

19. (a) H $1s^1$ (b) Na (Ne) $3s^1$ (c) S (Ne) $3s^2\ 3p^4$ (d) Ca (Ar) $4s^2$
(e) Mn (Ar) $3d^5\ 4s^2$ (f) Fe (Ar) $3d^6\ 4s^2$ (g) Kr (Ar) $3d^{10}\ 4s^2\ 4p^6$
(h) Sr (Kr) $5s^2$ (i) Ru (Kr) $4d^6\ 5s^2$ (j) Te (Kr) $4d^{10}\ 5s^2\ 5p^4$ (k) Ba (Xe) $6s^2$
(l) Po (Xe) $4f^{14}\ 5d^{10}\ 6s^2\ 6p^4$

20.

(a) Be

"number" of e^-	n	ℓ	m_ℓ	m_s
1,2	1	0	0	±1/2
3,4	2	0	0	±1/2

(b) B

"number" of e^-	n	ℓ	m_ℓ	m_s
1,2	1	0	0	±1/2
3,4	2	0	0	±1/2
5	2	1	-1	+1/2

(c) F

"number" of e^-	n	ℓ	m_ℓ	m_s
1,2	1	0	0	±1/2
3,4	2	0	0	±1/2
5,8	2	1	-1	±1/2
6,9	2	1	0	±1/2
7	2	1	+1	+1/2

(d) Ca

"number" of e^-	n	ℓ	m_ℓ	m_s
1,2	1	0	0	±1/2
3,4	2	0	0	±1/2
5,8	2	1	-1	±1/2
6,9	2	1	0	±1/2
7,10	2	1	+1	±1/2
11,12	3	0	0	±1/2
13,16	3	1	-1	±1/2
14,17	3	1	0	±1/2
15,18	3	1	+1	±1/2
19,20	4	0	0	±1/2

(e) Sc (first 20 electrons same as for Ca)

	n	ℓ	m_ℓ	m_s
21	3	2	-2	+1/2

(f) Fe (first 20 electrons same as for Ca)

	n	ℓ	m_ℓ	m_s
21,26	3	2	-2	±1/2
22	3	2	-1	+1/2
23	3	2	0	+1/2
24	3	2	+1	+1/2
25	3	2	+2	+1/2

20. (cont'd)

(g) Zn (first 20 electrons same as for Ca)

	n	ℓ	m_ℓ	m_s
21,26	3	2	-2	±1/2
22,27	3	2	-1	±1/2
23,28	3	2	0	±1/2
24,29	3	2	+1	±1/2
25,30	3	2	+2	±1/2

(h) Se (first 30 electrons same as for Zn)

	n	ℓ	m_ℓ	m_s
31,34	4	1	-1	±1/2
32	4	1	0	+1/2
33	4	1	+1	+1/2

21.

(e) impossible-maximum value of ℓ is (n-1) (f) impossible-maximum value of m_ℓ is 2 because $\ell = 2$. 22. (a) Be (b) N (c) impossible (d) impossible (e) impossible (f) P (g) impossible (for a neutral atom) (h) impossible (i) impossible (j) Cu

Chapter Four

CHEMICAL PERIODICITY AND IONIC BONDING

D, G, W: Sec. 6-1 through 6-10 and 6-12; Sec. 1-16 and 1-17

CHEMICAL PERIODICITY

4-1 The Periodic Table

Elements are arranged in the periodic table according to increasing atomic number. Similar electronic configurations and properties recur periodically among elements in the same groups (vertical columns). The horizontal rows are called periods. The "A" Groups, or representative elements, have valence electrons in s and p sublevels, while the "B" Groups, or d-transition elements, have their "last" electrons in d sublevels one shell inside the outermost occupied shell. The lanthanides and actinides, or f-transition elements, are filling f sublevels two shells inside the outermost shell. The metalloids are the elements along the stepwise division in the upper right corner of the periodic table. Their properties are intermediate between those of the metals, below and to the left, and the nonmetals, above and to the right.

Variations in properties are much more regular among the representative elements than among the d- or f-transition elements and we shall focus our attention on the A Groups. It is often convenient to describe the electronic structures of the representative elements in terms of dot formulas which show only the outermost s and p electrons. Dot formulas are identical for all representative elements of a given group. See Table 4-3 in the text.

4-2 Atomic Radii

The radii of neutral atoms increase from right-to-left across a period because the increasing positive nuclear charge attracts the outer shell electrons more strongly which reduces the sizes of the atoms as we move to the right in a given period. They also increase from top-to-bottom within groups as electrons occupy shells farther from the nucleus. See Figure 4-2 in the text.

Example 4-1: Arrange the following elements in order of increasing atomic radii: Be, Li, N, B, F

These are all elements of period 2, and their radii increase from right-to-left in the periodic table.

$$\xrightarrow{\text{Increasing Radius}}$$
$$F < N < B < Be < Li$$

Example 4-2: Arrange the following elements in order of increasing atomic radii: Al, Ga, B, Tl, In

These are all elements of Group IIIA, and their radii increase from top-to-bottom.

$$\xrightarrow{\text{Increasing Radius}}$$
$$B < Al < Ga < In < Tl$$

Example 4-3: Which of the following elements has the smallest atomic volume and which has the largest? Na, Cl, Xe, Rb, Te

The volume of a sphere is $(4/3)\,\pi r^3$ where r is the radius. Thus the atom with the smallest radius, Cl, is the smallest and that with the largest radius, Rb, is the largest.

4-3 Ionic Radii

Simple cations are always smaller than the neutral atoms from which they are derived and, for a given element, the size decreases with increasing positive charge. For example, the radii of Pb, Pb^{2+}, and Pb^{4+} are 1.75 Å, 1.21 Å, and 0.84 Å, respectively. This decrease is due to the increasing attraction of a constant positive charge for decreasing numbers of electrons.

Simple anions are always larger than the neutral atoms from which they are derived. For example, the radii of oxygen atoms and oxide ions, O^{2-}, are 0.66 Å and 1.40 Å, respectively. The increase is due to the greater number of electrons (which repel each other) and a constant number of protons.

Finally, within a series of isoelectronic species, such as N^{3-}, O^{2-}, F^-, Ne, Na^+, and Mg^{2+}, all of which have ten electrons, the one with the most negative charge is the largest, due to the smaller nuclear charge and the correspondingly smaller attraction of the nucleus for the electrons. Clearly the ion with the highest positive charge is the smallest ion within a series of monatomic isoelectronic species.

Example 4-4: Arrange the following ions in order of increasing ionic radii: K^+, Na^+, Rb^+, Cs^+, H^+

These are all simple cations (1+) of Group IA elements. Like the neutral atoms from which they are derived, their radii also increase from top-to-bottom.

$$\xrightarrow{\text{Increasing Radius}}$$
$$H^+ < Na^+ < K^+ < Rb^+ < Cs^+$$

Example 4-5: Arrange the following ions in order of increasing ionic radii: S^{2-}, Se^{2-}, Te^{2-}, O^{2-}

These are all simple anions (2-) of Group VIA elements, and so their radii increase as the group is descended.

$$O^{2-} < S^{2-} < Se^{2-} < Te^{2-} \quad \xrightarrow{\text{Increasing Radius}}$$

<u>Example 4-6:</u> Arrange the following species in order of increasing size: As^{3-}, Se^{2-}, Br^-, Kr, Rb^+, Sr^{2+}

These are all isoelectronic with Kr, and the one with most negative charge is the largest.

$$Sr^{2+} < Rb^+ < Kr < Br^- < Se^{2-} < As^{3-} \quad \xrightarrow{\text{Increasing Radius}}$$

4-4 Ionization Energy

The amount of energy required to ionize the most loosely held electron of one mole of neutral gaseous atoms is called the first ionization energy of the element. See Table 4-4 and Figure 4-3 in the text.

$$\text{Atom}^{\circ}\,(g) + \text{Ionization Energy} \longrightarrow \text{Ion}^{+}\,(g) + e^{-}$$

The higher the ionization energy is, the more difficult it is to remove an electron. In general, first ionization energies increase from left-to-right across a period and from bottom-to-top within a group. The trends of increasing ionization energies are approximately parallel with the trends of decreasing radii for the A Group elements. Obviously, the very stable electronic configurations of the rare gases are difficult to disrupt and therefore the rare gas has the highest ionization energy of any element of a given period. The Group IA metals have the most loosely held electrons (one each in the outermost shell) and therefore the lowest first ionization energies.

Exceptions to these trends occur at the IIIA and VIA elements of the first four periods. The IIIA and VIA elements have slightly lower ionization energies than the preceding elements from Group IIA and VA. The IIIA elements have only one electron in their outermost p sublevel which is more easily removed than one of the electrons in the filled s sublevel in the same shell. The VIA elements have two paired and two unpaired electrons in their outermost p sublevel. Removal of one of the paired electrons to produce a relatively stable half-filled p sublevel is easier than removal of an unpaired electron of a VA element, which destroys a half-filled p sublevel.

<u>Example 4-7:</u> Arrange the following atoms in order of increasing first ionization energies: Sb, N, As, P, Bi

These are the elements of Group VA.

$$\xrightarrow{\text{Increasing 1st I.E.}}$$
$$Bi < Sb < As < P < N$$

Example 4-8: Arrange the following in order of increasing first ionization energy: K, Ca, Ga, Ge, As, Se, Br, Kr

These are the elements of period 4.

$$\xrightarrow{\text{Increasing 1st I.E.}}$$
$$K < Ga < Ca < Ge < As < Se < Br < Kr$$

Note that Ga from Group IIIA has a lower first ionization energy than Ca from Group IIA, while Se from Group VIA has approximately the same value as As from Group VA. (See Table 4-4 in the text.)

4-5 Electron Affinity

Electron affinity is defined as the amount of energy involved when one mole of neutral gaseous atoms add one mole of electrons to form one mole of anions with 1- charges. It is assigned a negative value when energy is released.

$$\text{Atom}^{o}\ (g) \ + \ e^{-} \longrightarrow \text{Ion}^{-}\ (g) \ + \ \text{Electron Affinity}$$

Highly negative electron affinities are associated with atoms having the greatest tendencies to form simple anions. There are several irregular variations in trends, but in general electron affinities become more negative from left-to-right across a period and from bottom-to-top within a group. See Table 4-5 in the text.

4-6 Electronegativity

The electronegativity of an element is a measure of the ability of its atoms to attract electrons to themselves when chemically combined with another element. It can be thought of as a rough average of ionization energy and electron affinity. Elements with the greatest attraction for electrons in chemical bonds have the highest electronegativities, which in general increase from left-to-right across periods and from bottom-to-top within groups. See Table 4-6 in the text.

Example 4-9: Arrange the following elements in order of increasing electronegativity: Na, Li, H, K, Cs

These are elements of Group IA.

$$\xrightarrow{\text{Increasing Electronegativity}}$$
$$Cs < K < Na \approx Li < H$$

Example 4-10: Arrange the following elements in order of increasing electronegativity: Po, Pb, Tl, Cs, Ba

These are elements of the sixth period.

$$\xrightarrow{\text{Increasing Electronegativity}}$$
$$Cs < Ba < Tl < Pb < Po$$

<u>Example 4-11</u>: Arrange the following elements in order of increasing electronegativity: F, Br, Be, Rb, Li

These elements are from different parts of the periodic table, but we can still determine the correct order by inspection.

$$\xrightarrow{\text{Increasing Electronegativity}}$$
$$Rb < Li < Be < Br < F$$

CHEMICAL BONDING

The attractive forces between atoms and ions in compounds are called chemical bonds. There are two major classes of bonds, <u>ionic</u> and <u>covalent</u> bonds.

Ionic bonds or electrovalent bonds are formed by the transfer of one or more electrons from one atom or group of atoms to another. Covalent bonds are formed by two atoms sharing one or more electron pairs. Compounds containing predominantly ionic bonds are called ionic compounds and those containing predominantly covalent bonds are called covalent compounds.

<u>4-7 Ionic Bonding</u>

Elements at the extreme left of the periodic table, the metals, lose electrons easily to form positive ions (cations). Nonmetals located far to the right in the periodic table gain electrons easily to form negative ions (anions).

Potassium (Group IA) and fluorine (Group VIIA) react explosively to form the ionic solid, potassium fluoride.

$$2\,K\,(s) \quad + \quad F_2\,(g) \quad \longrightarrow \quad 2\,KF\,(s) \quad + \quad \text{heat}$$

The reaction involves molecular fluorine rather than atomic fluorine, but it is instructive to consider the reaction in terms of the dot formulas of the atoms. Although all electrons are identical they are represented differently here for "bookkeeping" purposes only.

$$K\cdot \quad + \quad \cdot\ddot{\underset{\cdot\cdot}{F}}: \quad \longrightarrow \quad K^{+}\ [:\ddot{\underset{\cdot\cdot}{F}}:]^{-}$$

Both potassium and fluorine achieve noble gas configurations in this reaction.

$_{19}K$	(Ar)			$\uparrow$ (4s)				$\longrightarrow$	$_{19}K^{+}$	(Ar)			__ (4s)			
$_{9}F$	$\uparrow\downarrow$ (1s)	$\uparrow\downarrow$ (2s)		$\uparrow\downarrow$	$\uparrow\downarrow$ (2p)	$\uparrow$			$_{9}F^{-}$	$\uparrow\downarrow$ (1s)	$\uparrow\downarrow$ (2s)		$\uparrow\downarrow$	$\uparrow\downarrow$ (2p)	$\uparrow\downarrow$	

Potassium ions are isoelectronic with Ar atoms ($18\ e^-$), while fluoride ions are isoelectronic with neon ($10\ e^-$). The alkali metals (Group IA) have low first ionization energies, low electronegativities and low electron affinities while the halogens (Group VIIA) have high first ionization energies, high electronegativities and high electron affinities. Similar reactions also occur between all the IA and VIIA elements. These reactions can be generalized as follows:

$$2\,M + X_2 \longrightarrow 2\,M^+X^- \qquad \begin{matrix} M = Li \longrightarrow Cs \\ X = F \longrightarrow I \end{matrix}$$

This general equation actually represents the twenty reactions that produce the IA metal halides. Francium and astatine have been omitted because they are artificial, radioactive elements. They react in the same way. Hydrogen has also been omitted because the hydrogen halides, such as HF and HCl, are covalent rather than ionic compounds.

Example 4-12: Write the balanced equation for the reaction of lithium with bromine. Show the dot formula and the electronic configuration for each species.

$$2\,Li + Br_2 \longrightarrow 2\,LiBr$$

$$\left.\begin{matrix} Li & 1s^22s^1 \\ Br & (Ar)\ 4s^23d^{10}4p^5 \end{matrix}\right] \longrightarrow \begin{matrix} Li^+ & 1s^2 & \text{(He structure)} \\ Br^- & (Ar)\ 4s^23d^{10}4p^6 & \text{(Kr structure)} \end{matrix}$$

$$Li\cdot + \cdot\ddot{\underset{..}{Br}}: \longrightarrow Li^+\ [:\ddot{\underset{..}{Br}}:]^-$$

The Group IIA metals (alkaline earths) react with lighter Group VA nonmetals to form binary ionic compounds of the general formula M_3X_2 as the next example shows.

Example 4-13: Write the balanced equation for the reaction of calcium with nitrogen. Write the dot formula for each species in the equation.

$$3\,Ca + N_2 \longrightarrow Ca_3N_2$$

$$3\,Ca: + 2\cdot\ddot{\underset{\cdot}{N}}\cdot \longrightarrow 3\,Ca^{2+},\ 2[:\ddot{\underset{..}{N}}:]^{3-}$$

The equation with nitrogen in its molecular form is shown below.

$$3\,Ca: + :N\vdots\vdots N: \longrightarrow 3\,Ca^{2+},\ 2[:\ddot{\underset{..}{N}}:]^{3-}$$

Example 4-14: Write the balanced equation for the reaction of strontium with phosphorus, P_4. Show dot formulas for the atoms and ions.

$$6\ Sr + P_4 \longrightarrow 2\ Sr_3P_2$$

$$6\ Sr\!: + 4\cdot\ddot{P}\cdot \longrightarrow 6\ Sr^{2+},\ 4[:\ddot{P}:]^{3-}$$

Since Group VIA nonmetals form simple anions by gaining two electrons, they react with IA metals to form ionic compounds (M_2X), with IIA metals to form ionic compounds (MX), and with IIIA metals to form ionic compounds (M_2X_3).

Table 4-10 in the text summarizes the general formulas of the simple binary ionic compounds resulting from combinations of elements from different groups in the periodic table.

Example 4-15: Write formulas for the following compounds.

Compound	Formula	Compound	Formula
calcium phosphide	Ca_3P_2	sodium bromide	NaBr
rubidium sulfide	Rb_2S	gallium(III) chloride	$GaCl_3$
aluminum selenide	Al_2Se_3	strontium sulfide	SrS
magnesium oxide	MgO	lithium nitride	Li_3N
barium fluoride	BaF_2	aluminum oxide	Al_2O_3

4-8 Oxidation Numbers

The oxidation number or oxidation state of an element can be thought of as the number of electrons gained or lost by an atom when it forms binary ionic compounds. For single atom ions it corresponds to the formal change on the ion. Some elements exhibit different oxidation states in different compounds. Oxidation numbers of elements in covalent compounds can be used effectively as a mechanical aid in writing formulas and balancing equations. The more electronegative element is assigned a negative oxidation number and the less electronegative element is assigned a positive oxidation number. The most common oxidation states exhibited by elements are shown in Table 4-11 in the text. Three general rules for assigning oxidation numbers are given below. A more extensive set of rules is given in Section 4-12 in the text.

1. The oxidation number of any free, uncombined element is zero.
2. In any ion the sum of the oxidation numbers of the constituent elements is equal to the charge on the ion.
3. In a compound, the sum of the oxidation numbers of all elements is zero.

Example 4-16: What is the oxidation number of chromium in each of the following compounds or ions? (a) $CrCl_3$, (b) $CrCl_2$, (c) CrO_3, (d) Na_2CrO_4, (e) $K_2Cr_2O_7$, (f) $Cr_2O_7^{2-}$, (g) $Cr(OH)_4^-$, (h) $KCr(OH)_4$, and (i) $MnCrO_4$

The sum of the oxidation numbers of all the atoms present must equal the formal charge on the species. The oxidation numbers are shown above the formulas and the sums of the oxidation numbers are shown below.

(a) $\overset{+3}{Cr}\overset{-1}{Cl_3}$

$x + 3(-1) = 0$

$x = +3$

According to rule 4 (text) the oxidation number of chlorine is -1 so the oxidation number of chromium is +3.

(b) $\overset{+2}{Cr}\overset{-1}{Cl_2}$

$x + 2(-1) = 0$

$x = +2$

The oxidation number of chromium is +2.

(c) $\overset{+6}{Cr}\overset{-2}{O_3}$

$x + 3(-2) = 0$

$x = +6$

According to rule 5 (text) the oxidation number of oxygen is -2 since this is neither a peroxide nor a superoxide.

(d) $\overset{+1}{Na_2}\overset{+6}{Cr}\overset{-2}{O_4}$

$2(+1) + x + 4(-2) = 0$

$x = +6$

The oxidation numbers of sodium and oxygen are +1 and -2, respectively.

(e) $\overset{+1}{K_2}\overset{+6}{Cr_2}\overset{-2}{O_7}$

$2(+1) + 2x + 7(-2) = 0$

$x = +6$

The oxidation numbers of potassium and oxygen are +1 and -2, respectively.

(f) $\overset{+6}{Cr_2}\overset{-2}{O_7}{}^{2-}$

$2x + 7(-2) = -2$

$x = +6$

The dichromate ion is present in $K_2Cr_2O_7$ (e); the oxidation number of chromium is the same in both cases, +6.

(g) $\overset{+3}{Cr}(\overset{-2}{O}\overset{+1}{H})_4{}^-$

$x + 4(-2) + 4(+1) = -1$

$x = +3$

The oxidation numbers of oxygen and hydrogen and -2 and +1, respectively.

(h) $\overset{+1+3-2+1}{KCr(OH)_4}$

$+1 + x + 4(-2) + 4(+1) = 0$

$x = +3$

Potassium tetrahydroxychromate contains the $Cr(OH)_4^-$ ion (g).

(i) $\overset{+2\ +6\ -2}{MnCrO_4}$

$x + 6 + 4(-2) = 0$

$x = +2$

Both manganese and chromium are transition metals and show variable oxidation numbers. The oxidation number of chromium is +6 in CrO_4^{2-} (d). It is the same here and the oxidation number of manganese is +2.

<u>Example 4-17:</u> What is the oxidation number of nitrogen in the following species? (a) $RbNO_3$, (b) N_2, (c) HNO_2, (d) NH_3, and (e) NF_3

(a) $\overset{+1+5-2}{RbNO_3}$

$1 + x + 3(-2) = 0$

$x = +5$

Rubidium and oxygen exhibit oxidation numbers of +1 and −2, respectively.

(b) $\overset{0}{N_2}$

$2(0) = 0$

Any element, free and uncombined, exhibits an oxidation number of zero.

(c) $\overset{+1+3-2}{HNO_2}$

$1 + x + 2(-2) = 0 \quad x = +3$

Hydrogen and oxygen exhibit oxidation numbers of +1 and −2, respectively.

(d) $\overset{-3+1}{NH_3}$

$x + 3(+1) = 0$

$x = -3$

Nitrogen is more electronegative than hydrogen and is assigned a negative oxidation number.

(e) $\overset{+3-1}{NF_3}$

$x + 3(-1) = 0$

$x = +3$

Fluorine is more electronegative than nitrogen, so nitrogen is assigned a positive oxidation number.

<u>Example 4-18:</u> What is the oxidation number of the underlined element in the following species? (a) calcium hydride, $Ca\underline{H}_2$, (b) sodium hydride, $Na\underline{H}$, (c) sodium peroxide, $Na_2\underline{O}_2$, (d) the peroxide ion, $\underline{O}_2^{2-}$, (e) potassium superoxide, $K\underline{O}_2$, and (f) the superoxide ion, $\underline{O}_2^{-}$. These examples are somewhat unusual.

(a) $\overset{+2\,-1}{CaH_2}$

$2 + 2(x) = 0$

$x = -1$

The metal exhibit positive oxidation numbers in metal hydrides. Thus, the oxidation number of hydrogen is −1 in metal hydrides.

(b) $\overset{+1\,-1}{NaH}$

$1 + x = 0$

$x = -1$

(c) $\overset{+1\;\;-1}{Na_2O_2}$

$2(+1) + 2(x) = 0$

$x = -1$

The representative metals exhibit their common positive oxidation numbers in peroxides. The oxidation number of oxygen is −1 in peroxides.

(d) $\overset{-1}{O_2^{2-}}$

$2(x) = -1$

$x = -1$

(e) $\overset{+1\,-1/2}{KO_2}$

$1 + 2(x) = 0$

$x = -1/2$

Group IA and IIA elements exhibit oxidation numbers +1 and +2, respectively, in all their common compounds, and so oxygen must exhibit an oxidation number of −1/2 in superoxides.

(f) $\overset{-1/2}{O_2^{-}}$

$2(x) = -1$

$x = -1/2$

EXERCISES

1. Arrange the members of the following groups in order of increasing atomic radius.

   (a) Na, Mg, Al, Si, S
   (b) Si, Sn, C, Pb, Ge
   (c) H, O, Cl, F, I
   (d) K, Ca, Sc, Ti, V, Cr

2. Arrange the members of the following groups in order of increasing ionic radii.

   (a) F^-, Cl^-, Br^-, I^-, At^-
   (b) Tl^{3+}, Ga^{3+}, Al^{3+}, In^{3+}
   (c) Fe^{2+}, Fe^{3+}
   (d) Se^{2-}, Sr^{2+}, Rb^+, Br^-

3. Arrange the following species in order of increasing atomic or ionic volume.

   (a) S, Se, Se^{2-}, S^{2-} (b) Cu, Cu^+, Cu^{2+} (c) F^-, Ne, N^{3-}, Al^{3+}, Na^+

4. Arrange the members of the following groups of elements in order of increasing first ionization energy.

   (a) N, P, As, Sb, Bi
   (b) Ca, Ba, Mg, Be, Sr
   (c) Li, Be, B, C, N, O, F
   (d) Ba, Cs, Pb, Po, Tl

5. Explain any deviations in general trends observed in problem 4.

6. Arrange the members of the following groups of elements in order of increasing electronegativity.

   (a) Na, Mg, Al, Si, S, Cl
   (b) Ge, Ga, Ca, K, Br
   (c) C, Si, Ge, Sn, Pb
   (d) I, Sn, Te, Sb, Sr

7. Show the transfer of electrons ($\underline{\uparrow\downarrow}$ notation) in the reactions between the following pairs of elements.

   (a) Sr and Cl (c) Na and S_8 (e) K and Cl_2 (g) Al and O_2
   (b) Li and F_2 (d) Ba and P_4 (f) Li and O_2 (h) Ca and Br_2

8. Write condensed representations of the electron distributions for atoms and ions of each of the reactants and products in problem 7.

9. Write formulas for the binary ionic compounds of the following elements.

(a) barium and sulfur
(b) sodium and nitrogen
(c) aluminum and nitrogen
(d) calcium and fluorine
(e) potassium and sulfur
(f) rubidium and selenium
(g) magnesium and iodine
(h) lithium and bromine
(i) cesium and tellurium
(j) barium and nitrogen

10. Assign oxidation numbers to all elements in the ionic compounds in problem 9.

11. Assign oxidation numbers of all elements in the compounds below.

(a) $KMnO_4$
(b) H_2SO_4
(c) $(NH_4)_2SO_4$
(d) $CaSO_3$
(e) $KHSO_4$
(f) $NaBrO_3$
(g) $Na_2Cr_2O_7$
(h) N_2O_4
(i) Cl_2
(j) NaH
(k) $PbCl_4$
(l) As_4O_6
(m) Na_2O_2
(n) H_2CO_3
(o) C_2H_6

12. Assign oxidation numbers to all elements in the ions below.

(a) ClO_4^-
(b) $S_2O_3^{2-}$
(c) NH_4^+
(d) $Ca(H_2O)_6^{2+}$
(e) $N_2H_5^+$
(f) ClO_2^-
(g) SO_3^{2-}
(h) MnO_4^-
(i) $Cu(NH_3)_4^{2+}$
(j) NO_3^-
(k) NO_2^-
(l) HCO_3^-
(m) CO_3^{2-}
(n) HSO_3^-
(o) SiF_6^{2-}

ANSWERS TO EXERCISES

1. (a) $S < Si < Al < Mg < Na$
(b) $C < Si < Ge < Sb < Pb$
(c) $H < F < O < Cl < I$
(d) $Cr < V < Ti < Sc < Ca < K$

2. (a) $F^- < Cl^- < Br^- < I^- < At^-$
(b) $Al^{3+} < Ga^{3+} < In^{3+} < Tl^{3+}$
(c) $Fe^{3+} < Fe^{2+}$
(d) $Sr^{2+} < Rb^+ < Br^- < Se^{2-}$

3. (a) $S < Se < S^{2-} < Se^{2-}$
(b) $Cu^{2+} < Cu^+ < Cu$
(c) $Al^{3+} < Na^+ < Ne < F^- < N^{3-}$

4. (a) Bi < Sb < As < P < N
(b) Ba < Sr < Ca < Mg < Be
(c) Li < B < Be < C < O < N < F
(d) Cs < Ba < Tl < Pb < Po

5. Generally, first ionization energies increase from left-to-right across a period and from bottom-to-top within a group. In 4(c) the ionization energy of B is less than that of Be. This is because the energy required to remove the most loosely held electron, the first one in a set of p orbitals, in the Group IIIA elements is less than that for the IIB elements, which must lose an electron from a filled s orbital. Completely filled sets of orbitals are relatively stable configurations. Another reversal in the expected trend occurs at the VIA elements where pairing of electrons in p orbitals begins.

6. (a) Na < Mg < Al < Si < S < Cl
(b) K < Ca < Ga < Ge < Br
(c) Pb < Sn ≈ Si < Ge < C (This is not the expected order; Ge is out of place.)
(d) Sr < Sn < Sb < Te < I

7. (a) Sr [Kr] 5s: ↑↓ ; 2 Cl [Ne] 3s: ↑↓ 3p: ↑↓ ↑↓ ↑
⟶ Sr^{2+} [Kr] 5s: — ; 2 Cl^- [Ne] 3s: ↑↓ 3p: ↑↓ ↑↓ ↑↓

(b) Li 1s: ↑↓ 2s: ↑ ; F 1s: ↑↓ 2s: ↑↓ 2p: ↑↓ ↑↓ ↑
⟶ Li^+ 1s: ↑↓ 2s: — ; F^- 1s: ↑↓ 2s: ↑↓ 2p: ↑↓ ↑↓ ↑↓

(c) 2 Na [Ne] 3s: ↑ ; S [Ne] 3s: ↑↓ 3p: ↑↓ ↑ ↑
⟶ 2 Na^+ [Ne] 3s: — ; S^{2-} [Ne] 3s: ↑↓ 3p: ↑↓ ↑↓ ↑↓

(d) 3 Ba [Xe] 6s: ↑↓ ; 2 P [Ne] 3s: ↑↓ 3p: ↑ ↑ ↑
⟶ 3 Ba^{2+} [Xe] 6s: — ; 2 P^{3-} [Ne] 3s: ↑↓ 3p: ↑↓ ↑↓ ↑↓

7. (e) K [Ar] ↑ (4s) → K^+ [Ar] __ (4s)

Cl [Ne] ↑↓ (3s) ↑↓ ↑↓ ↑ (3p) → Cl^- [Ne] ↑↓ (3s) ↑↓ ↑↓ ↑↓ (3p)

(f) 2 Li ↑↓ (1s) ↑ (2s) → 2 Li^+ ↑↓ (1s) __ (2s)

O ↑↓ (1s) ↑↓ (2s) ↑↓ ↑ ↑ (2p) → O^{2-} ↑↓ (1s) ↑↓ (2s) ↑↓ ↑↓ ↑↓ (2p)

(g) 2 Al [Ne] ↑↓ (3s) ↑ __ __ (3p) → 2 Al^{3+} [Ne] __ (3s) __ __ __ (3p)

3 O ↑↓ (1s) ↑↓ (2s) ↑↓ ↑ ↑ (2p) → 3 O^{2-} ↑↓ (1s) ↑↓ (2s) ↑↓ ↑↓ ↑↓ (2p)

(h) Ca [Ar] ↑↓ (4s) → Ca^{2+} [Ar] __ (4s)

2 Br^- [Ar] $3d^{10}$ ↑↓ (4s) ↑↓ ↑↓ ↑ (4p) → 2 Br^- [Ar] $3d^{10}$ ↑↓ (4s) ↑↓ ↑↓ ↑↓

8. (a) Sr $1s^2 2s^2 2p^6 3s^2 3p^6 4s^2 3d^{10} 4p^6 5s^2$

2 Cl $1s^2 2s^2 2p^6 3s^2 3p^5$

↓

Sr^{2+} $1s^2 2s^2 2p^6 3s^2 3p^6 4s^2 3d^{10} 4p^6$

2 Cl^- $1s^2 2s^2 2p^6 3s^2 3p^6$

(b) Li $1s^2 2s^1$ → Li^+ $1s^2$

F $1s^2 2s^2 2p^5$ → F^- $1s^2 2s^2 2p^6$

(c) 2 Na $1s^2 2s^2 2p^6 3s^1$ → 2 Na^+ $1s^2 2s^2 2p^6$

S $1s^2 2s^2 2p^6 3s^2 3p^4$ → S^{2-} $1s^2 2s^2 2p^6 3s^2 3p^6$

(d) 3 Ba $1s^2 2s^2 2p^6 3s^2 3p^6 4s^2 3d^{10} 4p^6 5s^2 4d^{10} 5p^6 6s^2$

2 P $1s^2 2s^2 2p^6 3s^2 3p^3$

↓

3 Ba^{2+} $1s^2 2s^2 2p^6 3s^2 3p^6 4s^2 3d^{10} 4p^6 5s^2 4d^{10} 5p^6$

2 P^{3-} $1s^2 2s^2 2p^6 3s^2 3p^6$

8. (e) K $1s^2 2s^2 2p^6 3s^2 3p^6 4s^1$ $\quad$ Cl $1s^2 2s^2 2p^6 3s^2 3p^5$ $\longrightarrow$ K^+ $1s^2 2s^2 2p^6 3s^2 3p^6$ $\quad$ Cl^- $1s^2 2s^2 2p^6 3s^2 3p^6$

(f) 2 Li $1s^2 2s^1$ $\quad$ O $1s^2 2s^2 2p^4$ $\longrightarrow$ 2 Li^+ $1s^2$ $\quad$ O^{2-} $1s^2 2s^2 2p^6$

(g) 2 Al $1s^2 2s^2 2p^6 3s^2 3p^1$ $\quad$ 3 O $1s^2 2s^2 2p^4$ $\longrightarrow$ 2 Al^{3+} $1s^2 2s^2 2p^6$ $\quad$ 3 O^{2-} $1s^2 2s^2 2p^6$

(h) Ca $1s^2 2s^2 2p^6 3s^2 3p^6 4s^2$ $\quad$ 2 Br $1s^2 2s^2 2p^6 3s^2 3p^6 4s^2 3d^{10} 4p^5$ $\longrightarrow$ Ca^{2+} $1s^2 2s^2 2p^6 3s^2 3p^6$ $\quad$ 2 Br^- $1s^2 2s^2 2p^6 3s^2 3p^6 4s^2 3d^{10} 4p^6$

9. (a) BaS (b) Na_3N (c) AlN (d) CaF_2 (e) K_2S (f) Rb_2Se
(g) MgI_2 (h) $LiBr$ (i) Cs_2Te (j) Ba_3N_2

10. (a) $\overset{+2\,-2}{BaS}$ (b) $\overset{+1\;-3}{Na_3N}$ (c) $\overset{+3\,-3}{AlN}$ (d) $\overset{+2\,-1}{CaF_2}$ (e) $\overset{+1\,-2}{K_2S}$ (f) $\overset{+1\;-2}{Rb_2Se}$
(g) $\overset{+2\;-1}{MgI_2}$ (h) $\overset{+1\,-1}{LiBr}$ (i) $\overset{+1\;-2}{Cs_2Te}$ (j) $\overset{+2\;-3}{Ba_3N_2}$

11. (a) $\overset{+1+7-2}{KMnO_4}$ (b) $\overset{+1+6-2}{H_2SO_4}$ (c) $\overset{-3+1\;\;+6-2}{(NH_4)_2SO_4}$ (d) $\overset{+2+4-2}{CaSO_3}$ (e) $\overset{+1+1+6-2}{KHSO_4}$
(f) $\overset{+1\;+5-2}{NaBrO_3}$ (g) $\overset{+1\;\;+6\;-2}{Na_2Cr_2O_7}$ (h) $\overset{+4\;-2}{N_2O_4}$ (i) $\overset{0}{Cl_2}$ (j) $\overset{+1\;-1}{NaH}$
(k) $\overset{+4\,-1}{PbCl_4}$ (l) $\overset{+3\;-2}{As_4O_6}$ (m) $\overset{+1\;\;-1}{Na_2O_2}$ (n) $\overset{+1+4-2}{H_2CO_3}$ (o) $\overset{-3\;+1}{C_2H_6}$

12. (a) $\overset{+7\,-2}{ClO_4}^-$ (b) $\overset{+2\,-2}{S_2O_3}^{2-}$ (c) $\overset{-3+1}{NH_4}^+$ (d) $\overset{+2\;\;+1-2}{Ca(H_2O)_6}^{2+}$
(e) $\overset{-2+1}{N_2H_5}^+$ (f) $\overset{+3-2}{ClO_2}^-$ (g) $\overset{+4-2}{SO_3}^{2-}$ (h) $\overset{+7\;-2}{MnO_4}^-$
(i) $\overset{+2\;\;-3+1}{Cu(NH_3)_4}^{2+}$ (j) $\overset{+5-2}{NO_3}^-$ (k) $\overset{+3-2}{NO_2}^-$ (l) $\overset{+1+4-2}{HCO_3}^-$
(m) $\overset{+4\;-2}{CO_3}^{2-}$ (n) $\overset{+1+4-2}{HSO_3}^-$ (o) $\overset{+4-1}{SiF_6}^{2-}$

Chapter Five

COVALENT BONDING AND INORGANIC NOMENCLATURE

D,G,W: Sec. 6-11 through 6-14, Sec. 7-1 through 7-10, Sec. 8-1 through 8-8

Covalent bonds are formed when one or more pairs of electrons are shared between atoms. If two electrons (one pair) are shared between two atoms the bond is a single bond, while four and six electrons shared between two atoms give double and triple bonds, respectively.

5-1 Dot Formulas of Molecules and Complex Ions

Recall that dot formulas of atoms show only the electrons in the outermost shells. The following rule-of-thumb is used to determine the number of electrons shared in covalent bonds involving representative (A-group) elements in molecules or polyatomic ions. Most atoms attain a share in eight electrons, an octet, or noble gas configuration (only two electrons for hydrogen). The total number of electrons shared (S) equals the sum of the number of outer shell electrons needed by all the atoms to attain noble gas configuration (N) minus the total number available in outermost shells (A).

$$S = N - A$$

$$\text{shared} = (\text{needed}) - (\text{available})$$

The periodic group number gives the number of electrons available in the outermost shell of a representative element. Eight are needed to attain a noble gas configuration (two for hydrogen). This very useful relationship is restricted to those cases in which the central atom achieves a noble gas electronic configuration. The relationship may not be used for the following cases.

1. most covalent compounds of beryllium
2. most covalent compounds of Group IIIA elements
3. compounds in which the central element must have a share of more than eight valence electrons to accommodate all substituents
4. compounds containing d- or f-transition metals
5. species containing an odd number of electrons

The central atom of a molecule or complex ion is always the one needing the greatest number of electrons to attain a noble gas configuration.

Example 5-1: Write the dot formula for carbon dioxide.

Each carbon and oxygen atom needs a share of eight electrons, so N = 8 x 3 = 24. Carbon (the central atom) has 4 available electrons and the oxygen atoms have 6 each, so A = 4 + 2(6) = 16.

N - A = S

24 - 16 = 8, or 8 e^- must be shared

:Ö::C::Ö: or :Ö = C = Ö: (double bonds between C and O)

Example 5-2: Write the dot formula for NaCN, an ionic compound. The charge on the cyanide ion is 1-.

For CN^-: N - A = S

2(8) - (4+5+1) = S, 16 - 10 = 6 e^- shared.

Na^+, :C⋮⋮N:$^-$ or Na^+, :C≡N:$^-$

Example 5-3: Write the dot formula for ammonium sulfate, $(NH_4)_2SO_4$, an ionic compound.

For NH_4^+: N - A = S

[8+4(2)] - [5+4(1) - 1] = S, 16 - 8 = 8 e^- shared

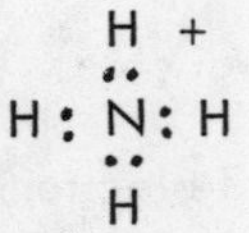

or

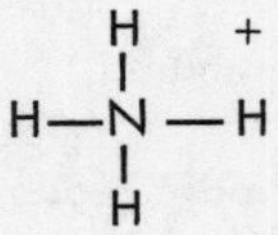

For SO_4^{2-}: N - A = S

5(8) - [6+4(6)+2] = S, 40 - 32 = 8 e^- shared

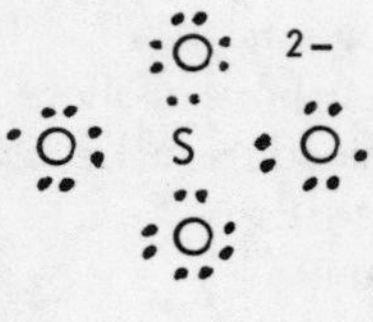

or

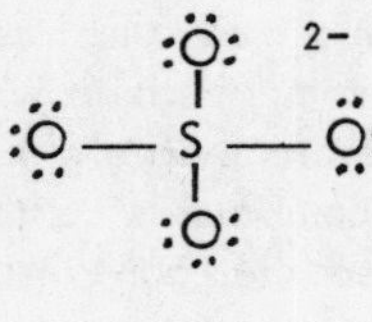

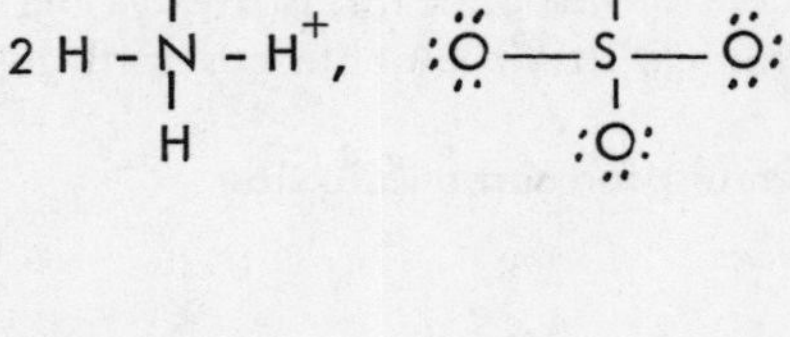

Example 5-4: Write the dot formula for the hydrocarbon, ethane, C_2H_6. Ethane is a covalent compound that contains a C–C bond.

$$N \quad - \quad A \quad = S$$

$$[2(8)+6(2)] - [2(4)+6(1)] = S, \quad 28 - 14 = 14 \text{ e}^- \text{ shared}$$

```
H  H                              H   H
..  ..                            |   |
H:C: C:H          or          H - C - C - H
..  ..                            |   |
H  H                              H   H
```

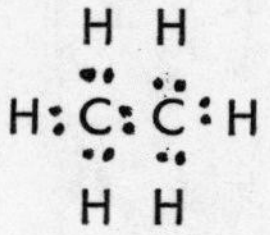

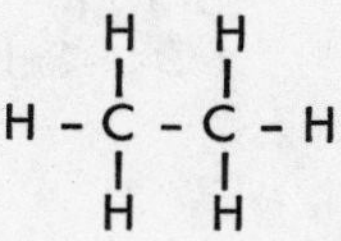

Example 5-5: Write the dot formula for the hydrocarbon, ethylene (or ethene), C_2H_4, a covalent compound that contains a C=C bond.

$$N \quad - \quad A \quad = S$$

$$[2(8)+4(2)] - [2(4)+4(1)] = S, \quad 24 - 12 = 12 \text{ e}^- \text{ shared}$$

```
H  H                              H   H
..  ..                            |   |
H:C::C:H          or          H - C = C - H
```

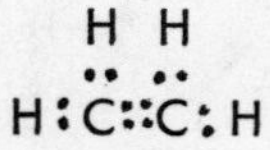

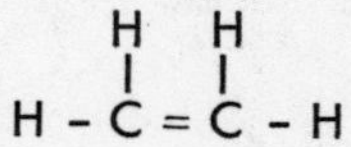

Example 5-6: Write the dot formula for the hydrocarbon, acetylene (or ethyne), C_2H_2.

$$N \quad - \quad A \quad = S$$

$$[2(8)+2(2)] - [2(4)+2(1)] = S \quad 20 - 10 = 10 \text{ e}^- \text{ shared}$$

```
         ..
H:C::C:H          or          H - C ≡ C - H
         ..
```

Example 5-7: Write the dot formula for phosphoric acid, H_3PO_4. Phosphorus is the central atom and all H atoms are bound to O atoms.

$$N \quad - \quad A \quad = S$$

$$[3(2)+8+4(8)] - [3(1)+5+4(6)] = S, \quad 46 - 32 = 14 \text{ e}^- \text{ shared}$$

```
   :O:                              :O:
   ..                               |
H:O: P:O:H        or          H - O - P - O - H
   ..                               |
   :O:                             :O:
   ..                               |
   H                                H
```

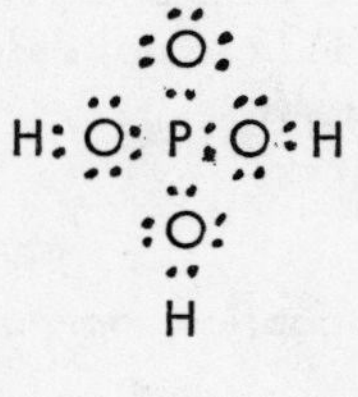

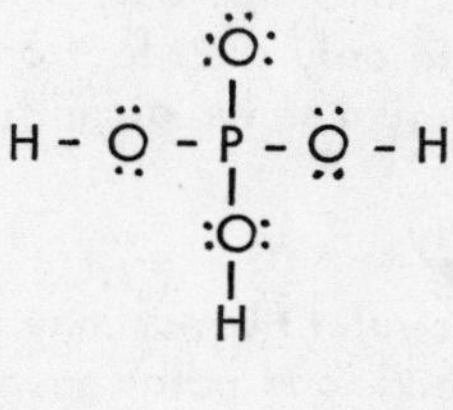

5-2 Resonance

Some molecules and complex ions cannot be represented by a single dot formula that reflects accurately their properties. For such species we draw several formulas, and such species are said to exhibit resonance. The different formulas that we draw are called resonance structures.

Example 5-8: Draw the three equivalent dot structures for sulfur trioxide, SO_3. Sulfur is the central atom.

$$N \quad - \quad A \quad = S$$

$$[8+3(8)] - [6+3(6)] = S, \qquad 32 - 24 = 8\ e^- \text{ shared}$$

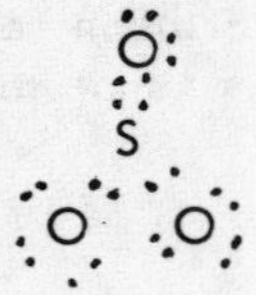

or

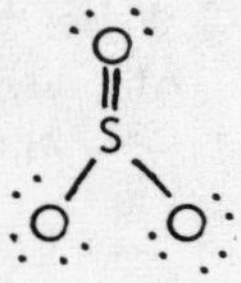

The three equivalent resonance structures are:

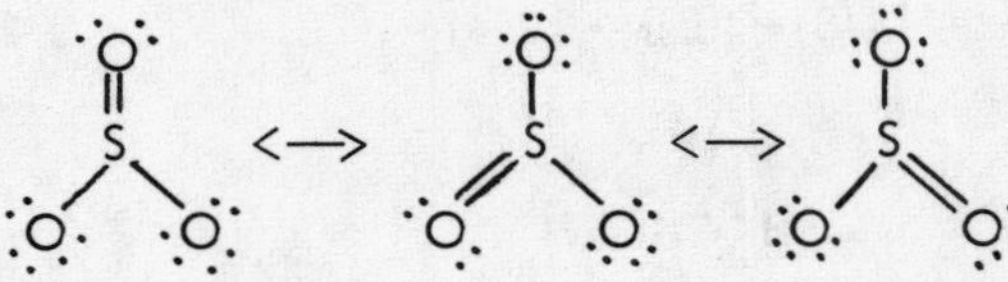

Experimental data indicate that there are no formal single or double bonds in SO_3 molecules. All bonds are equivalent (identical) and have one-third double bond character and two-thirds single bond character.

Example 5-9: Draw one resonance structure for a nitrate ion, NO_3^-. Nitrogen is the central element.

$$N \quad - \quad A \quad = S$$

$$[8+3(8)] - [5+3(6)+1) = S, \qquad 32 - 24 = 8\ e^- \text{ shared}$$

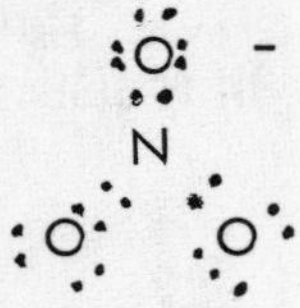

or

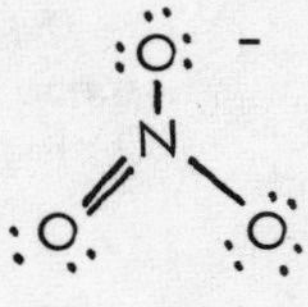

(There are two other equivalent structures.)

Please keep in mind the fact that dot formulas for molecules and polyatomic ions indicate only the electrons involved in bonding and are not designed to show the actual three-dimensional arrangements of atoms.

5-3 Polarity

Covalent bonds may be classified in two categories, nonpolar covalent bonds and polar covalent bonds.

1. Nonpolar Covalent Bonds

Nonpolar covalent bonds result from the equal sharing of electrons between two atoms of equal electronegativities. Strictly speaking, both atoms must be identical to form nonpolar bonds. These bonds occur mainly in homonuclear diatomic molecules such as H_2, N_2, O_2, F_2, Cl_2, Br_2, I_2, but are also present in some more complex molecules such as P_4 and S_8.

Nitrogen molecules have six shared electrons (triple bond). The electron density is symmetrically distributed about the two nuclei (as drawn below in only two dimensions).

:N⋮⋮N: or :N ≡ N:

Example 5-10: Draw a representation of the bromine, Br_2, molecule.

N - A = S

2(8) - 2(7) = 2 e^- shared

:Br:Br: or :Br — Br:

Bromine molecules contain single bonds with a symmetrical distribution of electron density about the two nuclei.

2. Polar Covalent Bonds

Bonds in which electron pairs are shared unequally between two atoms with different electronegativities are called polar covalent bonds. They are intermediate between nonpolar covalent bonds and ionic bonds. The greater the difference in electronegativities of the atoms is, the greater the charge separation and the more polar the bonds. Electronegativities are tabulated in Table 4-6 in the text.

	$\longmapsto$ H — Cl	$\longmapsto$ H — Br		$\delta+$ H — $\delta-$ Cl	$\delta+$ H — $\delta-$ Br
EN	2.1 3.0	2.1 2.8	or		
Δ(EN)	0.9	0.7			

The $\delta+$ and $\delta-$ designations and the crossed arrows both indicate the dipolar character of the bonds. It is not necessary to use both types of notation since they indicate the same thing. The $\delta-$ atom has greater than an equal share while the $\delta+$ atom has less than an equal share of bonding electrons. The atoms are said to possess partial charges, not formal charges as ions do. When the crossed arrow designation is used, the head of the arrow is directed toward the more electronegative atom. The electronegativity difference is slightly greater for HCl than for HBr, and so HCl molecules are slightly more polar.

Example 5-11:   Consider the following covalent compounds, CH_4, CF_4, CCl_4, CBr_4 and CI_4. Carbon is the central atom in each, and each has four single bonds. Draw the dot formula for each, and then determine the order of bond polarity.

For CH_4:   N   –   A   = S

$$[8+4(2)] - [4+4(1)] = S, \qquad 16 - 8 = 8 \; e^- \text{ shared}$$

```
        H                              H
        ..                             |
     H : C : H          or         H - C - H
        ..                             |
        H                              H
```

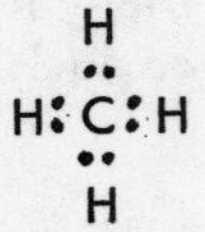

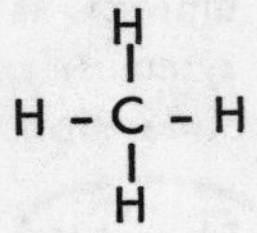

Since CI_4, CF_4, CCl_4, and CBr_4 molecules all involve carbon bonded to a Group VIIA element, all can be represented by the following (where X = F, Cl, Br, I):

CS_4:   N   –   A   = S

$$[8+4(8)] - [4+4(7)] = S \qquad 40 - 32 = 8 \; e^- \text{ shared}$$

```
      ..             ..              ..              ..
     :I:            :F:            :Cl:            :Br:
      |              |               |               |
  :I - C - I:    :F - C - F:    :Cl - C - Cl:    :Br - C - Br:
      |              |               |               |
     :I:            :F:            :Cl:            :Br:
      ..             ..              ..              ..
```

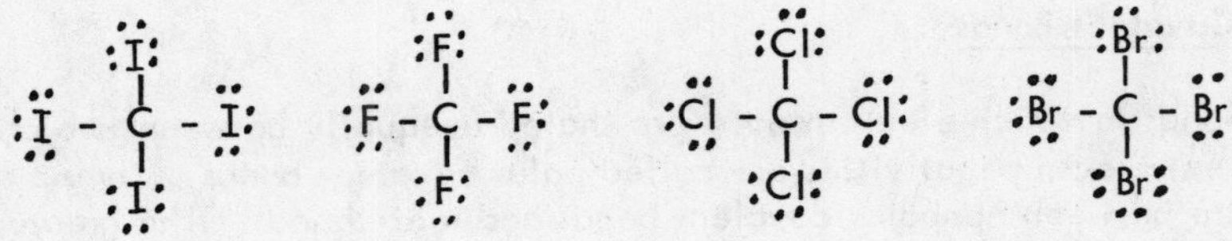

The electronegativities are C (2.5), H (2.1), I (2.5), F (4.0), Cl (3.0), Br (2.8). The differences in electronegativities, Δ(EN), are:

	C-H	C-F	C-Cl	C-Br	C-I
Δ(EN)	0.4	1.5	0.5	0.3	~ 0

Note that C is more electronegative than H but less electronegative than F, Cl, and Br, which accounts for the directional differences in polarity. To two significant figures, the electronegativities of C and I are the same, and the Cl-I bond is essentially nonpolar. The magnitude of bond polarity increases in the order.

$$\underline{\text{C-I} < \text{C-Br} < \text{C-H} < \text{C-Cl} < \text{C-F}}$$
Increasing bond polarity

5-4 Polar and Nonpolar Molecules

The fact that molecules like CF_4 have very polar bonds does not mean that they are polar molecules. In fact CF_4, as well as all other molecules in Example 5-11, are nonpolar because the individual bond dipoles are symmetrically located, and therefore they cancel. On the other hand molecules like HF, HCl, HBr, and HI all contain polar covalent bonds and are also polar molecules.

The polarities of molecules (not individual bonds) are described in terms of dipolemoments which measure the tendencies of molecules to align themselves with an applied external electric field. Dipole moments (μ) must be experimentally measured and are usually expressed in Debyes (D). Dipole moment is the magnitude of charge separated times the distance of separation. In diatomic molecules, the charge separation in the bonds is the same as the charge separation in molecules. But in more complex molecules, the individual bond dipoles may interact constructively or destructively. The dipole moment of a molecule is determined by the magntiudes and the directions of the bond dipoles. The representations of CH_4, CF_4, CCl_4, CBr_4, and CI_4 in Example 5-11 do not show three-dimensional shapes; the four bonds in each molecule are directed toward the corners of a regular tetrahedron, so that their effects add vectorially to zero. Dipole moments give valuable information about the geometry of molecules.

<u>Example 5-12:</u> Draw the dot formulas of CO_2 and SO_2. The dipole moment of CO_2 is zero, while for SO_2 μ = 1.6 D. Explain.

For CO_2: N - A = S, [8+2(8)] - [4+2(6)] = S, 24 - 16 = 8 e^- shared in two double bonds

:Ö = C = Ö: , the molecule must be linear because μ = 0 D

For SO_2: N - A = S

[8+2(8)] - [6+2(6)] = S, 24 - 18 = 6 e^- shared

:Ö::S̈:Ö: or :Ö = S̈ − Ö: <—> :Ö − S̈ = Ö: (resonance structures)

Sulfur dioxide molecules contain two equivalent bonds that may be represented as

O ═ S̈ ═ O (unshared e^- on O's not shown)

However the dipole moment of SO_2, μ = 1.6 D, precludes the possibility of a linear structure and the molecule is a bent (or angular) molecule. (The bond angle is 119.5^o.)

, $\mu = 1.6$ D (unshared e^- on O's not shown)

bent

5-5 Geometry of Covalent Molecules and Polyatomic Ions

The most important factor in determining the geometry of molecules and complex ions is the number of sets of electrons (regions of high electron density) around the central element. Table 5-1 indicates the electronic geometries and (bond) angles corresponding to different numbers of electron sets. For these purposes, a set of electrons may consist of: a single bond, a double bond, a triple bond, or a lone pair of electrons (unshared pair) in the outermost electronic shell of the central atom.

Table 5-1

Number of Electron Sets About the Central Atom and Electronic Geometry

Number of Electron Sets	Electronic Geometry*		Angles**
2	linear		180°
3	trigonal planar	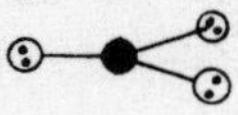	120°
4	tetrahedral	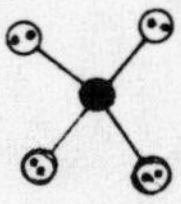	109° 28'
5	trigonal bipyramidal	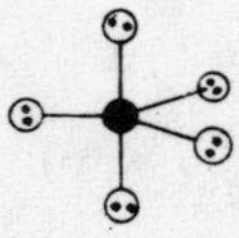	90°, 120° 180°
6	octahedral	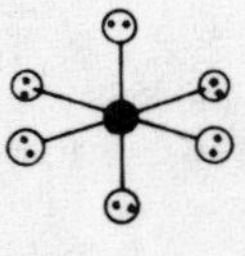	90°, 180°

*Electronic geometries are illustrated using only single pairs of electrons as electron groups.

**Angles made by imaginary lines through the nucleus and the centers of regions of high electron density.

The angles in molecules may be distorted by different substituents associated with electron sets. The electronic geometry describes the arrangement of the electron groups while molecular geometry describes the arrangment of atoms surrounding the central element. Electronic and molecular geometries are identical if and only if there are no lone (unshared) pairs of electrons on the central atom.

These electronic geometries are predicted (and experimentally observed) from the Valence Shell Electron Pair Repulsion Theory (VSEPR). It states: groups of electrons arrange themselves around the central atom in the way that minimizes repulsions among sets of electrons. Molecular geometries are described in terms of the positions of atoms, not electron pairs.

Example 5-13: Describe the electronic and molecular geometries of methane, CH_4. What are the bond angles?

In Example 5-11 we demonstrated that the dot formula of CH_4 is:

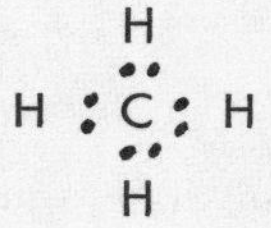

There are four sets of electrons, each associated with a hydrogen atom. Thus, both the electronic geometry and molecular geometry are tetrahedral. All H-C-H bond angles are 109°28'.

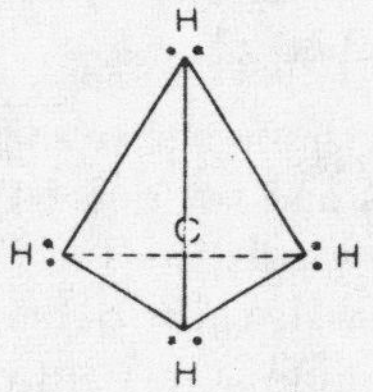

Example 5-14: Describe the electronic and molecular geometries and bond angles in boron trichloride molecules, BCl_3. (Note: This is a compound in which the central element is not surrounded by eight outer shell electrons.)

Its dot formula is:

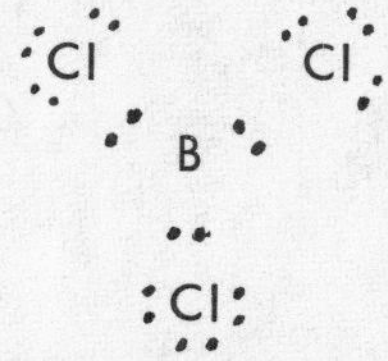

Since there are no lone pairs of electrons on boron, the electronic and molecular geometries must be the same, trigonal planar. The bond angles are 120°.

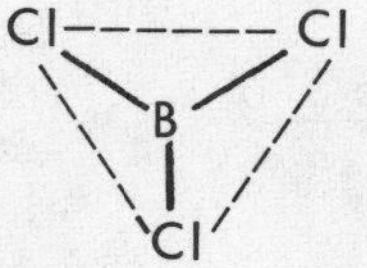

If the substituents (atoms) associated with the pairs of electrons are different, distortions of bond angles result. The distortions depend upon the relative magnitudes of the repulsions of the electron groups. The following generalization can be made. Repulsions between two lone pairs (lp) are greater than the repulsions between a lone pair and a bonded pair (bp) which, in turn, are greater than those between bonded pairs.

$$lp\text{-}lp > lp\text{-}bp > bp\text{-}bp$$

Example 5-15: Describe the electronic geometry, molecular geometry and bond angle of H_2O molecules. Are the molecules polar or nonpolar?

First write the dot formula.

$$N \quad - \quad A \quad = S$$

$$[8+2(2)] - [6+2(1)] = S, \qquad 12 - 8 = 4\ e^- \text{ shared}$$

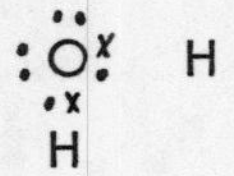

There are four regions of high electron density so the electronic geometry is tetrahedral.

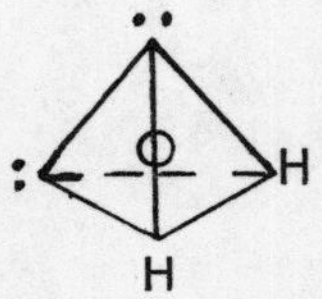

Repulsions between the lone pairs are greater than repulsions between lone pairs and bonded pairs. The smallest repulsion is between the bonded pairs. As a result the H-O-H bond angle is 104.5°, somewhat less than 109°28'. The molecular geometry is defined by the positions of the atoms, and the molecules are bent or angular. The bond dipoles are directed from the hydrogen atoms toward the more electronegative oxygen. The lone pairs give dipoles away from the oxygen toward the lone pairs. The net dipole (the vector sum of the four individual dipoles) is shown below.

individual dipoles

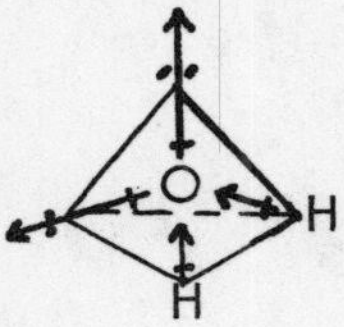

net dipole

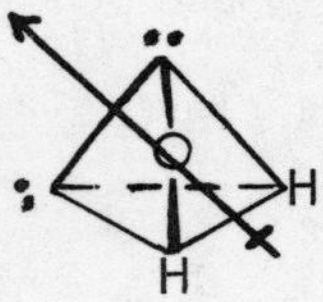

Example 5-16: Predict geometry, bond angles, and polarity of beryllium iodide, BeI_2, a covalent compound with fewer than eight electrons around the central element.

Since the rule-of-thumb for determining dot formulas cannot be used we must use chemical intuition. Beryllium contains two electrons in its outermost shell, and each one is shared with an iodine atom. Thus, there are two electron groups around the beryllium atom and both electronic and molecular geometries are linear. The I-Be-I angle is 180°.

Iodine is more electronegative than beryllium and the two bond dipoles, equal in magnitude but opposite in direction, cancel. The molecule is nonpolar.

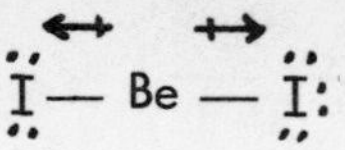

Example 5-17: Predict the electronic and molecular geometries, bond angles, and polarity of sulfur dioxide molecules.

The dot formula and shape of SO_2 were described in Example 5-12.

(unshared e^- on O's not shown)

There are three electron groups around the sulfur atom. The electronic geometry is trigonal planar and the molecular geometry is bent or angular. Due to the presence of the lone pair of electrons on the sulfur atoms the O-S-O bond angle is slightly less than 120°. It is approximately 119.5°. The molecule is polar since it is unsymmetrical.

Example 5-18: Predict the geometries, bond angles, and polarity of SO_3 molecules.

The dot formula of SO_3 was determined in Example 5-8. It has three resonance structures, which can be summarized as

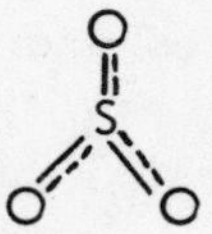

(unshared e^- on O's not shown)

There are three electron groups around the central atom and both the electronic and molecular geometries are trigonal planar, with 120° bond angles. Since the molecule is symmetrical it is nonpolar even though it has three polar bonds.

Example 5-19: Predict the geometry and bond angles of the sulfite ion, SO_3^{2-}.

$$N \quad - \quad A \quad = S$$

$$[8+3(8)] - [6+3(6)+2] = S, \quad 32 - 26 = 6\ e^- \text{ shared}$$

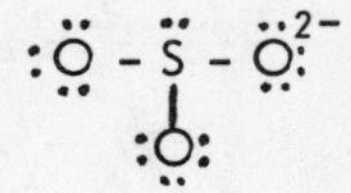

There are four regions of electron density about the sulfur atom and so the electronic geometry is tetrahedral. The one lone pair is not included in the ionic geometry, and the ion is shaped like a tripod (pyramidal) with the sulfur atom at the apex. The O-S-O bond angles are all slightly less than 109°28' due to the presence of the lone pair on the sulfur atom.

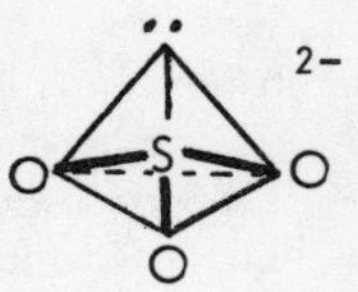

Example 5-20: Predict the electronic and molecular geometries, bond angles, and polarity of the ammonia molecule, NH_3.

$$N \quad - \quad A \quad = S$$

$$[8 + 3(2)] - [5+3(1)] = S, \qquad 14 - 8 = 6\ e^- \text{ shared}$$

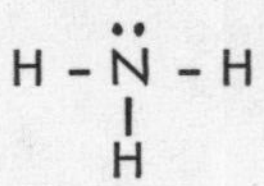

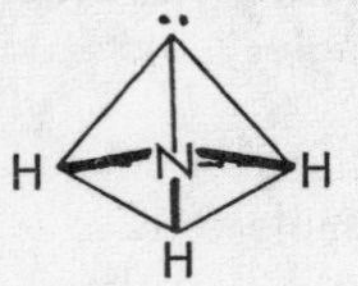

The electronic geometry is tetrahedral and the molecular geometry is pyramidal (tripod-like). The H-N-H bond angles are slightly less than 109°28' (107.3°), and the molecule is polar with a net dipole directed toward the lone pair.

Example 5-21: Determine the electronic and molecular geometries, and the bond angles of the ammonium ion, NH_4^+.

A coordinate covalent bond is a covalent bond in which both of the shared electrons are furnished by the same atom. As an example, the formation of the ammonium ion by addition of H^+ from an acid to the lone pair of electrons of ammonia results in a coordinate covalent bond.

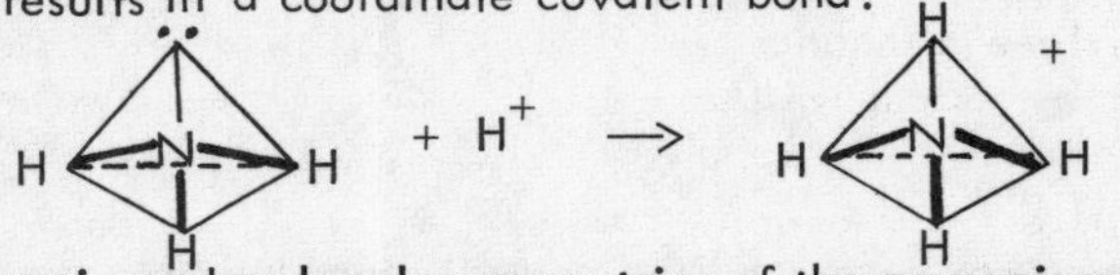

The electronic and molecular geometries of the ammonium ion are tetrahedral, and the bond angles are 109°28'.

Some nonmetals in period three (and below) of the periodic table are capable of accommodating more than eight electrons in their outermost shells by utilizing empty d orbitals in the same shell as their valence electrons.

Example 5-22: Describe the electronic and molecular geometries, bond angles, and polarity of phosphorus pentachloride, PCl_5, a covalent compound in the gas phase.

The dot formula is derived from chemical intuition since there are ten electrons involved in bonding. Phosphorus has five electrons in its outermost electron shell and each chlorine atom shares one of its electrons with phosphorus.

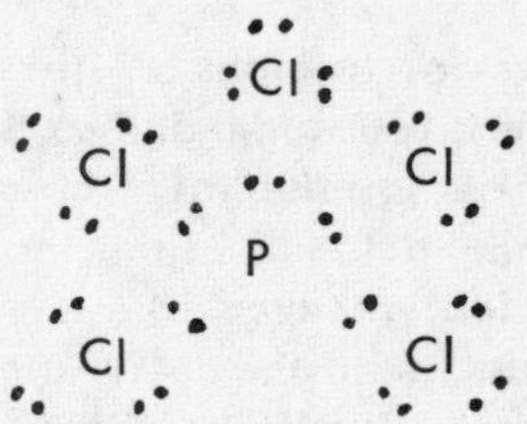

There are five electron groups in the outer shell of phosphorus, and both the electronic and molecular geometries are trigonal bipyramidal. The bond angles are 90°, 120°, and 180°. Since the molecule is symmetrical, all

P–Cl bond dipoles cancel and the molecule is nonpolar.

Example 5–23: Determine the geometries, bond angles, and polarity of IF_5 molecules.

Iodine has seven electrons in its outermost shell. Five of them are involved in bonds to the five fluorine atoms, leaving one lone pair. The dot formula is:

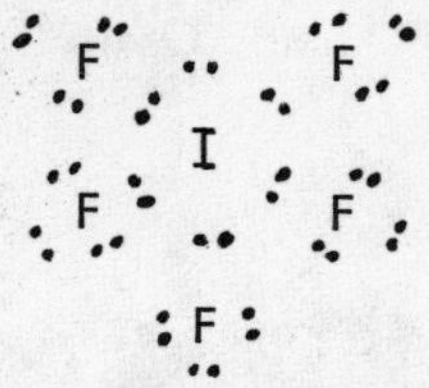

The electronic geometry is octahedral, and the molecular geometry is a distorted pyramid with a square base.

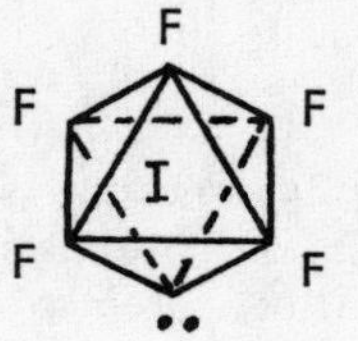

or

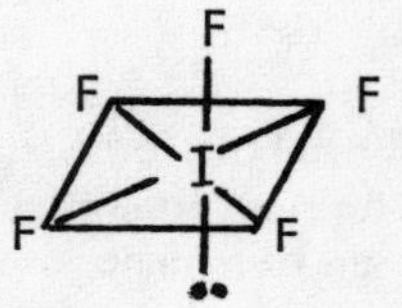

(pyramid with square base)

The I is somewhat "below" the plane of the 4 "in-plane" F atoms. Due to the presence of the lone pair of electrons on the iodine atom, the bond angles are slightly less than 90° and 180° and the molecule is polar.

5–6 Valence Bond Approach to Covalent Bonding

Valence Bond theory describes covalent bonding in terms of overlap of orbitals on separate atoms which form bonds as electrons in these orbitals pair up with each other. Such bonding often involves hybridization of atomic orbitals of the central element. Hybridized orbitals have some of the characterisitics of the component atomic orbitals, yet they are not the same as atomic orbitals.

Consider the case of boron trichloride, BCl_3, which has trigonal planar electronic and molecular geometries with 120° bond angles (Example 5–14). Hybridization involves the mixing of one s orbital and two p orbitals in boron's outermost (second) shell, and is therefore called sp^2 hybridization. This sp^2 designation does not indicate the number of electrons occupying the orbitals, but rather the number and kind of atomic orbitals mixed to form hybrid orbitals.

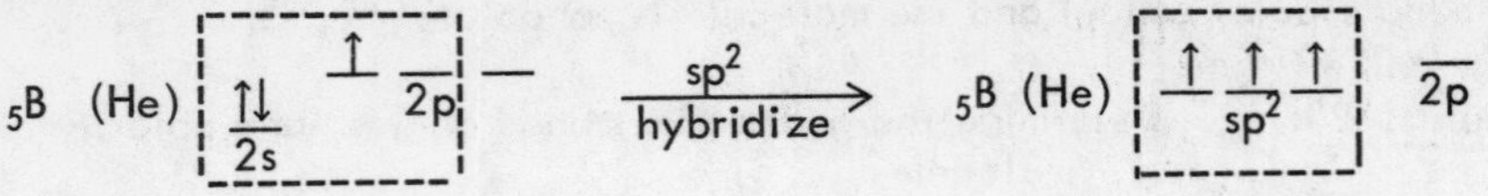

Atomic Orbitals Hybridized Orbitals

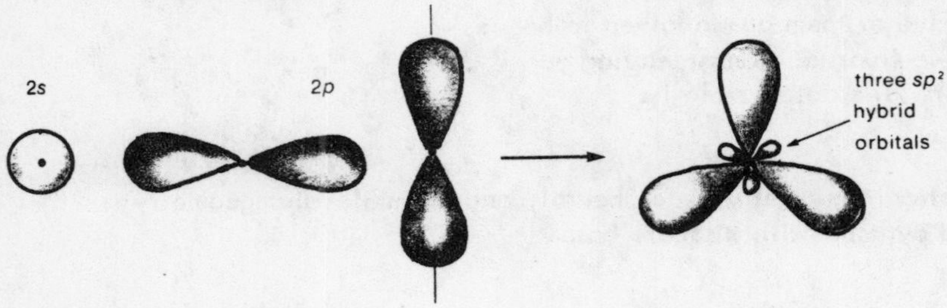

As the sp^2 hybrid orbitals form, one of the original 2s electrons is promoted to one of the three equivalent sp^2 orbitals, each of which is represented by a single lobe and can accommodate two electrons. Thus, three sp^2 orbitals can hold a maximum of six electrons, the same number as the individual unhybridized s and two p orbitals could accommodate. Sets of hybrid orbitals have the same total electronic capacity as their component atomic orbitals, even though the number of lobes may be different.

The three unpaired electrons (from boron) in sp^2 orbitals pair up with unpaired electrons from chlorine to form the three B-Cl bonds. Only the sp^2 hybrid orbitals of boron and the 3 p_z orbitals of chlorine are shown here.

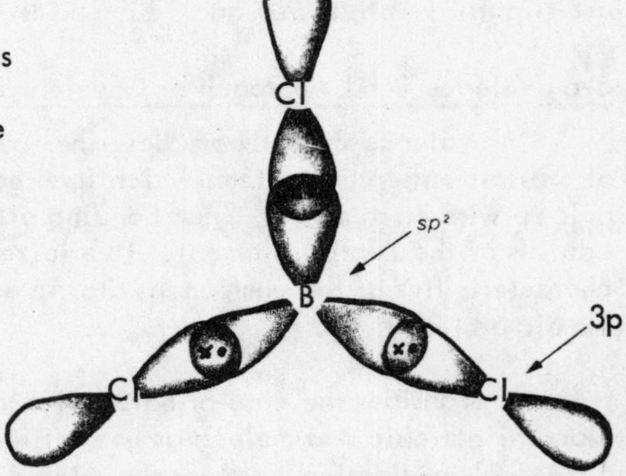

Consider silane, SiH_4, which has the same dot formula and geometry (tetrahedral) as methane, CH_4, as another example. The hybridization of the outermost atomic orbitals of the silicon atom involves the mixing of one s orbital with all three p orbitals in the third energy level to form a set of four equivalent sp^3 orbitals separated from each other by angles of 109°28'. One of the original 3s electrons is promoted into one of the sp^3 orbitals.

$_{14}$Si (Ne) | $\uparrow\downarrow$ $\underline{3s}$ | $\underline{\uparrow}$ $\underline{\uparrow}$ $\overline{3p}$ — | $\xrightarrow[\text{hybridize}]{sp^3}$ | $_{14}$Si | $\uparrow$ $\uparrow$ $\uparrow$ $\uparrow$ $\overline{sp^3}$

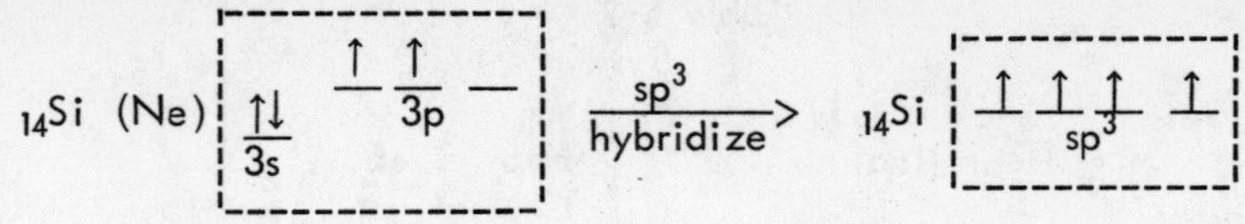

Atomic Orbitals   **Hybrid Orbitals**

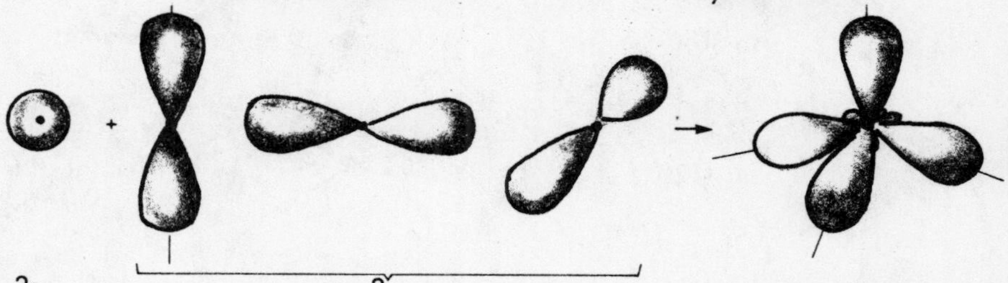

3s   3p   four *sp³* hybrid orbitals

Each of the four sp³ hybrid orbitals on silicon contains one electron that can pair with a 1s electron of a hydrogen atom to fill the sp³ orbitals to capacity. All bond angles are 109°28'.

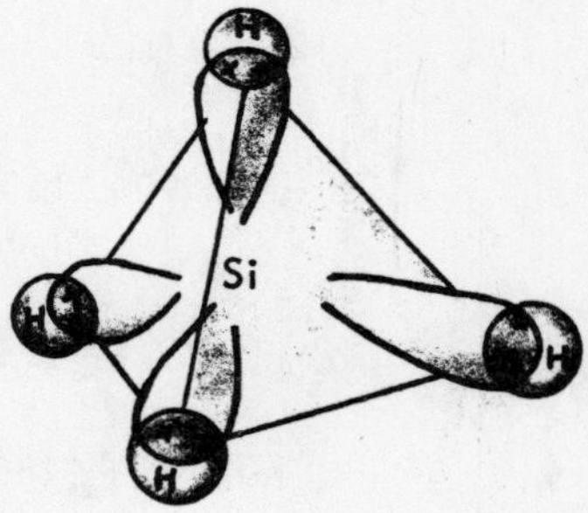

Table 5-2 (next page) shows some common kinds of hybridization, the number of electron groups, and the resulting hybrid orbital orientation.

Example 5-24:   Describe the electronic and molecular geometries, bond angles and hybridization with respect to carbon and nitrogen in the methylamine molecule, CH_3NH_2. Show the hybridization of atomic orbitals.

$N - A = S$, $[2(8) + 5(2)] - [4 + 5 + 5(1)] = 26 - 14 = 12$ e⁻ shared

$$H:C:N:H$$

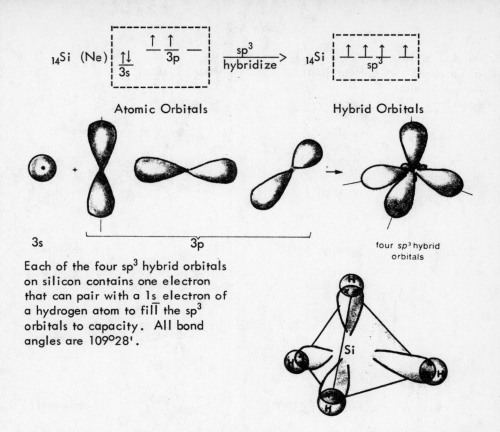

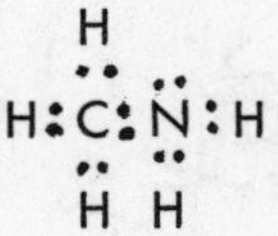

There are four regions of high electron density around both carbon and nitrogen and both have tetrahedral electronic geometry and sp³ hybridization.

This example continues after Table 5-2.

Table 5-2

Number of Sets of Electrons Around Central Atom	Hybridization of Central Atom	Hybridized Orbital Orientation	
2	sp (180°)		linear
3	sp^2 (120°)	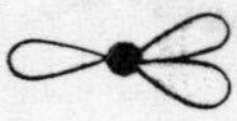	trigonal planar
4	sp^3 (109°28')		tetrahedral
5	sp^3d (90°, 120°, 180°)		trigonal bipyramidal
6	sp^3d^2 (90°, 180°)		octahedral

Although the molecule cannot actually be synthesized in this way, we may visualize its formation by removal of one hydrogen and one electron each from one molecule of methane, CH_4, and from one molecule of ammonia, NH_3, and the subsequent overlap of the sp^3 orbitals, with the pairing of the two remaining electrons to form a covalent bond. The carbon is at the center of a tetrahedron and the nitrogen at the center of pyramid.

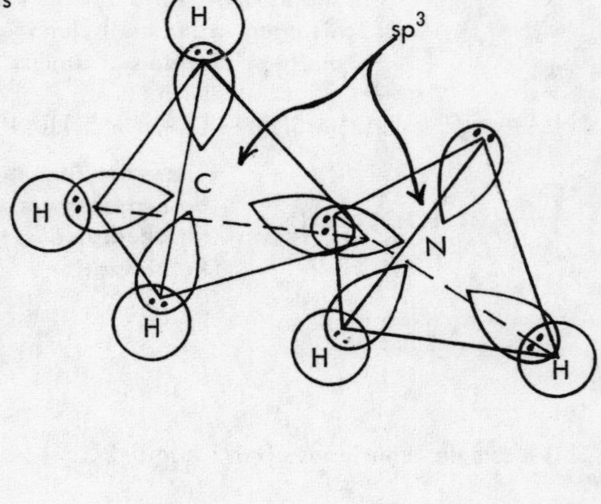

The H - N - H angles = 106°, while the C - N - H angles = 112°.

Because of the lone pair of electrons on the nitrogen atom, and the fact that the atoms bonded to the nitrogen are not identical, the angles around nitrogen are distorted slightly from the 109°28' angles of a regular tetrahedron.

Example 5-25: Show the hybridization of sulfur in SO_2 molecules.

In Example 5-12 we found that one of the two equivalent resonance structures of SO_2 (angular molecules) is

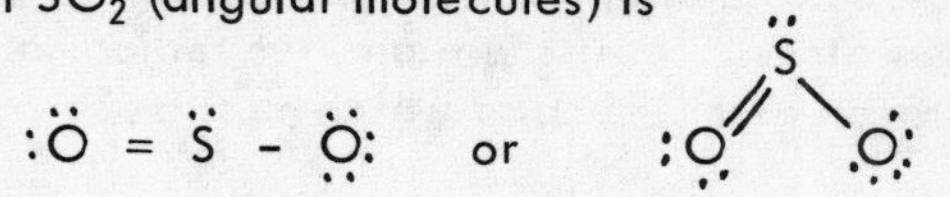

(unshared pairs on O's not shown)

dot formula shape

Since there are three electron groups around the sulfur atom, the hybridization is sp^2.

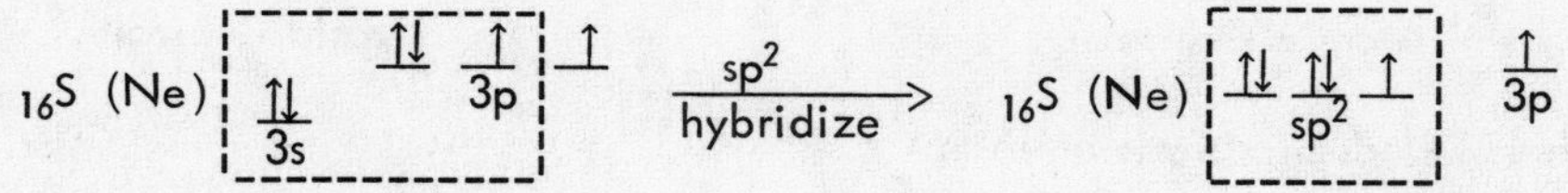

Atomic Orbitals Hybrid Orbitals

One of the resonance structures could be visualized as follows. (In actuality, both sulfur-oxygen bonds are equivalent and intermediate between single and double bonds. One resonance structure treats the molecules as if formal localized single and double bonds were present. They are not.) One pair of electrons in the sp^2 orbitals of sulfur remains as the lone pair. A second pair is donated to the singly-bonded oxygen to form a coordinate covalent bond. The only unpaired electron in an sp^2 orbital pairs with an electron in one orbital on the doubly-bonded oxygen atom. The remaining electron from the sulfur, in the unhybridized 3p orbital, pairs with an electron in another p orbital of the same oxygen atom to form the second bond between the two. The orbitals holding all the outer shell electrons of sulfur and the bonding electrons of the oxygen atoms are shownat the right.

One Resonance Form of SO_2

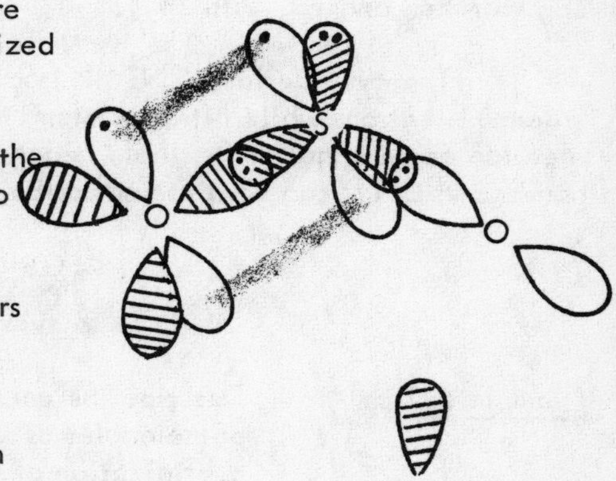

sp^2 hybrid orbitals

The second bond of the double bond is called a pi (π) bond; it involves side-to-side overlap of both top and bottom lobes of an oxygen p orbital with a sulfur p orbital and involves a total of two electrons. The other two bonds (sp-p) involve only single head-to-head overlap of lobes and are called sigma (σ) bonds.

Example 5-26: How are nitrogen atoms hybridized in the nitrogen molecule, N_2?

We saw in Section 5-3 that the N_2 molecule involves a triple bond and one unshared electron pair on each nitrogen atom. Thus, there are two regions of high electron density around each nitrogen atom, each of which must be sp hybridized.

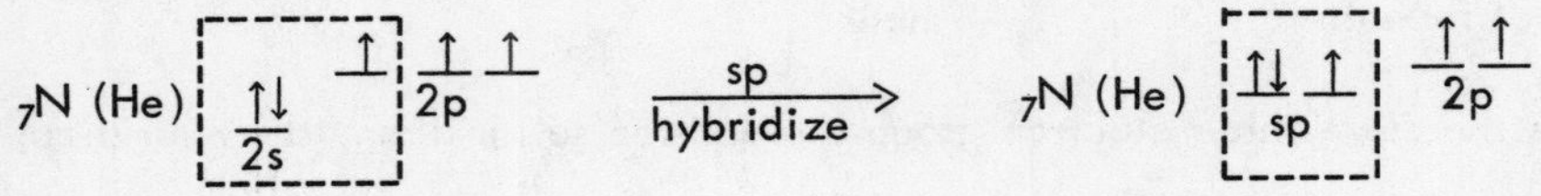

Atomic Orbitals　　　　Hybrid Orbitals

:N≡N:

The half filled sp orbitals of each nitrogen atom overlap with each other and the unpaired electrons in them pair up to form a sigma bond. Two pi bonds result from the side-to-side overlap of two sets of corresponding parallel unhybridized 2p orbitals on the two atoms (say $2p_y$ with $2p_y$ and $2p_z$ with $2p_z$).

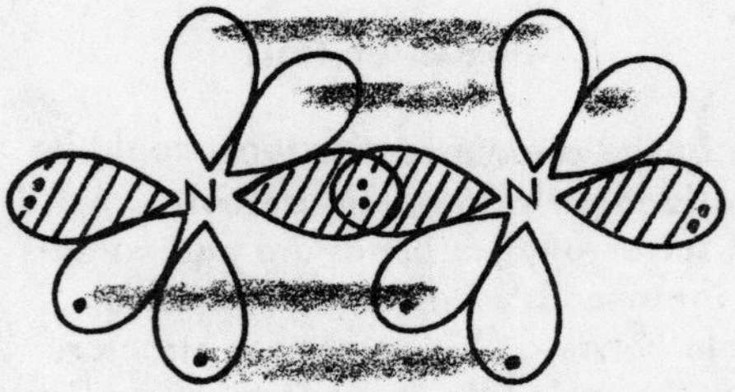

The cyanide ion, CN^-, is isoelectronic with N_2. Carbon has four valence electrons while nitrogen atoms have five, but the ion has an extra electron accounting for its single negative charge. Thus the hybridization and bonding in CN^- can be pictured in the same way as that in N_2.

$$[:C \equiv N:]^-$$

sp hybridized

Example 5-27: Describe the geometry, bond angles and the polarity of molecules of xenon tetrafluoride, XeF_4. Show the hybridization of the xenon atom.

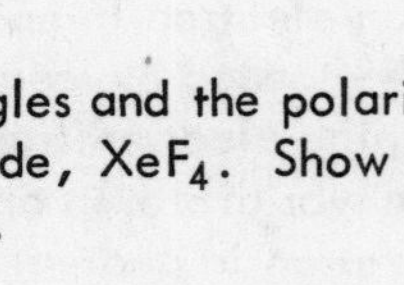

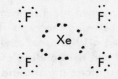

The xenon atom has an expanded coordination sphere and the octet rule does not apply. We must generate the dot formula using chemical intuition. Atomic xenon has eight valence electrons and each fluorine contributes four, for a total of twelve, or six pairs around the xenon (octahedral electronic geometry).

The presence of six regions of high electron density around the xenon requires sp^3d^2 hybridization.

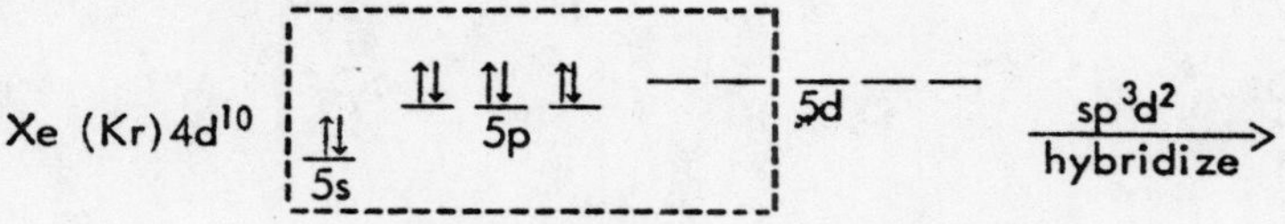

Atomic Orbitals

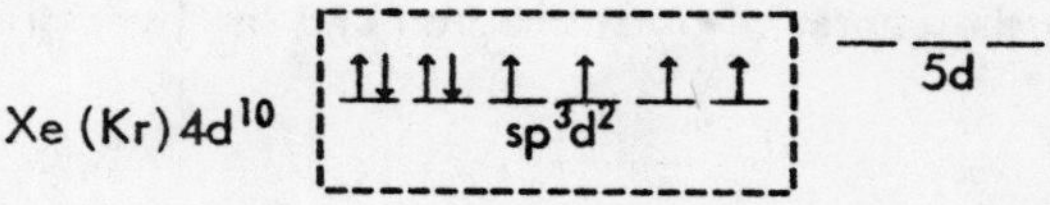

Hybrid Orbitals

The resulting four unpaired electrons in sp^3d^2 hybrids of Xe pair up with the unpaired electrons in the overlapping 2p orbitals of each of four fluorine atoms to form four covalent bonds. The other four electrons of xenon remain in sp^3d^2 orbitals as unshared pairs. Since lone pair-lone pair repulsions are so great the lone pairs are as far apart as possible (180°). This leaves the four fluorines and the xenon in the same plane and the molecular geometry is said to be square planar.

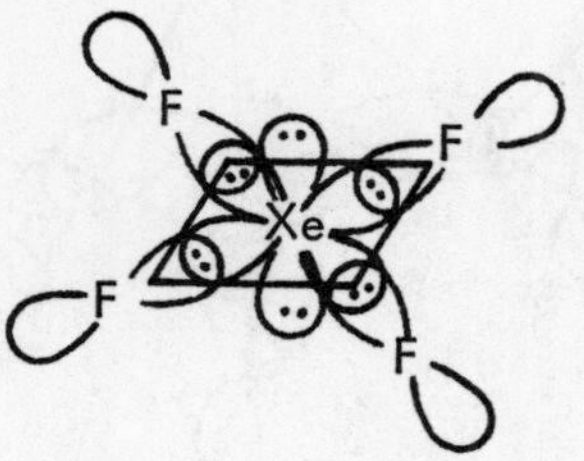

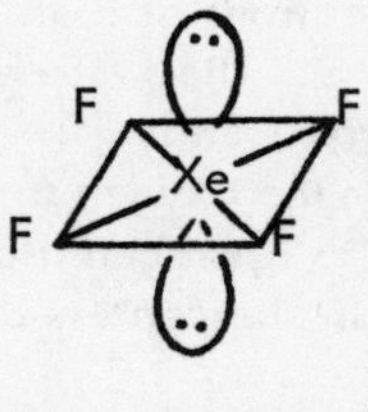

F-Xe-F bond angles = 90° and 180°

<u>Example 5-28:</u> Describe the hybridization at the sulfur atom and the geometry relative to the sulfur atom in sulfuric acid molecules, H_2SO_4. All oxygen atoms are bonded to the sulfur atom and the hydrogen atoms are bonded to different oxygen atoms.

We are interested in the geometry with respect to sulfur and before anything can be determined in that respect we should determine the dot formula.

$$N - A = S$$

$$[2(2)+1(8)+4(8)] - [2(1)+1(6)+4(6)] = S, \qquad 44 - 32 = 12\ e^- \text{ shared}$$

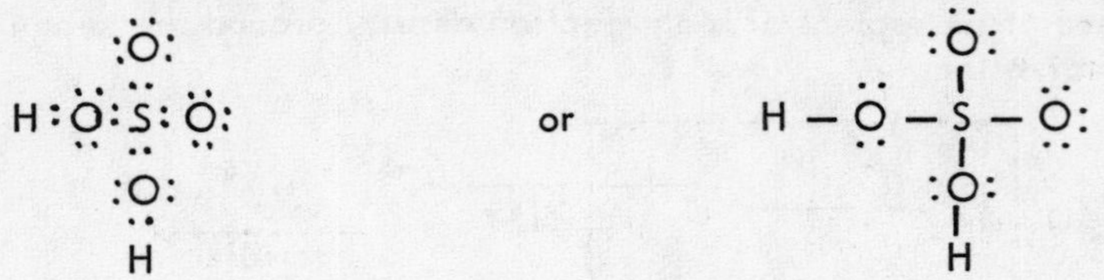

There are four groups of electrons surrounding the sulfur atom so the sulfur must be at the center of a tetrahedron and the hybridization is sp^3.

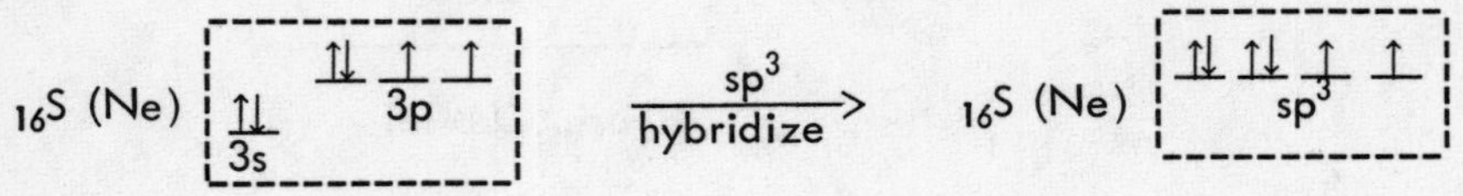

Atomic Orbitals

Hybrid Orbitals

Two of the pairs of electrons in the sp^3 orbitals are associated with the coordinate covalent bonds to the two oxygen atoms to which no hydrogen atoms are attached. The other two electrons are shared with unpaired electrons of the other two oxygen atoms. The O–S–O angles are all approximately 109°. Can you see why the oxygen atoms that are bonded to hydrogen atoms are also shown in the centers of tetrahedra?

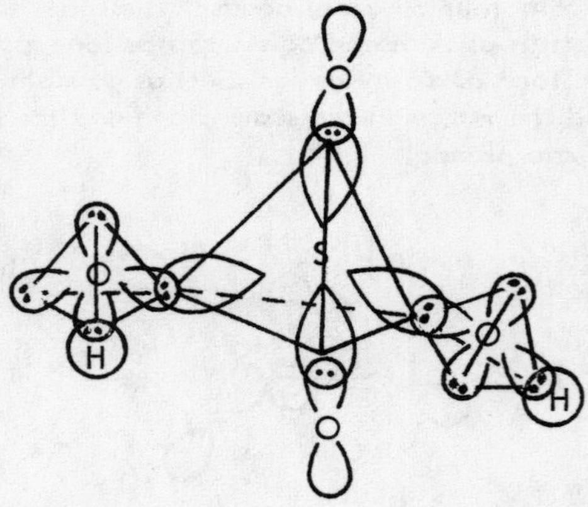

NAMING OF INORGANIC COMPOUNDS

For naming purposes, it is advantageous to classify simple compounds in two major categories. These are binary compounds, those consisting of two elements, and ternary compounds, those consisting of three elements.

5-7 Binary Compounds

Binary compounds may be either ionic or covalent. In both cases we name the less electronegative element first and the more electronegative element last. The more electronegative element is named by adding an "-ide" suffix to the element's characteristic stem. The stems to which the suffixes are added are given in Sec. 5-23 of the text.

Binary ionic compounds contain metal cations and nonmetal anions. Thus, the cation is named first and the anion second.

Formula	Name	Formula	Name
LiF	lithium fluoride	Al_2O_3	aluminum oxide
$MgBr_2$	magnesium bromide	BaS	barium sulfide
KH	potassium hydride	Rb_3N	rubidium nitride

This method is adequate for naming binary ionic compounds containing representative (A-group) cations. However, some transition (B-group) elements and a few of the more electronegative representative metals exhibit more than one oxidation state. These metals can form two or more binary compounds with nonmetals. In order to distinguish among all possibilities the oxidation state of the metal is indicated by a Roman numeral in parenthesis following its name. Roman numerals are not used for metals that exhibit only one nonzero oxidation state.

Formula	Name	Formula	Name
$CuCl_2$	copper(II) chloride	SnF_2	tin(II) fluoride
CuCl	copper(I) chloride	SnS_2	tin(IV) sulfide
Fe_2O_3	iron(III) oxide	PbO_2	lead(IV) oxide
FeO	iron(II) oxide	PbS	lead(II) sulfide

An older method, still in use but not recommended by the IUPAC, involves the use of "-ous" and "-ic" suffixes to indicate the lower and higher of two possible oxidation states, respectively. However, this sytem is capable of distinguishing between only two different oxidation states of a metal and, therefore, is not as generally useful as the Roman numeral system.

Formula	Name	Formula	Name
CuBr	cuprous bromide	FeS	ferrous sulfide
$CuBr_2$	cupric bromide	SnO	stannous oxide
FeF_3	ferric fluoride	SnO_2	stannic oxide

Pseudobinary ionic compounds are those whose names also end in "-ide" but contain more than two elements. In these compounds one or more of the ions consist of more than one element but behave as if they were simple ions. Three common examples of such ions are the ammonium ion, NH_4^+, the cyanide ion, CN^-, and the hydroxide ion, OH^-.

Formula	Name	Formula	Name
NH_4Br	ammonium bromide	$Ca(OH)_2$	calcium hydroxide
$Sr(CN)_2$	strontrium cyanide	NH_4CN	ammonium cyanide
$Sn(OH)_2$	stannous hydroxide or tin(II) hydroxide		
$Fe(OH)_3$	ferric hydroxide or iron(III) hydroxide		

Most binary covalent compounds involve two nonmetals. Many nonmetals can exhibit different oxidation states, but their oxidation states are not indicated by Roman numerals or suffixes. Rather, elemental proportions in binary covalent compounds are indicated by utilizing a prefix system for both elements. The first twelve Greek or Latin prefixes used (without hyphens) are mono, di, tri, tetra, penta, hexa, hepta, octa, nona, deca, undeca, and dodeca. The prefix mono is usually omitted; carbon monoxide, CO, is an exception.

Formula	Name	Formula	Name
N_2O	dinitrogen oxide	P_4O_{10}	tetraphosphorus decoxide
NO_2	nitrogen dioxide	IF_7	iodine heptafluoride
N_2O_4	dinitrogen tetroxide	S_4N_4	tetrasulfur tetranitride

Binary acids are compounds of hydrogen and the more electronegative nonmetals. These compounds act as acids when dissolved in aqueous solutions. The pure compounds are named as typical binary compounds. Their aqueous solutions are named by modifying the characteristic stem of the nonmetal with the prefix "hydro-" and the suffix "-ic" followed by the word "acid". The stem for sulfur, in this instance is "sulfur" than than "sulf".

Formula	Name of Compound	Name of Aqueous Solution
HBr	hydrogen bromide	hydrobromic acid, HBr (aq.)
HF	hydrogen fluoride	hydrofluoric acid, HF (aq.)
H_2Se	hydrogen selenide	hydroselenic acid, H_2Se (aq.)
HCN	hydrogen cyanide	hydrocyanic acid, HCN (aq.)

5-8 Ternary Compounds

Ternay acids (oxyacids) are compounds of hydrogen, oxygen, and (usually) another nonmetal. One common ternary acid of each nonmetal is (somewhat arbitrarily) designated as the "-ic acid". That is, it is named "stem-ic acid". The common ternary "-ic acids" are shown in Sec. 5-24 of the text. It is important to learn the names and formulas of these acids since the names of other ternary acids and their salts are derived from them.

Acids containing one fewer oxygen atom per central nonmetal atom are named in the same way except that the "-ic" suffix is changed to "-ous".

Formula	Name	Formula	Name
H_2SO_3	sulfurous acid	H_2SeO_3	selenous acid
HNO_2	nitrous acid	$HBrO_2$	bromous acid

Ternary acids having two fewer oxygen atoms per central nonmetal atom than the "-ic acid" have an "-ous" suffix instead of "-ic" and a "hypo-" prefix. Notice that $H_2N_2O_2$ has a 1:1 ratio of nitrogen to oxygen, as would the hypothetical HNO.

Formula	Name	Formula	Name
$HClO$	hypochlorous acid	$HBrO$	hypobromous acid
H_3PO_2	hypophosphorous acid	$H_2N_2O_2$	hyponitrous acid

Acids containing one more oxygen atom per central nonmetal atom than the "-ic acid" are named in the same way as the "-ic acids" with an additional prefix, "per-".

Formula	Name
$HClO_4$	perchloric acid
$HBrO_4$	perbromic acid
HIO_4	periodic acid

Ternary salts or salts of oxyacids (oxysalts) usually contain metal cations and oxyacid anions. As with binary compounds, the cation is named first. The name of the anion is based on the name of the ternary acid from which it is derived.

An anion derived from an acid with an "-ic" ending is named by dropping the "-ic acid" and replacing it with "-ate". An anion derived from an "-ous acid" is named by dropping the "-ous acid" and adding "-ite". The "per-" and "hypo-" prefixes are retained.

Formula	Name
$MgCO_3$	magnesium carbonate (CO_3^{2-} derived from H_2CO_3)
$NaBrO_3$	sodium bromate (BrO_3^- derived from $HBrO_3$)
$Al(NO_2)_3$	aluminum nitrite (NO_2^- derived from HNO_2)
Na_2SO_3	sodium sulfite (SO_3^{2-} derived from H_2SO_3)
$Cr(ClO_4)_3$	chromium(III) perchlorate (ClO_4^- derived from $HClO_4$)
$KClO$	potassium hypochlorite (ClO^- derived from $HClO$)

Acidic salts contain anions derived from ternary acids in which one or more acidic hydrogen atoms remain. These salts are named in the same way as they would be if they were the usual type of ternary salts except that the word hydrogen, with the appropriate prefix, is inserted after the name of the metal cation.

Formula	Name
$NaHSO_4$	sodium hydrogen sulfate
$NaHSO_3$	sodium hydrogen sulfite
NaH_2AsO_4	sodium dihydrogen arsenate
Na_2HAsO_4	sodium hydrogen arsenate
$NaHCO_3$	sodium hydrogen carbonate

An older, commonly used method (which is not suggested by the IUPAC) involves the use of the prefix "bi-" attached to the name of the anion to indicate the presence of an acidic hydrogen atom. According to this method, $NaHSO_4$ also could be called sodium bisulfate and $NaHCO_3$ could be named sodium bicarbonate.

EXERCISES

1. Draw an electron dot formula for each of the following covalent compounds. The central atom is underlined. (*indicates hydrogen atoms bound only to oxygen; **indicates nitrogen atoms bound to each other and each nitrogen atom is bound to two oxygen atoms)

   (a) N_2
   (b) $\underline{C}S_2$
   (c) $\overline{\underline{N}}_2O_4$**
   (d) $\overline{\underline{As}}Cl_3$
   (e) $\underline{P}Cl_5$
   (f) $\overline{\underline{N}}F_3$
   (g) $\overline{H}_3\underline{P}O_4$*
   (h) $H\overline{\underline{N}}\overline{O}_2$*
   (i) $H_2\underline{Se}$
   (j) $\underline{Si}\overline{Cl}_4$
   (k) $\overline{\underline{C}}_2H_6$
   (l) $\overline{\underline{C}}_2H_4$
   (m) $\underline{C}H_3\underline{O}H$

2. Show the hybridizations with arrow representations of electrons in atomic and hybridized orbitals of the central elements of problem 1. Show bonding electrons of substituent atoms with "x's."

3. Sketch three-dimensional representations of the molecules of problem 1 emphasizing the geometry with respect to the central element. Name the electronic geometry and molecular geometry relative to the central atom.

4. Draw an electron dot formula for each of the ions below. The central atoms are underlined.

   (a) $\underline{Cl}O_3^-$
   (b) $\overline{\underline{N}}O_3^-$
   (c) $\overline{\underline{S}}O_3^{2-}$
   (d) $H\underline{C}O_3^-$
   (e) $\underline{P}\overline{H}_4^+$
   (f) $\overline{\underline{C}}N^-$
   (g) $\underline{Se}O_4^{2-}$
   (h) $\overline{\underline{N}}O_2^-$
   (i) $\overline{\underline{Cl}}O_4^-$
   (j) $\underline{P}O_4^{3-}$

5. What is the hybridization at the central element in each of the ions of problem 4? What are the electronic and ionic geometries relative to these elements?

6. Sketch three-dimensional representations of each of the ions of problem 4.

7. Which of the molecules below are polar molecules? The central element is underlined.

   (a) HBr
   (b) $\underline{S}\overline{O}_2$
   (c) $\overline{H}_2\underline{Te}$
   (d) H_2
   (e) $\underline{C}S_2$
   (f) $\overline{\underline{Sb}}Cl_5$
   (g) $\underline{Se}O_3$
   (h) $\overline{\underline{N}}O_2$
   (i) $\overline{F}_2$
   (j) $\underline{C}Br_4$
   (k) $\overline{\underline{C}}HBr_3$
   (l) $\overline{\underline{N}}H_2Cl$

8. Draw resonance formulas for the following species. The central element is underlined.

   (a) $\underline{N}O_3^-$
   (b) $\overline{\underline{N}}O_2^-$
   (c) $H\underline{C}O_3^-$
   (d) $\underline{N}_2O_4$

9. Write formulas for the following compounds:

(a)________ sodium nitrite
(b)________ potassium sulfide
(c)________ ammonium thiocyanate
(d)________ cupric phosphate
(e)________ cobaltous dichromate
(f)________ chromium(III) oxide
(g)________ lead acetate
(h)________ ammonium sulfide
(i)________ aluminum nitrite
(j)________ iron(II) sulfide
(k)________ ammonium bisulfate
(l)________ copper(II) chlorate
(m)________ cobalt(II) chromate
(n)________ chromium(VI) oxide
(o)________ manganese(III) fluoride
(p)________ ammonium sulfite

10. Write names for the following compounds:

(a) $KHSO_3$
(b) $KHSO_4$
(c) $Ba(HCO_3)_2$
(d) $Ni(SCN)_2$
(e) $Co(SCN)_2$
(f) $LiHSO_3$
(g) Ag_3AsO_4
(h) $Mn(ClO_4)_3$
(i) $SnCl_2$
(j) $SnCl_4$
(k) $Sc(C_2H_3O_2)_3$
(l) $CdCr_2O_7$

11. Write formulas for the following compounds:

(a)________ manganic fluoride*
(b)________ chromic permanganate
(c)________ chromium(III) hypochlorite
(d)________ ferric oxalate
(e)________ cobaltic fluoride
(f)________ tetraphosphorus hexoxide
(g)________ dinitrogen trioxide
(h)________ scandium nitrate
(i)________ manganous dichromate
(j)________ chromous sulfate
(k)________ ferrous oxalate
(l)________ cobaltous iodide
(m)________ dichlorine heptoxide
(n)________ tetraphosphorus decoxide
(o)________ dinitrogen pentoxide
(p)________ titanium(III) sulfate

12. Write names for the following compounds:

(a) $Cr(C_2H_3O_2)_2$ *
(b) $Mn_2(Cr_2O_7)_3$
(c) $AgMnO_4$
(d) $Cr(ClO_3)_3$
(e) $Sc(ClO_2)_3$
(f) $Co(ClO_3)_2$
(g) $Pb(ClO_4)_2$
(h) $Ni(BrO)_2$
(i) $AgBrO_3$
(j) NH_4Br
(k) K_3AsO_4
(l) CoC_2O_4

13. Write formulas for the following acids:

(a) arsenous acid ________
(b) hydrosulfuric acid ________
(c) sulfuric acid ________
(d) nitrous acid ________
(e) hypobromous acid ________
(f) hydrocyanic acid ________
(g) hydrofluoric acid ________
(h) selenic acid ________
(i) phosphoric acid ________
(j) sulfurous acid ________

---

*Note that some salts have been named by the older system to provide familiarity with this system.

14. Write names for the following as acids in aqueous solution:

(a) HBr ________
(b) $HBrO_4$ ________
(c) HNO_3 ________
(d) H_3PO_3 ________
(e) $HClO_3$ ________
(f) H_2TeO_3 ________
(g) H_2CO_3 ________
(h) HIO ________
(i) $H_2N_2O_2$ ________

1. (a) N = 2(8) = 16 e^-
A = 2(5) = 10 e^-
N–A = S = 6 e^-

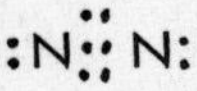

(b) N = 3(8) = 24
A = 4 + 2(6) = 16
S = 8 e^-

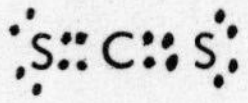

(c) N = 6(8) = 48
A = 2(5) + 4(6) = 34
S = 14 e^-

(d) N = 4(8) = 32
A = 5 + 3(7) = 26
S = 6 e^-

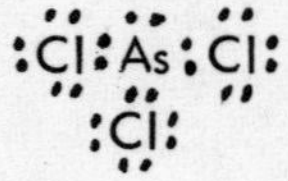

(e) N–A = S does not apply;
5 substituents around P

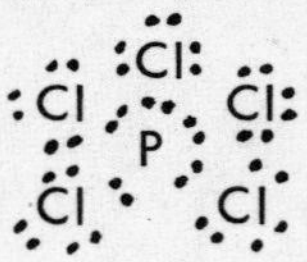

(f) N = 4(8) = 32
A = 5 + 3(7) = 26
S = 6 e^-

(g) N = 3(2) + 5(8) = 46
A = 3(1) + 5 + 4(6) = 32
S = 14 e^-

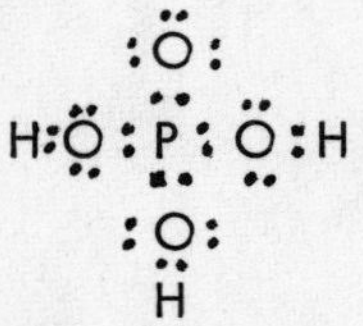

(h) N = 2 + 3(8) = 26
A = 1 + 5 + 2(6) = 18
S = 8 e^-

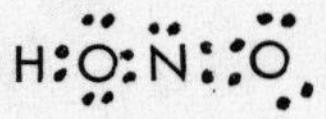

(i) N = 2(2) + 8 = 12
A = 2(1) + 6 = 8
S = 4 e^-

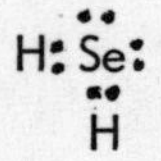

(j) N = 5(8) = 40
A = 4 + 4(7) = 32
S = 8 e^-

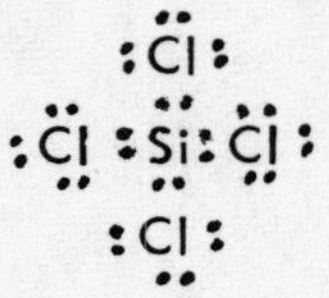

(k) $N = 2(8) + 6(2) = 28$
$A = 2(4) + 6(1) = 14$
$S = 14\ e^-$

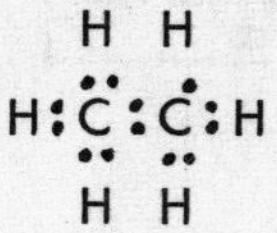

(l) $N = 2(8) + 4(2) = 24$
$A = 2(4) + 4(1) = 12$
$S = 12\ e^-$

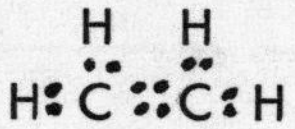

(m) $N = 2(8) + 4(2) = 24$
$A = 4 + 6 + 4(1) = 14$
$S = 10\ e^-$

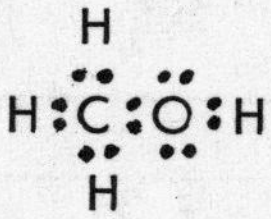

2. (a) N_2 — 2 e^- groups around each N ∴ sp hybridization

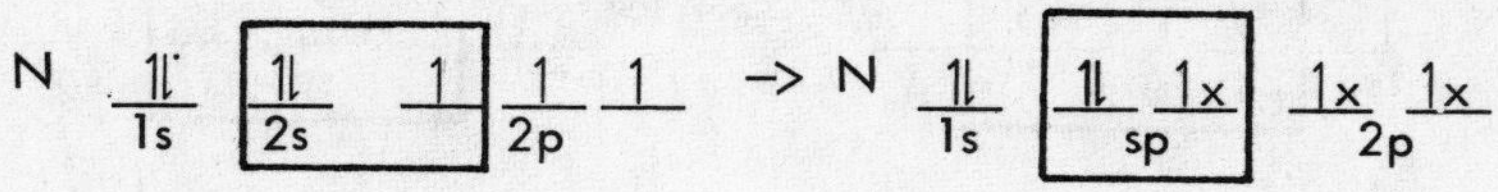

(b) CS_2 — 2 e^- groups around C ∴ sp hybridization

(c) N_2O_4 — 3 e^- groups around each N ∴ sp^2 hybridization

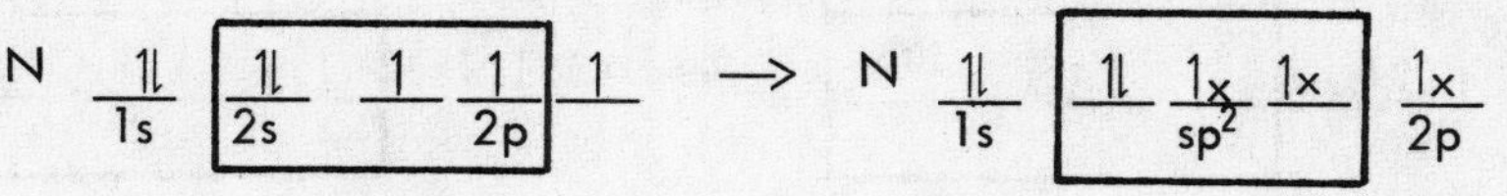

(d) $AsCl_3$ — 4 e^- groups around As ∴ sp^3 hybridization

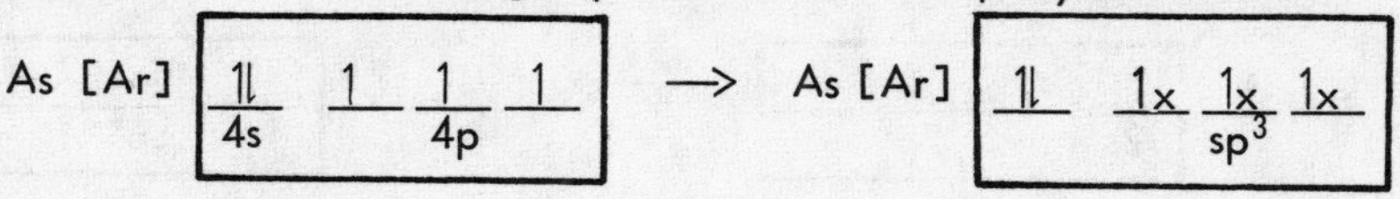

(e) PCl_5 — 5 e^- groups around P ∴ sp^3d hybridization

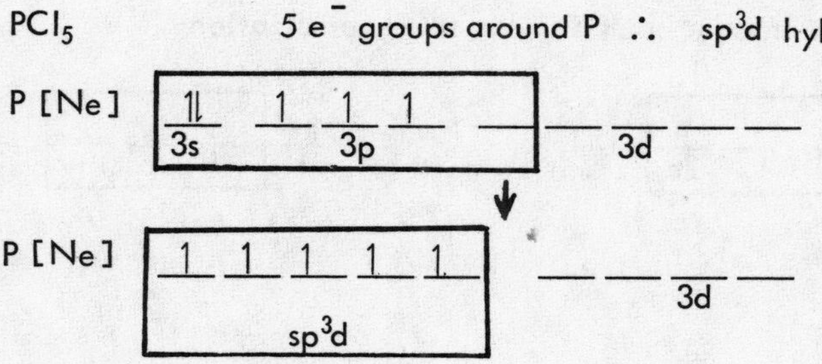

2. (f) 4 e^- groups around N ∴ sp^3 hybridization

N ⇅ (1s) [⇅ (2s) ↑ ↑ ↑ (2p)] → N ⇅ (1s) [⇅ ↑x ↑x ↑x (sp^3)]

(g) 4 e^- groups around P ∴ sp^3 hybridization

P [Ne] [⇅ (3s) ↑ ↑ ↑ (3p)] → P [Ne] [⇅ ↑x ↑x ↑x (sp)]

(h) 3 e^- groups around N ∴ sp^2 hybridization

N ⇅ (1s) [⇅ (2s) ↑ ↑ (2p)] ↑ → N ⇅ (1s) [⇅ ↑x ↑x (sp^2)] ↑x (2p)

(i) 4 e^- pairs around Se ∴ sp^3 hybridization

Se [Ar] $3d^{10}$ [⇅ (4s) ⇅ ↑ ↑ (4p)] → Se [Ne] $3d^{10}$ [⇅ ⇅ ↑x ↑x (sp^3)]

(j) 4 e^- pairs around Si ∴ sp^3 hybridization

Si [Ne] [⇅ (3s) ↑ ↑ __ (3p)] → Si [Ne] [↑x ↑x ↑x ↑x (sp^3)]

(k) 4 e^- pairs around each C ∴ sp^3 hybridization

C ⇅ (1s) [⇅ (2s) ↑ ↑ __ (2p)] → C ⇅ (1s) [↑x ↑x ↑x ↑x (sp^3)]

(l) 3 sets of e^- around each C ∴ sp^2 hybridization

C ⇅ (1s) [⇅ (2s) ↑ ↑ (2p)] __ → C ⇅ (1s) [↑x ↑x ↑x (sp^2)] ↑x (2p)

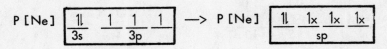

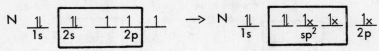

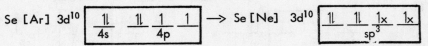

2. (m) 4 sets of e$^-$ around C and 4 e$^-$ sets of e$^-$ around O ∴ sp^3 hybridization at both C and O.

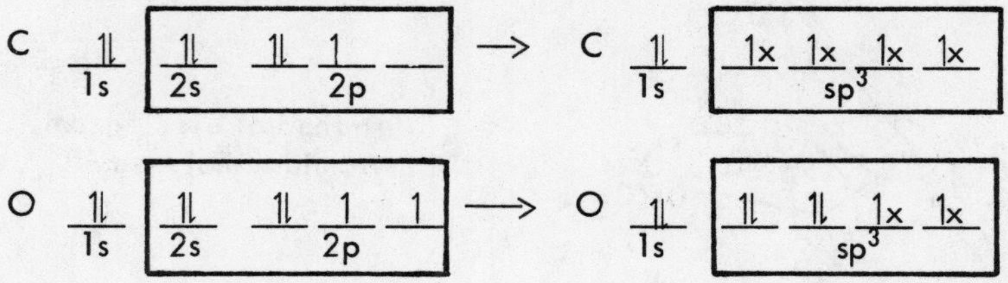

3. (a)

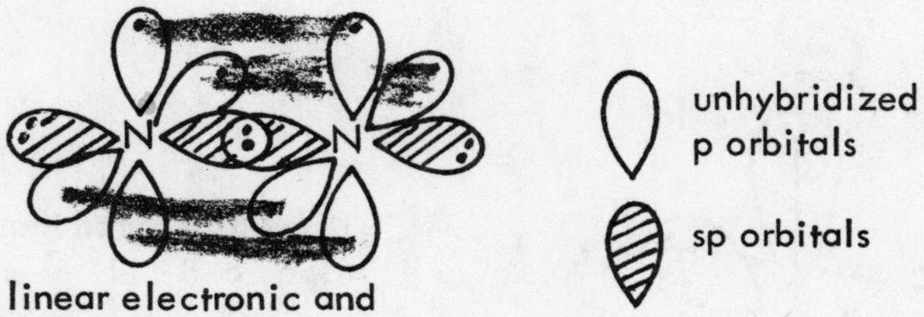

linear electronic and molecular geometry

(b)

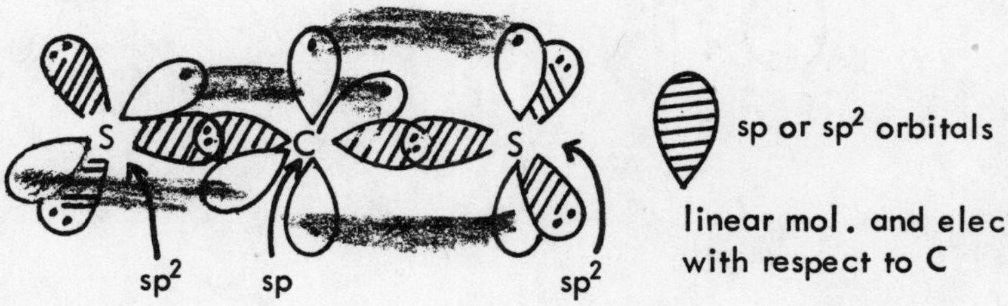

linear mol. and elec. geom. with respect to C

(c)

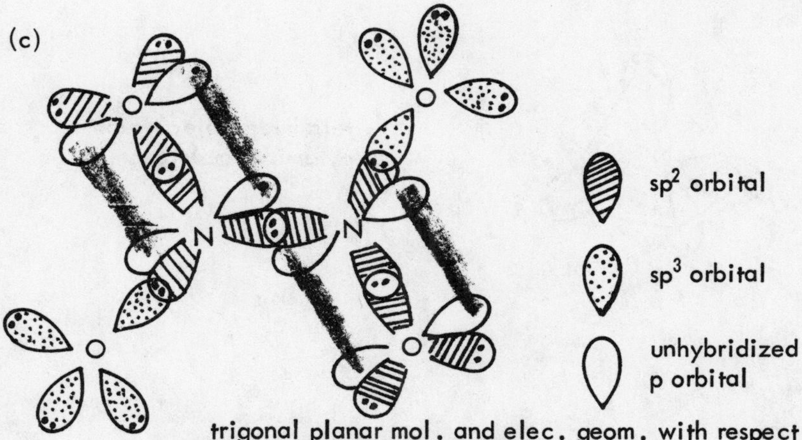

trigonal planar mol. and elec. geom. with respect to N
only one resonance form shown

(d)

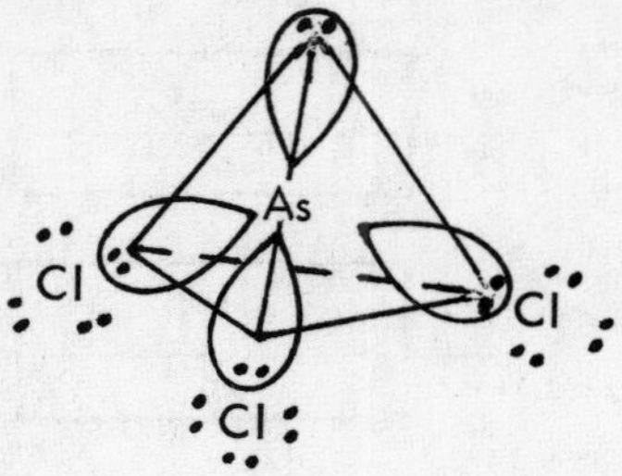

tetrahedral elec. geom.
pyramidal mol. geom.

(e)

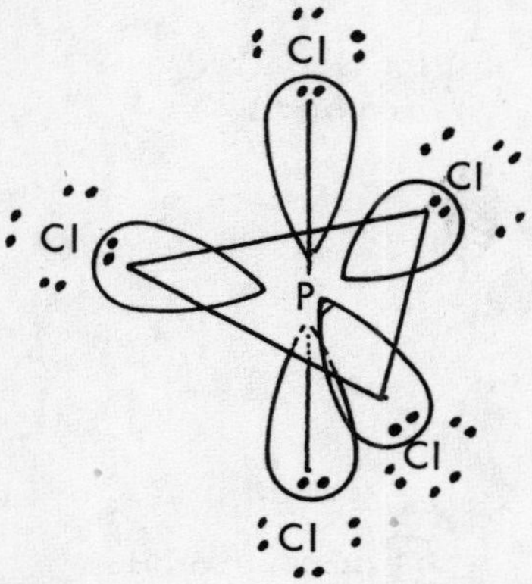

trigonal bipyramidal
elec. and mol. geom.

(f)

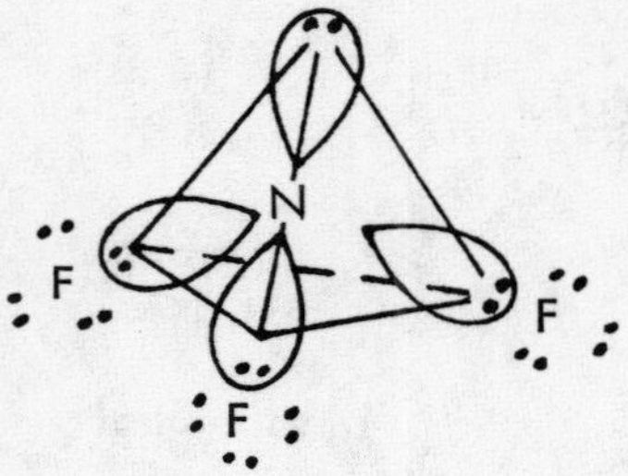

tetrahedral elec. geom.
pyramidal mol. geom.

(g)

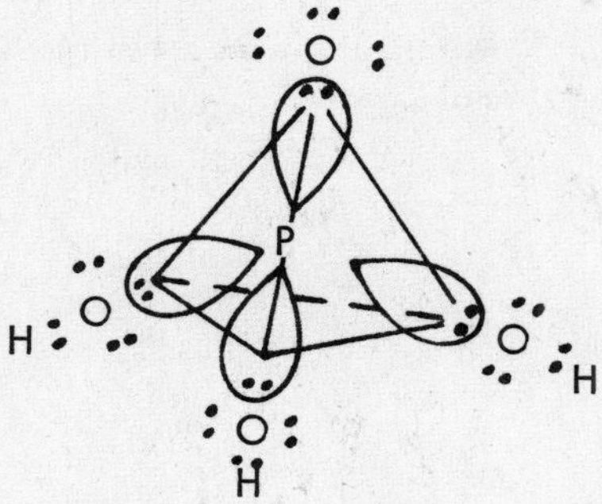

distorted tetrahedral
elec. and mol. geom.
with respect to P

(h)

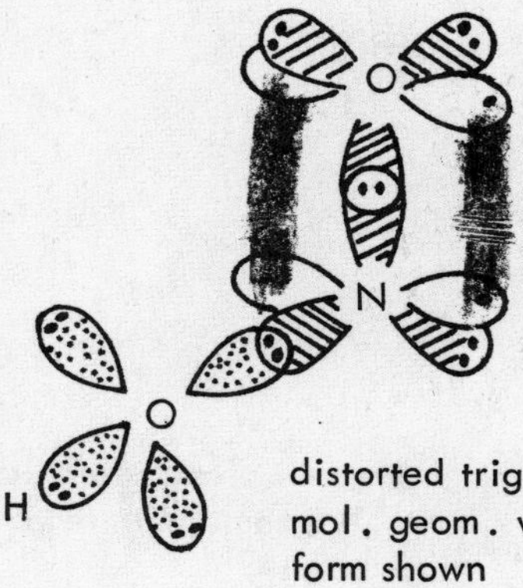

sp² hybrid orbitals

sp³ hybrid orbitals

unhybridized p orbitals

distorted trig. planar elec. geom. and angular
mol. geom. with respect to N; only one resonance
form shown

(j)

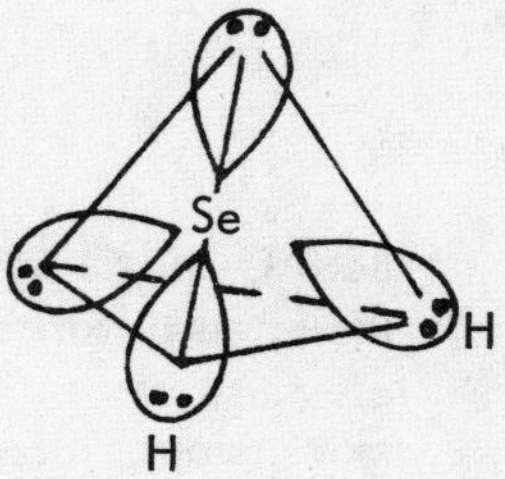

distorted tetrahedral
elec. geom.; angular
mol. geom.

(j)

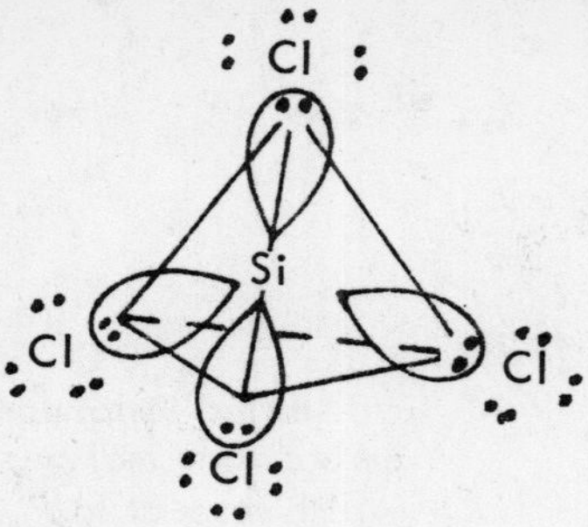

tetrahedral elec. and mol. geom.

(k)

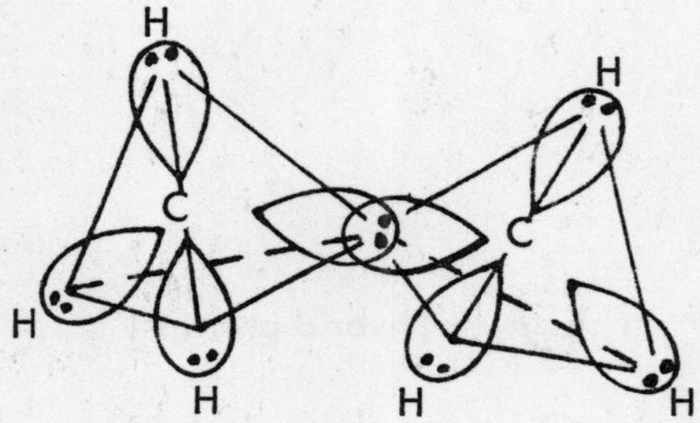

tetrahedral elec. geom. with respect to each C. mol. geom.: two tetrahedra sharing a corner

(l)

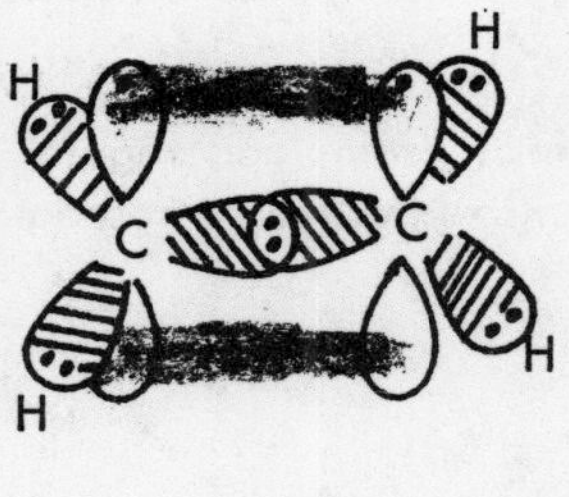

trigonal planar elec. and mol. geom. with respect to each C.

(m)

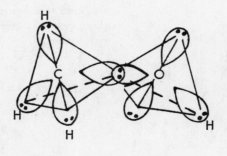

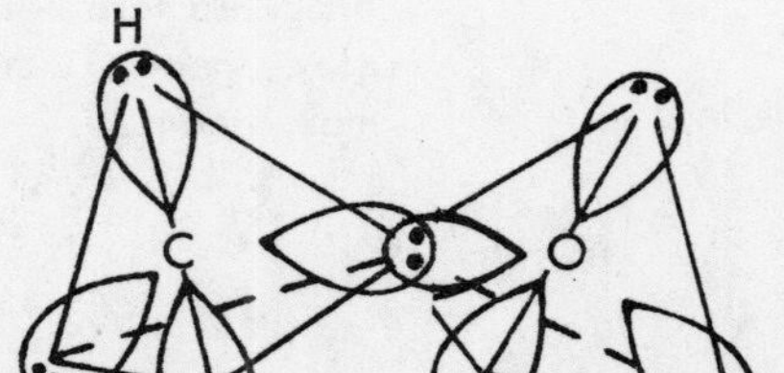

Both C and O have distorted tetrahedral elec. geom.
mol. geom.: distorted tetrahedron with respect to carbon, angular with respect to oxygen

4. (a) $N = 4(8) = 32$
$A = 7 + 3(6) + 1 = 26$
$S = 6\ e^-$

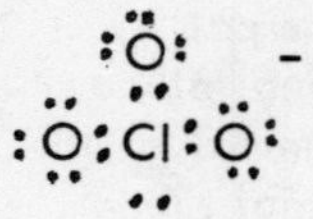

(b) $N = 4(8) = 32$
$A = 5 + 3(6) + 1 = 24$
$S = 8\ e^-$

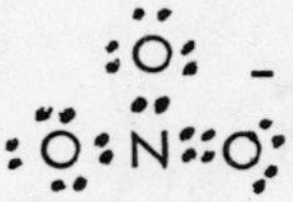

(c) $N = 4(8) = 32$
$A = 6 + 3(6) + 2 = 26$
$S = 6\ e^-$

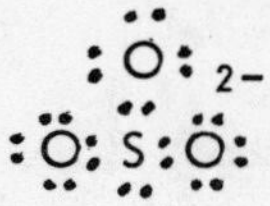

(d) $N = 2 + 4(8) = 34$
$A = 1 + 4 + 3(6) + 1 = 24$
$S = 10\ e^-$

(e) $N = 4(2) + 8 = 16$
$A = 5 + 4(1) - 1 = 8$
$S = 8\ e^-$

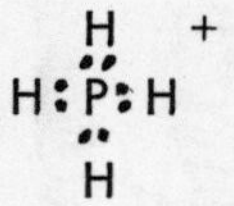

(f) $N = 2(8) = 16$
$A = 4 + 5 + 1 = 10$
$S = 6\ e^-$

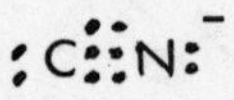

(g) $N = 5(8) = 40$
$A = 6 + 4(6) + 2 = 32$
$S = 8\ e^-$

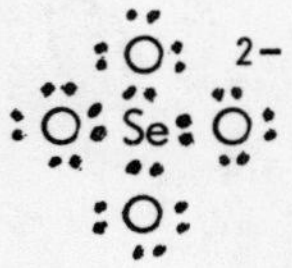

(h) $N = 3(8) = 24$
$A = 5 + 2(6) + 1 = 18$
$S = 6\ e^-$

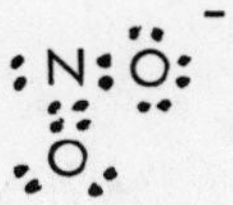

(i) $N = 5(8) = 40$
$A = 7 + 4(6) + 1 = 32$
$S = 8\ e^-$

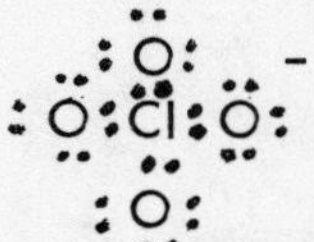

(j) $N = 5(8) = 40$
$A = 5 + 4(6) + 3 = 32$
$S = 8\ e^-$

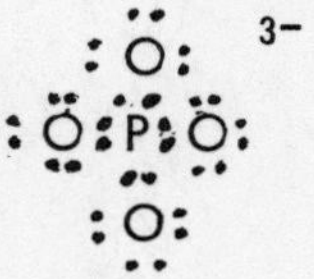

5.

	Hybridization at Central Element	Electronic Geometry	Ionic Geometry
(a)	sp^3	tetrahedral	pyramidal
(b)	sp^2	trigonal planar	trigonal planar
(c)	sp^3	tetrahedral	pyramidal
(d)	sp^2	trigonal planar	trigonal planar with respect to C
(e)	sp^3	tetrahedral	tetrahedral
(f)	sp	linear	linear
(g)	sp^3	tetrahedral	tetrahedral
(h)	sp^2	trigonal planar	angular
(i)	sp^3	tetrahedral	pyramidal
(j)	sp^3	tetrahedral	tetrahedral

6.(a)

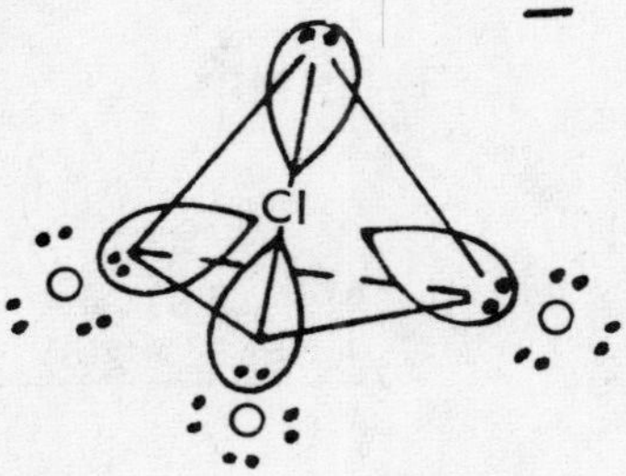

(b)

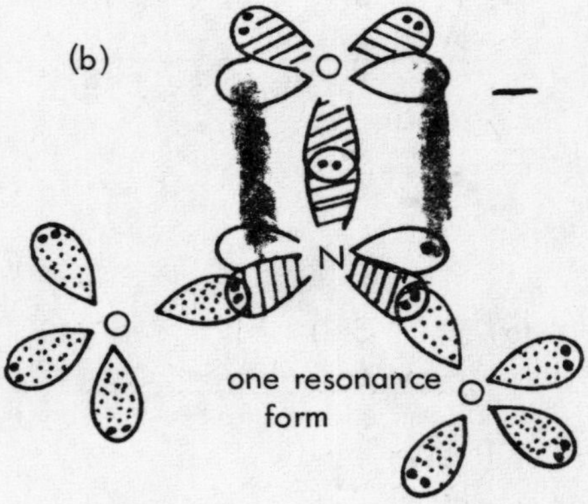

(c)

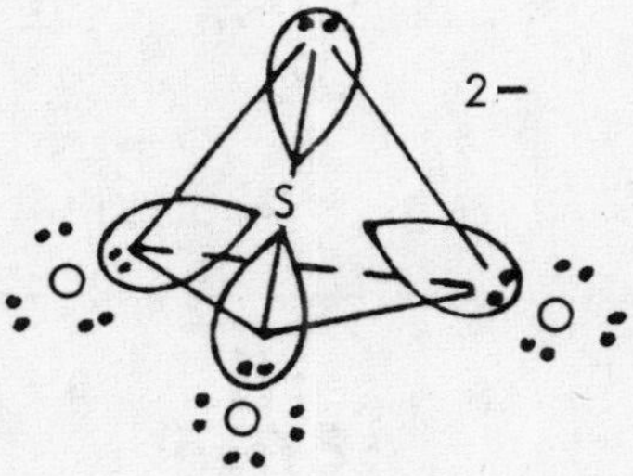

(d)

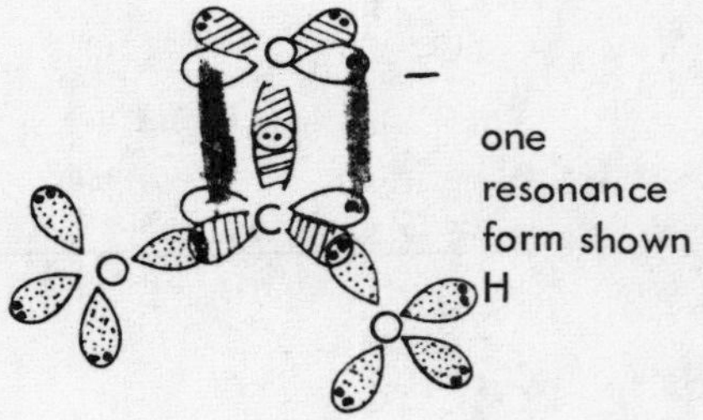

(e)

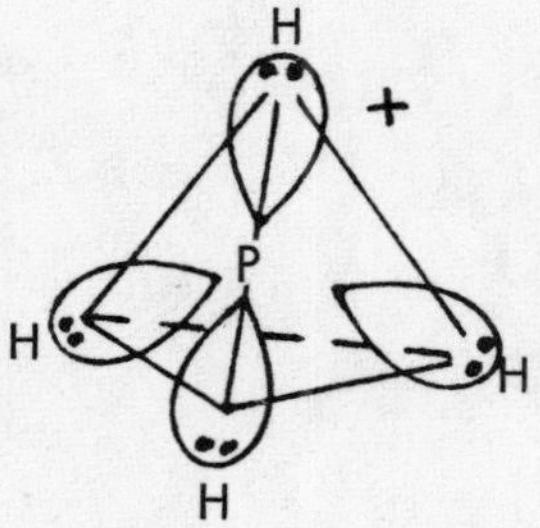

(f)

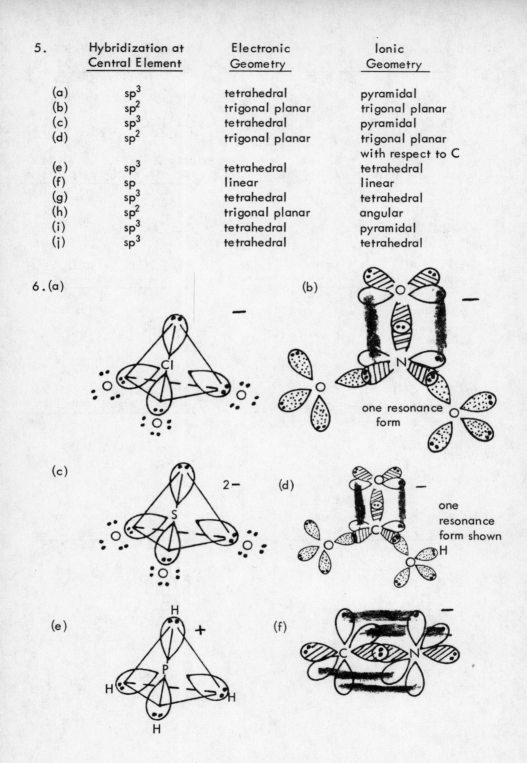

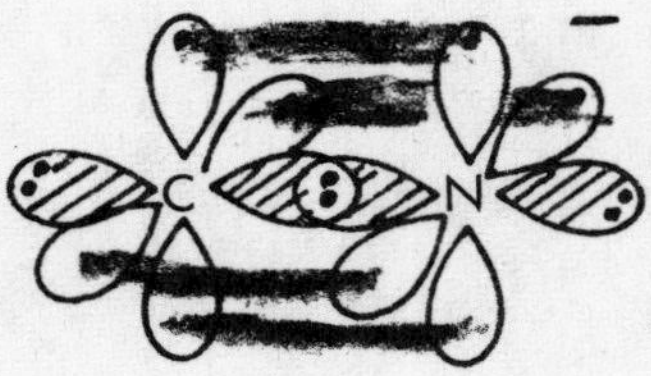

(g)

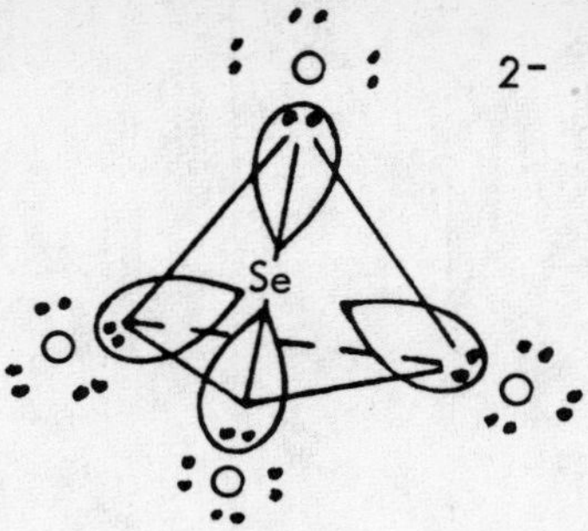

(h)

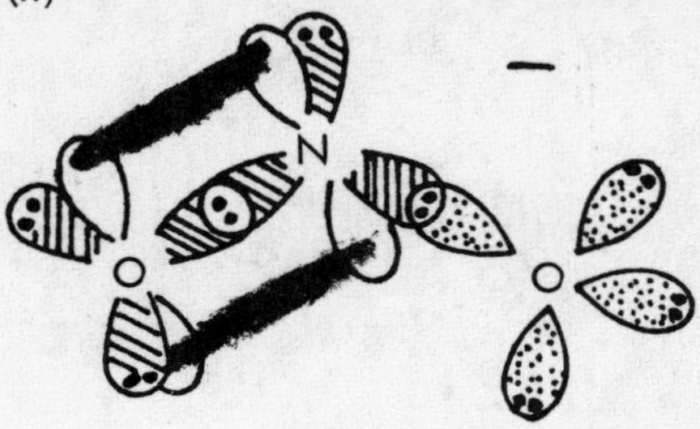

one resonance form shown

(i)

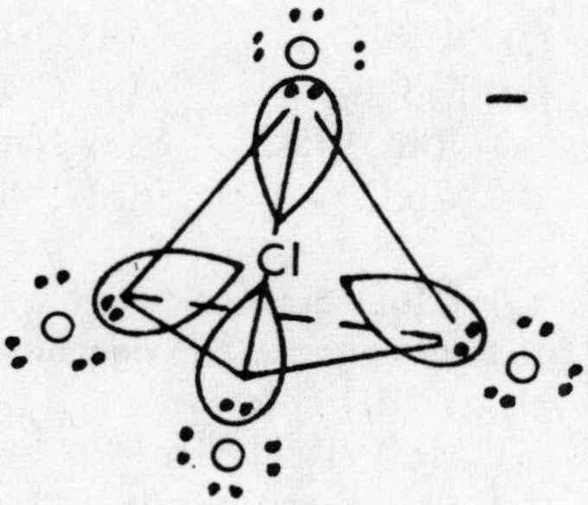

(j)

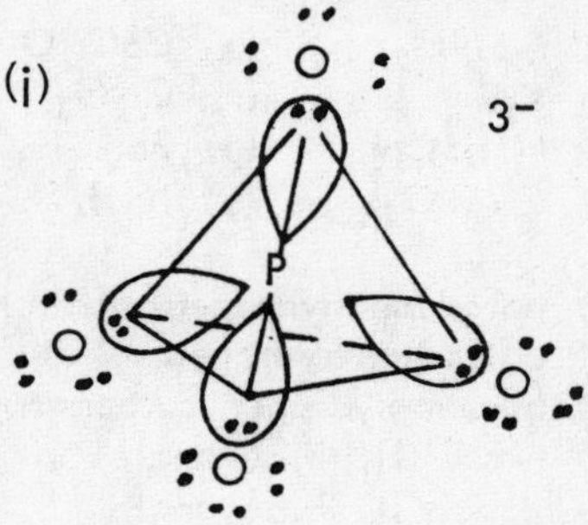

7. (a) polar (b) polar (c) polar (d) nonpolar (e) nonpolar
(f) nonpolar (g) nonpolar (h) polar (i) nonpolar (j) nonpolar
(k) polar (l) polar

8. (a)

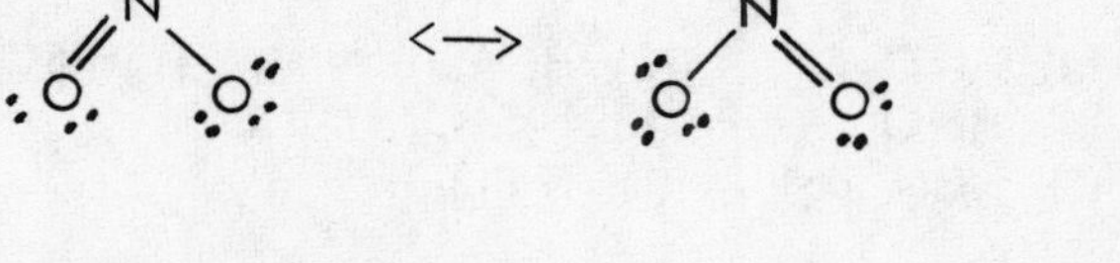

see 6(b)

(b)

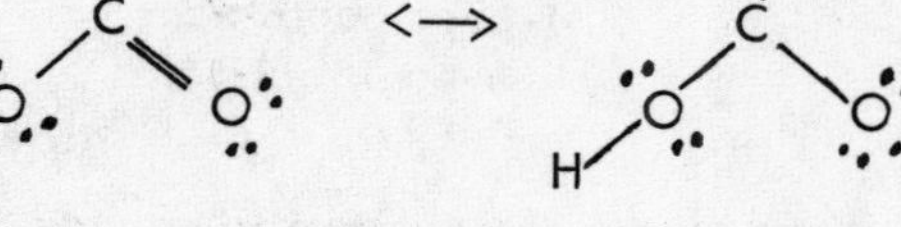

see 6(h)

(c)

see 6(d)

(d)

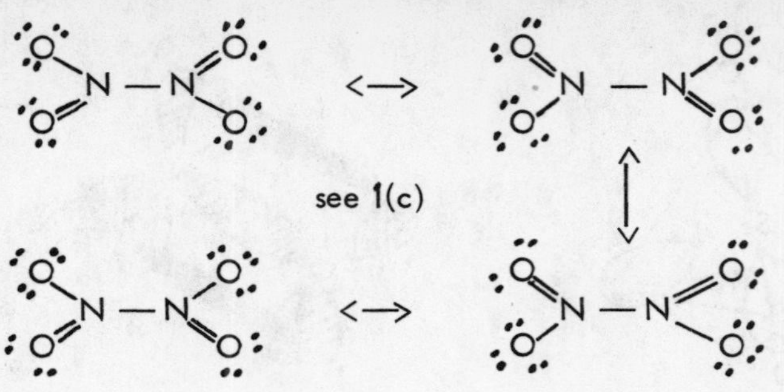

9. (a) $NaNO_2$ (b) K_2S (c) NH_4SCN (d) $Cu_3(PO_4)_2$
(e) $CoCr_2O_7$ (f) Cr_2O_3 (g) $Pb(C_2H_3O_2)_2$ (h) $(NH_4)_2S$
(i) $Al(NO_2)_3$ (j) FeS (k) $(NH_4)HSO_4$ (l) $Cu(ClO_3)_2$
(m) $CoCrO_4$ (n) CrO_3 (o) MnF_3 (p) $(NH_4)_2SO_3$

10. (a) potossium hydrogen sulfite
(b) potassium hydrogen sulfate
(c) barium hydrogen carbonate
(d) nickel(II) thiocyanate
(e) cobalt(II) thiocyanate
(f) lithium hydrogen sulfite
(g) silver arsenate
(h) manganese(III) perchlorate
(i) tin(II) chloride
(j) tin(IV) chloride
(k) scandium acetate
(l) cadmium dichromate

11. (a) MnF_3* (b) $Cr(MnO_4)_3$ (c) $Cr(ClO)_3$ (d) $Fe_2(C_2O_4)_3$
(e) CoF_3 (f) P_4O_6 (g) N_2O_3 (h) $Sc(NO_3)_3$
(i) $MnCr_2O_7$ (j) $CrSO_4$ (k) FeC_2O_4 (l) CoI_2
(m) Cl_2O_7 (n) P_4O_{10} (o) N_2O_5 (p) $Ti_2(SO_4)_3$

12. (a) chromous acetate*
(b) manganic dichromate
(c) silver permanganate
(d) chromic chlorate
(e) scandium chlorite
(f) cobaltous chlorate
(g) lead(II) perchlorate
(h) nickel(II) hypobromite
(i) silver bromate
(j) ammonium bromide
(k) potassium arsenate
(l) cobalt(II) oxalate

13. (a) H_3AsO_3 (b) H_2S (aq) (c) H_2SO_4
(d) HNO_2 (e) HBrO (f) HCN (aq)
(g) HF (aq) (h) H_2SeO_4 (i) H_3PO_4
(j) H_2SO_3

14. (a) hydrobromic acid
(b) perbromic acid
(c) nitric acid
(d) phosphorous acid
(e) chloric acid
(f) tellurous acid
(g) carbonic acid
(h) hypoiodous acid
(i) hyponitrous acid

---

*Note that some salts have been named by the older system to provide familiarity with this system.

Chapter Six

MOLECULAR ORBITALS IN CHEMICAL BONDING

D, G, W: Sec. 8-9 through 8-14

6-1 Molecular Orbitals

Molecular Orbital theory provides an alternative to Valence Bond theory with respect to describing covalent bonding. The basic notion of MO theory is that atomic orbitals on different atoms overlap to form molecular orbitals which belong to the molecule as a whole, or a portion of the molecule, rather than to individual atoms.

The overlap of two (appropriate) atomic orbitals results in the formation of two molecular orbitals. One, the bonding MO, is lower in energy than the combining atomic orbitals, while the other, the antibonding MO, is higher in energy than the combining orbitals. The overlap of an s orbital on one atom with an s orbital on another atom results in a bonding σ_s (read sigma - s) and an antibonding $\sigma_s{}^*$ orbital. The asterisk superscript signifies an antibonding orbital. In the diagram below the dots (·) represent nuclei.

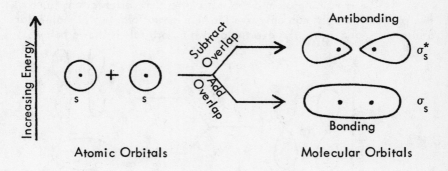

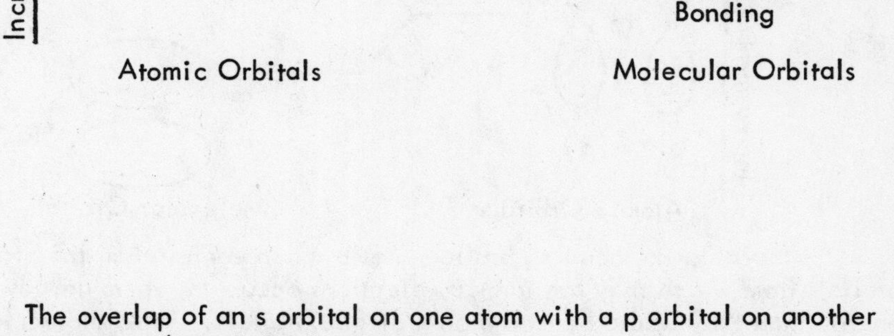

The overlap of an s orbital on one atom with a p orbital on another produces σ_{sp} and $\sigma_{sp}{}^*$ molecular orbitals.

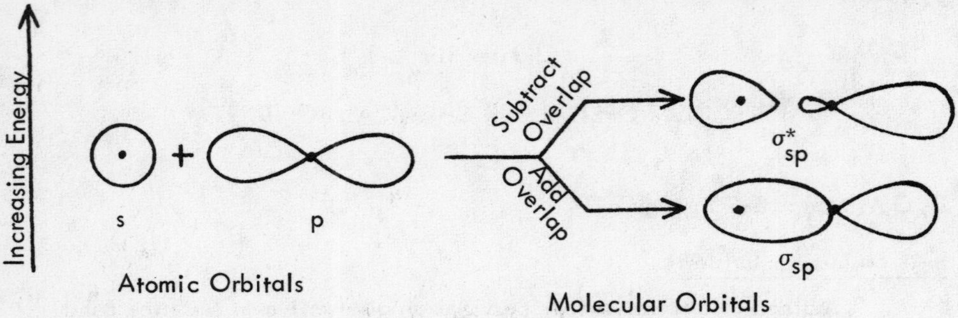

Two p orbitals of the same spatial orientation can overlap "head-on" to form σ_p and $\sigma_p{}^*$ molecular orbitals. For example two p_x orbitals could overlap as shown below.

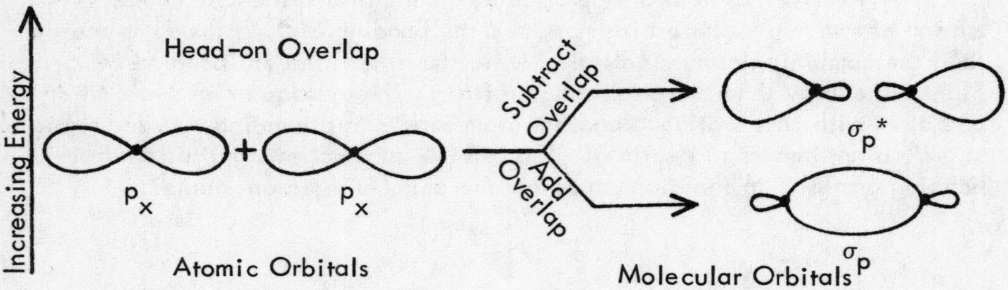

The remaining atomic orbitals of the same orientations (such as the two p_y's or two p_z's) can only overlap side-on to form two pi molecular orbitals, π_p and $\pi_p{}^*$. The overlap of two p_z orbitals is shown below.

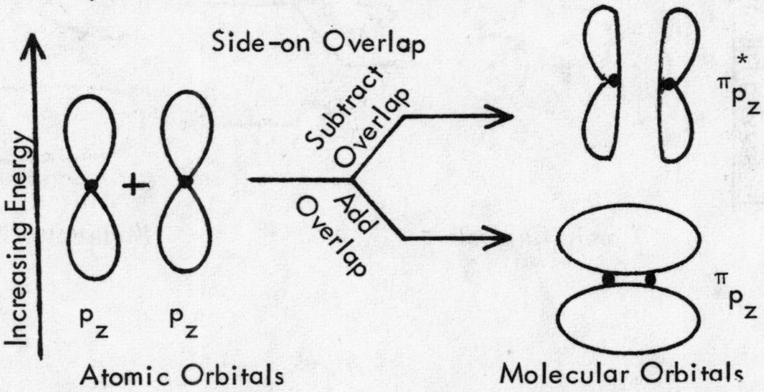

Because the bonding orbitals are lower in energy than the atomic orbitals from which they are formed, electrons occupying them are lower in energy than they would be in the pure atomic orbitals. Thus, molecules or complex ions are said to be stabilized by electrons in bonding orbitals.

Conversely, the occupation of the higher energy antibonding orbitals by electrons tends to destabilize species.

6-2 Molecular Orbital Energy-Level Diagrams

The diagram below shows the relative ordering of energies of molecular orbitals resulting from combinations of s and p orbitals of the first two energy levels for homonuclear diatomic molecules and ions from H_2 up through N_2.

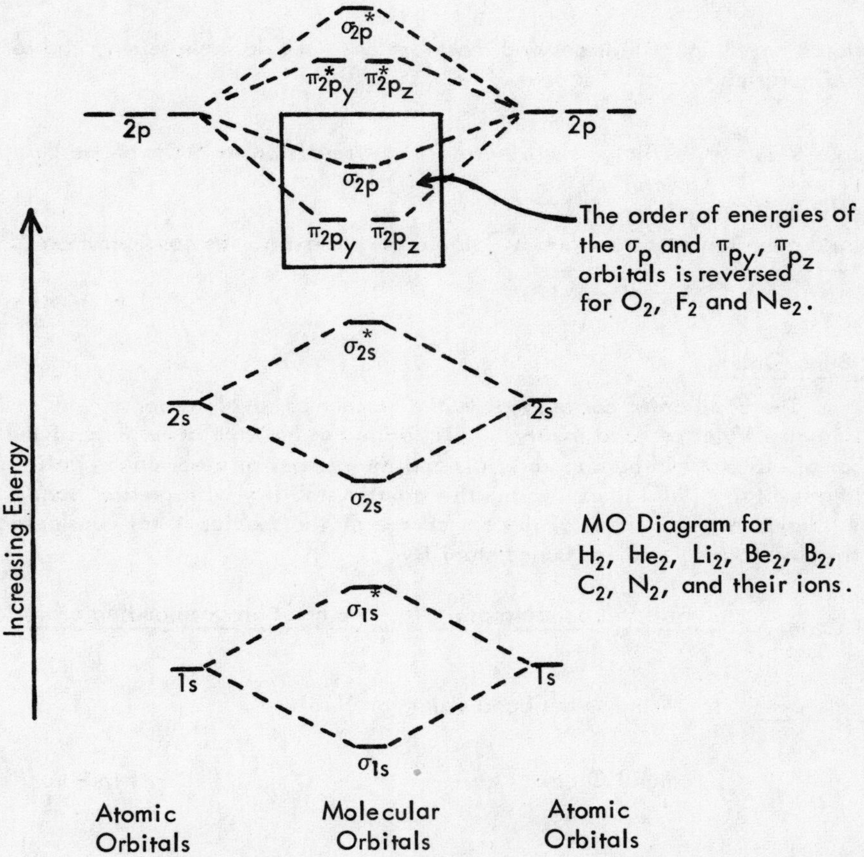

MO Diagram for H_2, He_2, Li_2, Be_2, B_2, C_2, N_2, and their ions.

The N_2 molecule consists of two N atoms, each having the $1s^2 2s^2 2p^3$ electronic configuration, so the N_2 molecule has fourteen electrons which fill MO's in accordance with Hund's Rule. That is, they fill the lowest energy MO's first and occupy a set of energetically equivalent orbitals singly before pairing up in any one orbital of the set. Thus N_2 has the electronic configuration: $\sigma_{1s}^{2}\ \sigma_{1s}^{*2}\ \sigma_{2s}^{2}\ \sigma_{2s}^{*2}\ \pi_{2p_y}^{2}\ \pi_{2p_z}^{2}\ \sigma_{2p}^{2}$.

Example 6-1: What is the electronic configuration in MO's of the N_2^+ ion?

We may visualize the formation of N_2^+ from N_2 by removal of one electron from the highest energy level of N_2 to produce N_2^+ which has the configuration: $\sigma_{1s}^2\ \sigma_{1s}^{*2}\ \sigma_{2s}^2\ \sigma_{2s}^{*2}\ \pi_{2p_y}^2\ \pi_{2p_z}^2\ \sigma_{2p}^1$.

Example 6-2: What is the electronic configuration in MO's of the Be_2^+ ion?

Be_2 would have eight electrons and therefore Be_2^+ would have seven, and so its configuration is: $\sigma_{1s}^2\ \sigma_{1s}^{*2}\ \sigma_{2s}^2\ \sigma_{2s}^{*1}$.

Example 6-3: What is the electronic configuration in MO's of the B_2^- ion?

B_2 would have ten electrons and B_2^- would have eleven. Its configuration is: $\sigma_{1s}^2\ \sigma_{1s}^{*2}\ \sigma_{2s}^2\ \sigma_{2s}^{*2}\ \pi_{2p_y}^2\ \pi_{2p_z}^1$.

6-3 Bond Order

The bond order corresponds to the number of covalent bonds according to Valence Bond theory. It is defined as half the difference of the number of electrons in bonding orbitals and the number of electrons in antibonding orbitals. MO theory associates greater stability with greater bond order. However, the effect of the net charge on the species is not considered and this, of course, will influence stability.

$$\text{Bond Order} = \frac{(\text{no of } e^- \text{ in bonding orbitals}) - (\text{no of } e^- \text{ in antibonding orbitals})}{2}$$

Example 6-4: What is the bond order for N_2?

$$\text{Bond Order} = \frac{10 - 4}{2} = 3 \qquad \text{(a triple bond)}$$

Example 6-5: What is the bond order for N_2^+? See Example 6-1.

$$\text{Bond Order} = \frac{9 - 4}{2} = 2\tfrac{1}{2}$$

Example 6-6: What is the bond order of Be_2^+? See Example 6-2.

$$\text{Bond Order} = \frac{4 - 3}{2} = \tfrac{1}{2}$$

Example 6-7: What is the bond order of B_2^-? See Example 6-3.

$$\text{Bond Order} = \frac{7 - 4}{2} = 1\tfrac{1}{2}$$

6-4 MO Diagram for Heteronuclear Diatomic Molecules

The MO diagrams of heteronuclear diatomic molecules are skewed because the atomic orbitals of the more electronegative element are lower in energy than those of the element with which it combines. The MO diagram for NO (nitrogen oxide or nitric oxide) is shown below.

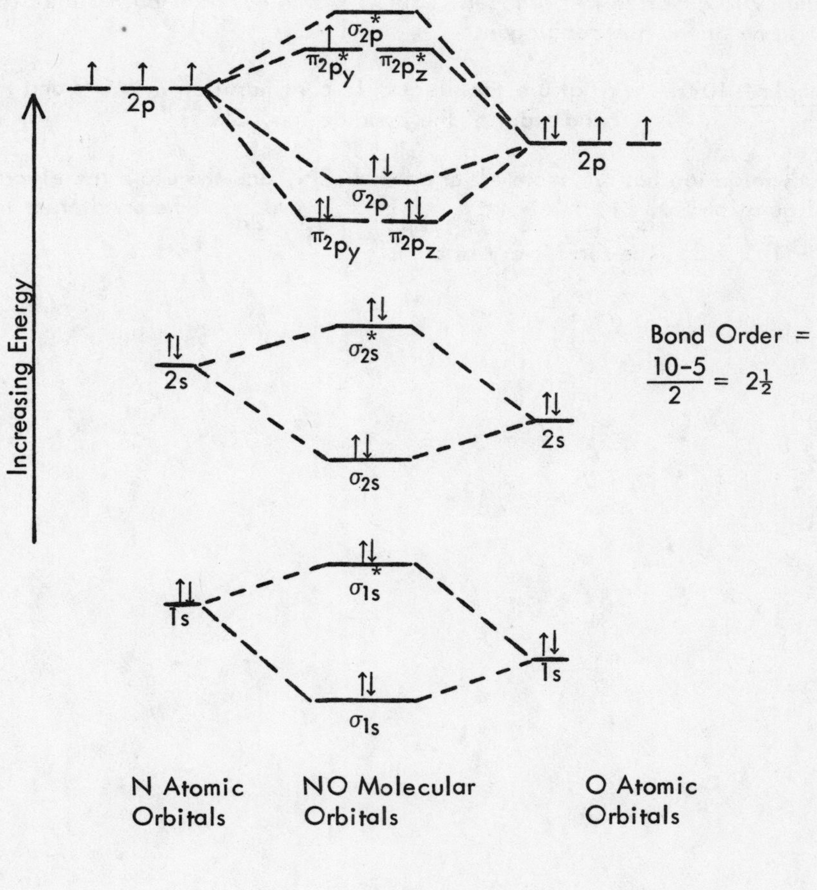

Example 6-8: What are the electronic configuration in MO's and the bond order of NO^+?

The electronic configuration involves one less electron than NO. It is: $\sigma_{1s}^2\ \sigma_{1s}^{*2}\ \sigma_{2s}^2\ \sigma_{2s}^{*2}\ \pi_{2p_y}^2\ \pi_{2p_z}^2\ \sigma_{2p}^2$. The bond order is $(10 - 4)/2 = 3$. This ion exists in a number of compounds.

Example 6-9: What are the electronic configuration in MO's and the bond order of the neutral (radical) molecule CN?

Carbon contributes six electrons and nitrogen contributes seven for a total of thirteen. The electron configuration is: $\sigma_{1s}^2\ \sigma_{1s}^{*2}\ \sigma_{2s}^2\ \sigma_{2s}^{*2}\ \pi_{2p_y}^2\ \pi_{2p_z}^2\ \sigma_{2p}^1$. The bond order is $(9 - 4)/2 = 2\frac{1}{2}$. This molecule is known but it has a strong tendency to dimerize to cyanogen, $(CN)_2$, which has no unpaired electrons and is one of the pseudohalogens.

Example 6-10: What are the electronic configuration in MO's and the bond order of the cyanide ion, CN^-?

The cyanide ion has one more electron than CN, and therefore the electronic configuration: $\sigma_{1s}^2\ \sigma_{1s}^{*2}\ \sigma_{2s}^2\ \sigma_{2s}^{*2}\ \pi_{2p_y}^2\ \pi_{2p_z}^2\ \sigma_{2p}^2$. The bond order is $(10 - 4)/2 = 3$. The ion is very stable.

EXERCISES

1. From memory, draw the MO energy level diagram for homonuclear diatomic molecules of the second period, showing the π_{2p_y} and π_{2p_z} orbitals below the σ_{2p} in energy.

2. Utilizing the sets of MO's you drew in problem 1, write out the electron configurations for the following molecules and ions in the form σ_{1s}^2 σ_{1s}^{*2} σ_{2s}^2 and so on: H_2, H_2^+, H_2^-, Be_2, Be_2^+, Be_2^-, B_2 and B_2^-.

3. (a) What is the bond order of each of the species of problem 2?
   (b) Which are diamagnetic (D) and which are paramagnetic (P)?
   (c) Apply MO theory to predict the relative stabilities of these species.
   (d) Comment on the validity of the predictions from (c). What else must be considered in addition to electron occupation of MO's?

4. Assuming that the σ_{2p} MO is lower in energy than the π_{2p_y} and π_{2p_z} MO's for the following species write out electron configurations for each one: O_2, O_2^+, Ne_2 and Ne_2^-.

5. (a) What is the bond order of each of the species of problem 4?
   (b) Which are diamagnetic (D) and which are paramagnetic (P)?
   (c) Apply MO theory to predict the relative stabilities of the species.

6. From memory, draw the MO energy level diagram for a heteronuclear diatomic molecule, XY, in which both elements are from the second period and Y is more electronegative than X. Show the σ_{2p} MO lower in energy than the π_{2p}'s.

7. Utilizing the sets of MO's you drew in problem 6, write out electron configurations for the following molecules and ions in the form σ_{1s}^2 σ_{1s}^{*2} σ_{2p}^2 and so on: NO^-, CN^+, NF^+, CF, CF^-, OF, and OF^-.

8. (a) What is the bond order of each of the species of problem 7?
   (b) Which are diamagnetic (D) and which are paramagnetic (P)?
   (c) Apply MO theory to predict the relative stabilities of the species.

ANSWERS TO EXERCISES

1. See page 121.

2. H_2 σ_{1s}^2

   H_2^+ σ_{1s}^1

   H_2^- σ_{1s}^2 σ_{1s}^{*1}

2. (cont'd)

Be_2 $\sigma_{1s}^{2}\ \sigma_{1s}^{*2}\ \sigma_{2s}^{2}\ \sigma_{2s}^{*2}$

Be_2^{+} $\sigma_{1s}^{2}\ \sigma_{1s}^{*2}\ \sigma_{2s}^{2}\ \sigma_{2s}^{*1}$

Be_2^{-} $\sigma_{1s}^{2}\ \sigma_{1s}^{*2}\ \sigma_{2s}^{2}\ \sigma_{2s}^{*2}\ \pi_{2p}^{1}$

B_2 $\sigma_{1s}^{2}\ \sigma_{1s}^{*2}\ \sigma_{2s}^{2}\ \sigma_{2s}^{*2}\ \pi_{2p_y}^{1}\ \pi_{2p_z}^{1}$

B_2^{-} $\sigma_{1s}^{2}\ \sigma_{1s}^{*2}\ \sigma_{2s}^{2}\ \sigma_{2s}^{*2}\ \pi_{2p_y}^{2}\ \pi_{2p_z}^{1}$

3.

Species	(a) Bond Order	(b) D or P	(c) Predicted Stability
H_2	1	D	stable
H_2^{+}	1/2	P	marginally stable
H_2^{-}	1/2	P	marginally stable
Be_2	0	D	unstable
Be_2^{+}	1/2	P	marginally stable
Be^{-}	1/2	P	marginally stable
B_2	1	P	stable
B_2^{-}	1 1/2	P	stable

(d) Some of the conclusions in (c) are not reliable because the effect of charges must be taken into account.

4. O_2 $\sigma_{1s}^{2}\ \sigma_{1s}^{*2}\ \sigma_{2s}^{2}\ \sigma_{2s}^{*2}\ \sigma_{2p}^{2}\ \pi_{2p_y}^{2}\ \pi_{2p_z}^{2}\ \pi_{2p_y}^{*1}\ \pi_{2p_z}^{*1}$

O_2^{+} $\sigma_{1s}^{2}\ \sigma_{1s}^{*2}\ \sigma_{2s}^{2}\ \sigma_{2s}^{*2}\ \sigma_{2p}^{2}\ \pi_{2p_y}^{2}\ \pi_{2p_z}^{2}\ \pi_{2p_y}^{*1}$

Ne_2 $\sigma_{1s}^{2}\ \sigma_{1s}^{*2}\ \sigma_{2s}^{2}\ \sigma_{2s}^{*2}\ \sigma_{2p}^{2}\ \pi_{2p_y}^{2}\ \pi_{2p_z}^{2}\ \pi_{2p_y}^{*2}\ \pi_{2p_z}^{*2}\ \sigma_{2p}^{*2}$

Ne_2^{-} $\sigma_{1s}^{2}\ \sigma_{1s}^{*2}\ \sigma_{2s}^{2}\ \sigma_{2s}^{*2}\ \sigma_{2p}^{2}\ \pi_{2p_y}^{2}\ \pi_{2p_z}^{2}\ \pi_{2p_y}^{*2}\ \pi_{2p_z}^{*2}\ \sigma_{2p}^{*2}\ \sigma_{3s}^{1}$

5.

Species	(a) Bond Order	(b) D or P	(c) Predicted Stability*
O_2	2	P	very stable
O_2^{+}	2 1/2	P	very stable
Ne_2	0	D	unstable
Ne_2^{+}	1/2	P	marginally stable

*See answer to 3(d).

6. See page 123.

7. NO^- same as O_2 in problem 4

CN^+ $\sigma_{1s}^{2}\ \sigma_{1s}^{*2}\ \sigma_{2s}^{2}\ \sigma_{2s}^{*2}\ \sigma_{2p}^{2}\ \pi_{2p_y}^{1}\ \pi_{2p_z}^{1}$

NF^+ same as O_2^+ in problem 4

CF same as O_2^+ in problem 4

CF^- same as O_2 in problem 4

OF $\sigma_{1s}^{2}\ \sigma_{1s}^{*2}\ \sigma_{2s}^{2}\ \sigma_{2s}^{*2}\ \sigma_{2p}^{2}\ \pi_{2p_y}^{2}\ \pi_{2p_z}^{2}\ \pi_{2p_y}^{*2}\ \pi_{2p_z}^{*1}$

OF^- $\sigma_{1s}^{2}\ \sigma_{1s}^{*2}\ \sigma_{2s}^{2}\ \sigma_{2s}^{*2}\ \sigma_{2p}^{2}\ \pi_{2p_y}^{2}\ \pi_{2p_z}^{2}\ \pi_{2p_y}^{*2}\ \pi_{2p_z}^{*2}$

8.

Species	(a) Bond Order	(b) D or P	(c) Predicted Stability*
NO^-	2	P	very stable
CN^+	2	P	very stable
NF^+	2 1/2	P	very stable
CF	2 1/2	P	very stable
CF^-	2	P	very stable
OF	1 1/2	P	stable
OF^-	1	D	stable

*See answer to 3(d)

Chapter Seven

CHEMICAL REACTIONS: A SYSTEMATIC STUDY

D, G, W: Sec. 9- 4 through 9-7

Chemical reactions can be classified in a variety of ways. The following categories are defined and illustrated below: (1) combination reactions, (2) decomposition reactions, (3) displacement reactions, (4) metathesis reactions, and (5) oxidation-reduction reactions. Some reactions fit into more than one catgeory and others do not fit neatly into any of these categories.

7-1 Combination Reactions

Combination reactions are those in which two or more substances combine to form a single compound. They may involve the combination of two elements to form a single new compound.

$$C\ (s) + O_2\ (g) \xrightarrow{\Delta} CO_2\ (g)$$

$$H_2\ (g) + Br_2\ (\ell) \longrightarrow 2\ HBr\ (g)$$

$$2\ Li\ (s) + Cl_2\ (g) \longrightarrow 2\ LiCl\ (s)$$

$$P_4\ (s) + 5\ O_2\ (g) \xrightarrow{\Delta} P_4O_{10}\ (s)$$

They may involve the combination of an element and a compound to form a single new compound.

$$4\ FeO\ (s) + O_2\ (g) \longrightarrow 2\ Fe_2O_3\ (s)$$

$$2\ CO\ (g) + O_2\ (g) \longrightarrow 2\ CO_2\ (g)$$

They may also involve the combination of two compounds to form a single new compound.

$$N_2O_5\ (s) + H_2O\ (\ell) \longrightarrow 2\ HNO_3\ (\ell)$$

$$SO_2\ (g) + CaO\ (s) \longrightarrow CaSO_3\ (s)$$

7-2 Displacement Reactions

Displacement reactions are those in which one element displaces another from a compound. For example, an active metal such as aluminum displaces hydrogen from hydrochloric acid to form aluminum chloride and gaseous hydrogen.

$$2\ Al\ (s) + 6\ HCl\ (aq) \longrightarrow 2\ AlCl_3\ (aq) + 3\ H_2\ (g)$$

Bromine is more electronegative than iodine. Therefore bromine will displace iodide ions from sodium iodide to form bromide ions and solid iodine.

$$Br_2\ (\ell) + 2\ NaI\ (aq) \longrightarrow 2\ NaBr\ (aq) + I_2\ (s)$$

7-3 Decomposition Reactions

Decomposition reactions involve the decomposition of a single compound. The products may be elements or compounds as the following examples show.

$$2\ HI\ (g) \xrightarrow{\Delta} H_2\ (g) + I_2\ (g)$$
$$2\ NaClO_3\ (s) \xrightarrow{\Delta} 2\ NaCl\ (s) + 3\ O_2\ (g)$$
$$BaCO_3\ (s) \xrightarrow{\Delta} BaO\ (s) + CO_2\ (g)$$

7-4 Metathesis Reactions

Metathesis reactions (sometimes called double displacement reactions) involve the reaction of two compounds to form two new compounds. Among the most common metathesis reactions are acid-base neutralization reactions in which H^+ ions (usually represented as H_3O^+) from the acid react with OH^- ions from the base to form water and a salt (cation of base + anion of acid). Two examples of acid-base metathesis reactions follow:

$$2\ HCl\ (aq) + Ca(OH)_2\ (aq) \longrightarrow CaCl_2\ (aq) + 2\ H_2O\ (\ell)$$
$$KOH\ (aq) + HC_2H_3O_2\ (aq) \longrightarrow KC_2H_3O_2\ (aq) + H_2O\ (\ell)$$

Other metathesis reactions (not classical acid-base) are given below:

$$BaCl_2\ (aq) + H_2SO_4\ (aq) \longrightarrow BaSO_4\ (s) + 2\ HCl\ (aq)$$

a precipitate

$$AgNO_3\ (aq) + NaCl\ (aq) \longrightarrow AgCl\ (s) + NaNO_3\ (aq)$$

a precipitate

$$2\ Bi(NO_3)_3\ (aq) + 3\ H_2S\ (aq) \longrightarrow Bi_2S_3\ (s) + 6\ HNO_3\ (aq)$$

a precipitate

7-5 Oxidation-Reduction Equations

Oxidation-reduction reactions (or redox reactions) are those in which changes in oxidation number are involved. They may also be called electron transfer reactions. Many redox reactions also fit into one of the four previously described categories of reactions. Examples are given below with oxidation numbers that change indicated above the elements. Recall that oxidation numbers are indicated by ±n while ionic charges are indicated by n± where n refers to the magnitude of the oxidation number or ionic charge, respectively.

$$2\,\overset{0}{Na}\,(s) + \overset{0}{Cl_2}\,(g) \longrightarrow 2\,\overset{+1}{Na}\overset{-1}{Cl}\,(s)$$ (also a combination reaction)

$$\overset{0}{Zn}\,(s) + 2\,\overset{+1}{H}Br\,(aq) \longrightarrow \overset{+2}{Zn}Br_2\,(aq) + \overset{0}{H_2}\,(g)$$ (also a displacement reaction)

$$2\,\overset{0}{Na}\,(s) + 2\,\overset{+1}{H_2}O\,(\ell) \longrightarrow 2\,\overset{+1}{Na}OH\,(aq) + \overset{0}{H_2}\,(g)$$ (also a displacement reaction)

Some examples of oxidation-reduction reactions that do not fit neatly into one of the other four categories are the following:

$$\overset{0}{Br_2}\,(\ell) + H_2\overset{+4}{S}O_3\,(aq) + H_2O\,(\ell) \longrightarrow 2\,H\overset{-1}{Br}\,(aq) + H_2\overset{+6}{S}O_4\,(aq)$$

$$2\,\overset{+4}{Cl}O_2\,(g) + 2\,NaOH\,(aq) \longrightarrow Na\overset{+1}{Cl}O\,(aq) + Na\overset{+7}{Cl}O_4\,(aq) + H_2O\,(\ell)$$

The following are not oxidation-reduction reactions because none of the elements change oxidation number.

$$\overset{+2}{Mg}\overset{+4}{C}\overset{-2}{O_3}\,(s) \xrightarrow{\Delta} \overset{+2}{Mg}\overset{-2}{O}\,(s) + \overset{+4}{C}\overset{-2}{O_2}\,(g)$$ (a decomposition reaction)

$$\overset{+1}{H}\overset{-1}{F}\,(g) + \overset{-3}{N}\overset{+1}{H_3}\,(g) \longrightarrow \overset{-3}{N}\overset{+1}{H_4}\overset{-1}{F}\,(s)$$ (a combination reaction)

7-6 Methods of Writing Equations

Equations such as the ones we have dealt with so far in this chapter are called molecular equations because the complete formulas for all substances are used (without regard to whether they are molecular or ionic). An example of a molecular equation for a reaction in aqueous solution is:

$$BaCl_2\,(aq) + Na_2SO_4\,(aq) \longrightarrow BaSO_4\,(s) + 2\,NaCl\,(aq)$$

$BaCl_2$, Na_2SO_4, and NaCl are all soluble in water and completely (or nearly completely) ionized in dilute aqueous solution. All such substances are called strong electrolytes because they are completely or nearly completely ionized in water, and therefore their aqueous solutions conduct electricity well. Most soluble ionic salts are strong electrolytes, as are strong acids (~100% ionized). The common strong acids and strong soluble bases (metal hydroxides of Group IA and heavier members of Group IIA) are listed below. These lists should be learned.

Strong Acids	Strong Soluble Bases	
HCl	LiOH	
HBr	NaOH	
HI	KOH	$Ca(OH)_2$
HNO_3	RbOH	$Sr(OH)_2$
H_2SO_4	CsOH	$Ba(OH)_2$
$HClO_4$		
$HClO_3$		

One may usually assume that other common acids (most acids are soluble in water) are weak acids, i.e., they are only slightly ionized in dilute aqueous solutions. Other common metal hydroxides are quite insoluble in water. The only common weak soluble bases (only slightly ionized) are ammonia, NH_3, and its derivatives, the amines, RNH_2, R_2NH, and R_3N (R = organic group).' Ammonia ionizes slightly in water to produce NH_4^+ and OH^- ions.

$$NH_3\ (g) + H_2O\ (\ell) \xrightleftharpoons{\text{slight ionization}} NH_4^+\ (aq) + OH^-\ (aq)$$

Total ionic equations are only important for reactions that occur in solution. They involve writing the formulas for all soluble strong electrolytes (soluble ionic salts*, strong acids, and strong soluble bases) in ionic form and those for weak electrolytes, covalent (non-ionic) compounds and insoluble compounds in molecular form. In other words total ionic equations show the predominant form of all species for reactions in aqueous solutions.

Consider again the molecular equation below.

$$BaCl_2\ (aq) + Na_2SO_4\ (aq) \longrightarrow BaSO_4\ (s) + 2\ NaCl\ (aq)$$

Since both the reactants and NaCl are soluble ionic salts they are strong electrolytes and are written in ionic form with square brackets ([ ]) used to indicate the sources of the ions. The total ionic equation is:

$$[Ba^{2+}\ (aq) + 2\ Cl^-\ (aq)] + [2\ Na^+\ (aq) + SO_4^{2-}\ (aq)] \longrightarrow \underset{\text{insoluble}}{BaSO_4\ (s)} + 2[Na^+\ (aq) + Cl^-\ (aq)]$$

Notice that Na^+ (aq) and Cl^- (aq) appear in the same form on both reactants and products sides of the equation. They do not actually participate in the

*The solubility rules are given in Sec. 7-1.5 in the text.

reaction and are called spectator ions. Cancellation of equal numbers of spectator ions from each side yields the net ionic equation.

$$Ba^{2+}\,(aq) \quad + \quad SO_4^{2-}\,(aq) \quad \longrightarrow \quad BaSO_4\,(s)$$

The net ionic equation shows only the essence of the reaction; in other words, the net reaction.

The reaction of the strong electrolytes HNO_3 (a strong acid) and NaOH (a strong base) in water to produce $NaNO_3$ (a soluble ionic salt) and H_2O (a covalent compound) provides another example.

$$HNO_3\,(aq) + NaOH\,(aq) \longrightarrow NaNO_3\,(aq) + H_2O\,(\ell) \qquad \underline{\text{molecular}}$$

$$[H_3O^+\,(aq) + NO_3^-\,(aq)] + [Na^+\,(aq) + OH^-\,(aq)] \longrightarrow$$
$$[Na^+\,(aq) + NO_3^-\,(aq)] + 2\,H_2O\,(\ell) \qquad \underline{\text{total ionic}}$$

$$H_3O^+\,(aq) + OH^-\,(aq) \longrightarrow 2\,H_2O\,(\ell) \qquad \underline{\text{net ionic}}$$

This is the net ionic equation for all reactions of strong soluble bases with strong acids to form soluble salts and water. The corrosive properties of H_3O^+ and OH^- are neutralized in forming harmless H_2O.

One further example is given. Note that HF is a weak acid.

$$2\,HF\,(aq) + Ba(OH)_2\,(aq) \longrightarrow BaF_2\,(s) + 2\,H_2O\,(\ell) \qquad \underline{\text{molecular}}$$

2 HF (aq)	+ Ba(OH)₂ (aq)	BaF₂ (s)	2 H₂O (ℓ)
weak acid	strong soluble base	insoluble salt	covalent compound

$$2\,HF\,(aq) + [Ba^{2+}\,(aq) + 2\,OH^-\,(aq)] \longrightarrow BaF_2\,(s) + 2\,H_2O \qquad \underline{\text{total ionic}}$$

There are no spectator ions and therefore the net ionic and total ionic equations are identical except that we do not include brackets in net ionic equations.

$$2\,HF\,(aq) + Ba^{2+}\,(aq) + 2\,OH^-\,(aq) \longrightarrow BaF_2\,(s) + 2\,H_2O\,(\ell)$$

net ionic

EXERCISES

Classify each of the reactions below into one of the following categories: (1) combination, (2) decomposition, (3) displacement, (4) metathesis, and (5) oxidation-reduction. For those that are oxidation-reduction reactions, place them into another category as well if possible.

1. $N_2O_4\ (g) \longrightarrow 2\ NO_2\ (g)$
2. $Mg\ (s) + H_2SO_4\ (aq) \longrightarrow MgSO_4\ (aq) + H_2\ (g)$
3. $3\ HCl\ (aq) + Al(OH)_3\ (s) \longrightarrow AlCl_3\ (aq) + 3\ H_2O\ (\ell)$
4. $H_2\ (g) + S\ (s) \xrightarrow{\Delta} H_2S\ (g)$
5. $Na_2S\ (aq) + NiCl_2\ (aq) \longrightarrow NiS\ (s) + 2\ NaCl\ (aq)$
6. $H_2O\ (\ell) + SO_2\ (g) \longrightarrow H_2SO_3\ (aq)$
7. $C\ (s) + 4\ HNO_3\ (aq) \xrightarrow{\Delta} 4\ NO_2\ (g) + CO_2\ (g) + 2\ H_2O\ (\ell)$
8. $H_2SO_4\ (aq) + Ba(OH)_2\ (aq) \longrightarrow BaSO_4\ (s) + 2\ H_2O\ (\ell)$
9. $2\ HBrO\ (aq) \longrightarrow 2\ HBr\ (g) + O_2\ (g)$
10. $CS_2\ (g) + 3\ Cl_2\ (g) \xrightarrow{\Delta} CCl_4\ (g) + S_2Cl_2\ (g)$
11. $Na_2CO_3\ (aq) + Cd(NO_3)_2\ (aq) \longrightarrow 2\ NaNO_3\ (aq) + CdCO_3\ (s)$
12. $CO\ (g) + H_2O\ (g) \xrightarrow{\Delta} CO_2\ (g) + H_2\ (g)$
13. $H_2\ (g) + CO_2\ (g) \xrightarrow{\Delta} CO\ (g) + H_2O\ (g)$
14. $2\ KClO_3\ (s) \xrightarrow{\Delta} 2\ KCl\ (s) + 3\ O_2\ (g)$
15. $BaSO_3\ (s) \xrightarrow{\Delta} BaO\ (s) + SO_2\ (g)$
16. $N_2O_3\ (g) \longrightarrow NO_2\ (g) + NO\ (g)$
17. $2\ SO_2\ (g) + O_2\ (g) \longrightarrow 2\ SO_3\ (\ell)$
18. $2\ NaOH\ (aq) + H_2CO_3\ (aq) \longrightarrow Na_2CO_3\ (aq) + 2\ H_2O\ (\ell)$
19. $KOH\ (aq) + HCN\ (aq) \longrightarrow KCN\ (aq) + H_2O\ (\ell)$
20. $C\ (s) + CO_2\ (g) \xrightarrow{\Delta} 2\ CO\ (g)$
21. $I_2\ (g) + Cl_2\ (g) \xrightarrow{\Delta} 2\ ICl\ (g)$
22. $Cd\ (s) + PbCl_2\ (aq) \xrightarrow{\Delta} CdCl_2\ (aq) + Pb\ (s)$
    (Note: $PbCl_2$ is slightly soluble in hot H_2O but does not ionize appreciably.)
23. $CH_4\ (g) + N_2\ (g) \xrightarrow{\Delta} NH_3\ (g) + HCN\ (g)$
24. $SbCl_5\ (g) \xrightarrow{\Delta} SbCl_3\ (g) + Cl_2\ (g)$
25. $2\ N_2O_5\ (s) \longrightarrow 4\ NO_2\ (g) + O_2\ (g)$
26. $2\ KCl\ (aq) + F_2\ (g) \longrightarrow 2\ KF\ (aq) + Cl_2\ (g)$
27. $Li_2O\ (s) + H_2O\ (\ell) \longrightarrow 2\ LiOH\ (aq)$
28. $CO_2\ (g) + 2\ Mg\ (s) \xrightarrow{\Delta} 2\ MgO\ (s) + C\ (s)$
29. $2\ KMnO_4\ (aq) + 6\ HCl\ (aq) + 5\ H_2O_2\ (aq) \longrightarrow 2\ MnCl_2\ (aq) + 2\ KCl\ (aq) + 5\ O_2\ (g) + 8\ H_2O\ (\ell)$
30. $CH_4\ (g) + 2\ O_2\ (g) \xrightarrow{\Delta} CO_2\ (g) + 2\ H_2O\ (g)$
31. $4\ NH_3\ (g) + 5\ O_2\ (g) \xrightarrow{\Delta} 4\ NO\ (g) + 6\ H_2O\ (g)$
32. $Sr_3(PO_4)_2\ (s) + 3\ H_2SO_4\ (aq) \longrightarrow 2\ H_3PO_4\ (aq) + 3\ SrSO_4\ (s)$
33. $H_2S\ (aq) + Co(NO_3)_2\ (aq) \longrightarrow CoS\ (s) + 2\ HNO_3\ (aq)$

34. NH_3 (g) + HCl (g) $\longrightarrow$ NH_4Cl (s)
35. 2 $Cu(NO_3)_2$ (s) $\xrightarrow{\Delta}$ 2 CuO (s) + 4 NO_2 (g) + O_2 (g)
36. $AgNO_3$ (aq) + LiCl (aq) $\longrightarrow$ AgCl (s) + $LiNO_3$ (aq)
37. 2 Cs (s) + I_2 (s) $\xrightarrow{\Delta}$ 2 CsI (s)
38. N_2H_4 (ℓ) + HNO_2 (aq) $\xrightarrow{\Delta}$ HN_3 (aq) + 2 H_2O (ℓ)
39. 2 RbOH (aq) + H_2CO_3 (aq) $\longrightarrow$ Rb_2CO_3 (aq) + 2 H_2O (ℓ)
40. 4 CuO (s) $\xrightarrow{\Delta}$ 2 Cu_2O (s) + O_2 (g)

41. Write total ionic and net ionic equations for each of the above reactions where appropriate.

ANSWERS TO EXERCISES

1. decomposition
2. displacement and redox
3. metathesis
4. combination and redox
5. metathesis
6. combination
7. redox
8. metathesis
9. decomposition and redox
10. redox
11. metathesis
12. redox
13. redox
14. decomposition and redox
15. decomposition
16. decomposition and redox
17. combination and redox
18. metathesis
19. metathesis
20. combination and redox
21. combination and redox
22. displacement and redox
23. redox
24. decomposition and redox
25. decomposition and redox
26. displacement and redox
27. combination
28. displacement and redox
29. redox
30. redox
31. redox
32. metathesis
33. metathesis
34. combination
35. decomposition and redox
36. metathesis
37. combination and redox
38. redox
39. metathesis
40. decomposition and redox

To conserve space the designation (aq) has been omitted from all ions in aqueous solution in the equations that follow. Please keep in mind the fact that <u>all ions are hydrated in aqueous solution</u>.

41. (2) Mg (s) + [2 H_3O^+ + SO_4^{2-}] $\longrightarrow$ [Mg^{2+} + SO_4^{2-}] + H_2 (g) + 2 H_2O (ℓ)

Mg (s) + 2 H_3O^+ $\longrightarrow$ Mg^{2+} + H_2 (g) + 2 H_2O (ℓ)

(3) 3 [H_3O^+ + Cl^-] + $Al(OH)_3$ (s) $\longrightarrow$ [Al^{3+} + 3 Cl^-] + 6 H_2O (ℓ)

3 H_3O^+ + $Al(OH)_3$ (s) $\longrightarrow$ Al^{3+} + 6 H_2O (ℓ)

1. (cont'd)

(5) $[2\ Na^+ + S^{2-}] + [Ni^{2+} + 2\ Cl^-] \rightarrow NiS\ (s) + 2\ [Na^+ + Cl^-]$

$S^{2-} + Ni^{2+} \longrightarrow NiS\ (s)$

(7) $C\ (s) + 4\ [H_3O^+ + NO_3^-] \longrightarrow 4\ NO_2\ (g) + CO_2\ (g) + 6\ H_2O\ (\ell)$

$C\ (s) + 4\ H_3O^+ + 4\ NO_3^- \longrightarrow 4\ NO_2\ (g) + CO_2\ (g) + 6\ H_2O\ (\ell)$

(8) $[2\ H_3O^+ + SO_4^{2-}] + [Ba^{2+} + 2\ OH^-] \longrightarrow BaSO_4\ (s) + 4\ H_2O\ (\ell)$

$2\ H_3O^+ + SO_4^{2-} + Ba^{2+} + 2\ OH^- \longrightarrow BaSO_4\ (s) + 4\ H_2O\ (\ell)$

(9) $2\ HBrO + 2\ H_2O\ (\ell) \longrightarrow 2\ [H_3O^+ + Br^-] + O_2\ (g)$

$2\ HBrO + 2\ H_2O\ (\ell) \longrightarrow 2\ H_3O^+ + 2\ Br^- + O_2\ (g)$

(11) $[2\ Na^+ + CO_3^{2-}] + [Cd^{2+} + 2\ NO_3^-] \longrightarrow 2[Na^+ + NO_3^-] + CdCO_3\ (s)$

$CO_3^{2-} + Cd^{2+} \longrightarrow CdCO_3\ (s)$

(18) $2\ [Na^+ + OH^-] + H_2CO_3 \longrightarrow [2\ Na^+ + CO_3^{2-}] + 2\ H_2O\ (\ell)$

$2\ OH^- + H_2CO_3 \rightarrow CO_3^{2-} + 2\ H_2O\ (\ell)$

(19) $[K^+ + OH^-] + HCN \longrightarrow [K^+ + CN^-] + H_2O\ (\ell)$

$OH^- + HCN \longrightarrow CN^- + H_2O\ (\ell)$

(22) $Cd\ (s) + PbCl_2\ (aq) \longrightarrow [Cd^{2+} + 2\ Cl^-] + Pb\ (s)$

$Cd\ (s) + PbCl_2\ (aq) \longrightarrow Cd^{2+} + 2\ Cl^- + Pb\ (s)$

(26) $2\ [K^+ + Cl^-] + F_2\ (g) \longrightarrow 2\ [K^+ + F^-] + Cl_2\ (g)$

$2\ Cl^- + F_2\ (g) \longrightarrow 2\ F^- + Cl_2\ (g)$

(29) $2\ [K^+ + MnO_4^-] + 6\ [H_3O^+ + Cl^-] + 5\ H_2O_2 \longrightarrow$
$2[Mn^{2+} + 2\ Cl^-] + 5\ O_2\ (g) + 14\ H_2O\ (\ell)$

$2\ MnO_4^- + 6\ H_3O^+ + 5\ H_2O \longrightarrow 2\ Mn^{2+} + 5\ O_2\ (g) + 14\ H_2O\ (\ell)$

(32) $Sr_3(PO_4)_2\,(s) + 3[2\,H_3O^+ + SO_4^{2-}] \longrightarrow 2\,H_3PO_4 + SrSO_4\,(s) + 6\,H_2O\,(\ell)$

$Sr_3(PO_4)_2\,(s) + 6\,H_3O^+ + 3\,SO_4^{2-} \longrightarrow 2\,H_3PO_4 + 3\,SrSO_4\,(s) + 6\,H_2O\,(\ell)$

(33) $H_2S + [Co^{2+} + 2\,NO_3^-] + 2\,H_2O\,(\ell) \longrightarrow CoS\,(s) + 2\,[H_3O^+ + NO_3^-]$

$H_2S + Co^{2+} + 2\,H_2O\,(\ell) \longrightarrow CoS\,(s) + 2\,H_3O^+$

(36) $[Ag^+ + NO_3^-] + [Li^+ + Cl^-] \longrightarrow AgCl\,(s) + [Li^+ + NO_3^-]$

$Ag^+ + Cl^- \longrightarrow AgCl\,(s)$

(39) $2\,[Rb^+ + OH^-] + H_2CO_3 \longrightarrow [2\,Rb^+ + CO_3^{2-}] + H_2O\,(\ell)$

$2\,OH^- + H_2CO_3 \longrightarrow CO_3^{2-} + 2\,H_2O\,(\ell)$

Chapter Eight

GASES AND THE KINETIC-MOLECULAR THEORY

D,G,W: Chap. 11

The variables used to describe a gas are the number of moles of gas, its pressure, temperature, and volume.

8-1 Pressure

The pressure of a gas in a container is defined as the force it exerts per unit area on the walls of the container. Pressure can be, and often is, measured in terms of the height to which it forces a column of mercury to rise in an evacuated capillary tube inverted in a pool of mercury. Since the pressure exerted by the gas is directly proportional to the height of the column, pressure is often expressed in millimeters of mercury. The reference point for pressure is set at exactly 760 mm Hg which is defined to be one atmosphere of pressure. The units mm Hg and torr are interchangeable.

1 mm Hg = 1 torr, 1 atmosphere (atm) = 760 mm Hg = 101.3 kPa

8-2 Boyle's Law

Boyle's Law states that the volume of a given mass of a particular gas, at constant temperature, is inversely proportional to its pressure. In other words, as volume increases, pressure decreases and vice-versa. This can be stated algebraically as follows, where the subscripts refer to different pressure-volume conditions.

$$P_1V_1 = P_2V_2 \qquad T = \text{constant, mass} = \text{constant}$$

Example 8-1: A sample of O_2 occupies a volume of 60.0 liters at 25°C and standard pressure. What volume does it occupy at 25°C under a pressure of 0.300 atmosphere?

The problem can be solved by either of two methods. The first involves substitution into the above equation.

$P_1 = 1.00$ atm $\qquad P_2 = 0.300$ atm $\qquad P_1V_1 = P_2V_2$

$V_1 = 60.0$ L $\qquad V_2 = ?$

$$V_2 = \frac{P_1V_1}{P_2} = \frac{(1.00\ \text{atm})(60.0\ \text{L})}{0.300\ \text{atm}} = \underline{\underline{200\ \text{L}}}$$

The alternate solution involves only the application of chemical intuition in multiplying the original volume by the ratio of pressures. Since the pressure decreases we know the volume must increase, so the larger pressure must be in the numerator.

$$\underline{?}\ \text{L} = 60.0\ \text{L} \times \frac{1.00\ \text{atm}^*}{0.300\ \text{atm}} = \underline{\underline{200\ \text{L}}}$$

Example 8-2: Calculate the pressure change accompanying the compression of 500 mL of a gas at 0°C and 420 torr to a volume of 200 mL with no change in temperature.

In order to calculate the pressure change we must first calculate the final pressure, P_2.

$$P_1 = 420\ \text{torr} \qquad P_2 = ? \qquad P_1V_1 = P_2V_2$$

$$V_1 = 500\ \text{mL} \qquad V_2 = 200\ \text{mL}$$

$$P_2 = \frac{P_1V_1}{V_2} = \frac{(420\ \text{torr})(500\ \text{mL})}{200\ \text{mL}} = 1050\ \text{torr} = 1.05 \times 10^3\ \text{torr}$$

Alternatively, we can apply chemical intuition and multiply the original pressure, P_1, by a "volume factor" with the large volume in the numerator because the pressure increases as the volume decreases.

$$\underline{?}\ \text{torr} = 420\ \text{torr} \times \frac{500\ \text{mL}}{200\ \text{mL}} = 1050\ \text{torr} = 1.05 \times 10^3\ \text{torr}$$

The pressure increases by 630 torr (1050-420), to 2.5 times its original value, as the volume decreases from 500 mL to 200 mL at constant temperature.

8-3 Charles' Law

Charles' Law summarizes the relationship between the volume and temperature of a sample of gas at constant pressure. It states that for a given weight of a given gas at constant pressure, the volume is directly proportional to the absolute temperature. Stated in another way, as the absolute temperature increases the volume increases, and vice-versa.

$$V \alpha \frac{1}{T} \quad \text{or} \quad V = k(\frac{1}{T}) \qquad (\text{T in K})$$

---

*1.00 atm/0.300 atm is not a unit factor; it is a correction factor.

Charles' Law is often stated in a more useful form:

$$\frac{V_1}{T_1} = \frac{V_2}{T_2} \qquad P = \text{constant, mass} = \text{constant}$$

where the subscripts refer to different volume-temperature conditions.

Standard temperature is defined to be $0^\circ C$ or 273 K. The term STP (or SC) refers to standard temperature and pressure, 273 K and one atmosphere pressure ?

Example 8-3: A sample of methane at standard pressure occupies a volume of 16.2 liters at $20^\circ C$. What will be its volume at standard temperature and pressure.

This problem can also be solved by two methods. We can substitute into the useful form of Charles' Law.

$V_1 = 16.2\ L \qquad V_2 = ? \qquad \frac{V_1}{T_1} = \frac{V_2}{T_2}$

$T_1 = 20^\circ + 273^\circ = 293\ K \qquad T_2 = 273\ K$

$$V_2 = \frac{V_1 T_2}{T_1} = \frac{(16.2\ L)(273\ K)}{293\ K} = \underline{\underline{15.1\ L}}$$

Alternatively, we may multiply the initial volume by a temperature "correction factor" with the lower temperature in the numerator because the volume must decrease as the temperature decreases.

$$\underline{?}\ L = 16.2\ L \times \frac{273\ K}{293\ K} = \underline{\underline{15.1\ L}}$$

Example 8-4: A sample of helium occupies a volume of 50.0 L under 2.50 atm pressure at $10.0^\circ C$. What must the temperature be changed to ($^\circ C$) in order to maintain the same pressure if the volume is reduced to 15.0 L?

$V_1 = 50.0\ L \qquad V_2 = 15.0\ L \qquad \frac{V_1}{T_1} = \frac{V_2}{T_2}$

$T_1 = 283\ K \qquad T_2 = ?$

$$T_2 = \frac{V_2 T_1}{V_1} = \frac{(15.0\ L)(283\ K)}{50.0\ L} = 84.9\ K$$

$$K = {}^{\circ}C + 273^{\circ}$$

$${}^{\circ}C = K - 273^{\circ} = 84.9^{\circ} - 273^{\circ} = \underline{-188{}^{\circ}C}$$

The volume decreases and so the temperature must also decrease. Therefore the volume correction factor must be less than one.

$$\underline{?}\ K = 283\ K \times \frac{15.0\ L}{50.0\ L} = 84.9\ K = \underline{-188{}^{\circ}C}$$

8-4 Combined Gas Laws

The essence of <u>Boyle's Law</u> and <u>Charles' Law</u> can be incorporated into a single algebraic expression called the <u>Combined Gas Laws</u>.

Boyle's Law $P_1 V_1 = P_2 V_2$

Charles' Law $\dfrac{V_1}{T_1} = \dfrac{V_2}{T_2}$

$$\frac{P_1 V_1}{T_1} = \frac{P_2 V_2}{T_2}$$

Combined Gas Laws

This relationship is useful in cases in which five of the variables are known.

Example 8-5: A sample of nitrogen occupies 1200 mL and exerts a pressure of 540 torr at 30°C. What volume would it occupy at STP?

$$V_1 = 1200\ mL \qquad V_2 = ? \qquad \frac{P_1 V_1}{T_1} = \frac{P_2 V_2}{T_2}$$

$$P_1 = 540\ torr \qquad P_2 = 760\ torr$$

$$T_1 = 303\ K \qquad T_2 = 273\ K$$

$$V_2 = \frac{P_1 V_1 T_2}{T_1 P_2} = \frac{(540\ torr)(1200\ mL)(273\ K)}{(303\ K)(760\ torr)} = \underline{768\ mL}$$

Alternatively, we may use a pressure correction factor less than one and a volume correction factor less than one.

$$\underline{?}\ mL = 1200\ mL \times \frac{540\ torr}{760\ torr} \times \frac{273\ K}{303\ K} = \underline{768\ mL}$$

volume decreases as pressure increases

volume decreases as temperature decreases

Example 8-6: A certain weight of methane occupies a volume of 10.0 liters at STP. What temperature would be required for the same weight of methane to exert a pressure of 600 torr in an 8.00 liter container?

$V_1 = 10.0\ L$ $\quad V_2 = 8.00\ L$ $\quad \frac{P_1V_1}{T_1} = \frac{P_2V_2}{T_2}$

$P_1 = 760\ torr$ $\quad P_2 = 600\ torr$

$T_1 = 273\ K$ $\quad T_2 = ?$

$$T_2 = \frac{P_2V_2T_1}{P_1V_1} = \frac{(600\ torr)(8.00\ L)(273\ K)}{(760\ torr)(10.0\ L)} = 172\ K = \underline{\underline{-101\ ^\circ C}}$$

We may also use a volume correction factor less than one and a pressure correction factor less than one.

$$\underline{?}\ K = 273\ K \times \frac{8.00\ L}{10.0\ L} \times \frac{600\ torr}{760\ torr} = 172\ K = \underline{\underline{-101^\circ C}}$$

temperature decreases as volume decreases (arrow to $\frac{8.00\ L}{10.0\ L}$)

temperature decreases as pressure decreases (arrow to $\frac{600\ torr}{760\ torr}$)

8-5 The Standard Molar Volume

The volume occupied by one mole of an ideal gas under conditions of standard temperature and pressure is 22.4 L, the standard molar volume. Since no gases are ideal this relationship is only approximate for real gases. However, in calculations involving real gases, the use of this relationship provides entirely adequate results in most cases.

Example 8-7: A 50.0 gram sample of an unknown gas occupies 28.0 L at STP. Assume that the gas behaves ideally and determine the weight of one mole.

$$\frac{?\ g}{mol} = \frac{50.0\ g}{28.0\ L} \times \frac{22.4\ L}{1\ mol} = \underline{\underline{40.0\ g/mol}}$$

Example 8-8: A 0.765 gram sample of an unknown gaseous substance occupies 72.4 mL at 3.20 atmospheres pressure and 50.0°C. Assume that the gas behaves ideally and calculate its molecular weight.

We first convert the volume to STP so that we can use the standard molar volume, 22.4 L/mol. This can be done by the two methods illustrated in previous examples. Only one will be shown here.

$V_1 = 72.4\ mL$ $\quad V_2 = ?$ $\quad \frac{P_1V_1}{T_1} = \frac{P_2V_2}{T_2}$

$P_1 = 3.20\ atm$ $\quad P_2 = 1.00\ atm$

$T_1 = 323\ K$ $\quad T_2 = 273\ K$

$$V_2 = \frac{P_1V_1T_2}{T_1P_2} = \frac{(3.20 \text{ atm})(72.4 \text{ mL})(273 \text{ K})}{(323 \text{ K})(1.00 \text{ atm})} = 196 \text{ mL}$$

$$\frac{? \text{ g}}{\text{mol}} = \frac{0.765 \text{ g}}{0.196 \text{ L}} \times \frac{22.4 \text{ L}}{1 \text{ mol}} = \underline{\underline{87.4 \text{ g/mol}}}$$

8-6 Ideal Gas Law

The Combined Gas Law is often stated algebraically as shown below:

$$\frac{P_1V_1}{T_1} = \frac{P_2V_2}{T_2} \qquad \text{(mass = constant)}$$

For a given mass of a gas, it may also be stated as:

$$\frac{PV}{T} = \text{constant}$$

Since the equation involves a given mass, i.e., a given number of moles of a gas, the equation may be rewritten as

$$\frac{PV}{T} = nR$$

where n represents number of moles of the gas, a constant, and R is another constant, called the universal gas constant. The value of R may be determined quite easily because one mole of an ideal gas occupies 22.4 liters at STP. Making these substitutions produces the result:

$$R = \frac{PV}{nT} = \frac{(1 \text{ atm})(22.4 \text{ L})}{(1 \text{ mol})(273 \text{ K})} = 0.0821 \frac{\text{L}\cdot\text{atm}}{\text{mol}\cdot\text{K}}$$

If different units are used, R has a different numerical value. The Ideal Gas Law is usually written as,

$$PV = nRT$$

an equation that is often more easily utilized than the long form of the Combined Gas Laws.

Example 8-9: How many moles of gaseous chlorine are present in a 15.0 liter container of chlorine that has a pressure of 2650 torr at 430°C?

$$PV = nRT$$

$$n = \frac{PV}{RT} = \frac{(\frac{2650}{760} \text{ atm})(15.0 \text{ L})}{(0.0821 \frac{\text{L}\cdot\text{atm}}{\text{mol}\cdot\text{K}})(703 \text{ K})} = \underline{0.906 \text{ mol } Cl_2}$$

Note than in order to use R = 0.0821 L·atm/mol·K we must express pressure in atmospheres, volume in liters, and temperature in K.

Example 8-10: A 14.6 gram sample of a gas is placed inside a 2.00 liter glass bulb. The pressure exerted by the gas is measured at 1,200 torr when the temperature is 220°C. What is the molecular weight of the gas?

Since, PV = nRT, we can solve for n, the number of moles of gas.

$$n = \frac{PV}{RT} = \frac{(\frac{1200}{760}\text{ atm})(2.00\text{ L})}{(0.0821\ \frac{\text{L}\cdot\text{atm}}{\text{mol}\cdot\text{K}})(493\text{ K})} = 0.0780\text{ mol}$$

Molecular weights are often expressed in grams per mole.

$$\frac{?\text{ g}}{\text{mol}} = \frac{14.6\text{ g}}{0.0780\text{ mol}} = \underline{\underline{187\text{ g/mol}}}$$

Example 8-11: The molecular weight of a gas is 96.0 grams per mole. How many grams must be present in a 5.00 liter container to have conditions of standard temperature and pressure within the container?

$$PV = nRT = (\frac{wt}{MW})RT$$

$$wt = \frac{(MW)PV}{RT} = \frac{(\frac{96.0\text{ g}}{\text{mol}})(1.00\text{ atm})(5.00\text{ L})}{(0.0821\ \frac{\text{L}\cdot\text{atm}}{\text{mol}\cdot\text{K}})(273\text{ K})} = \underline{\underline{21.4\text{ g}}}$$

Example 8-12: What pressure is exerted by 10.0 grams of gaseous sulfur trioxide, SO_3, enclosed in a liter container at 1000°C?

$$PV = nRT \qquad n = \frac{10.0\text{ g}}{80.1\text{ g/mol}} = 0.125\text{ mol }SO_3$$

$$P = \frac{nRT}{V} = \frac{(0.125\text{ mol})(0.0821\ \frac{\text{L}\cdot\text{atm}}{\text{mol}\cdot\text{K}})(1273\text{ K})}{1.00\text{ L}} = \underline{\underline{13.1\text{ atm}}}$$

Example 8-13: Analysis of a sample of a volatile liquid showed that it contained 40.0% carbon, 6.67% hydrogen, and 53.3% oxygen by mass. A 1.00 gram sample of the gaseous compound occupied 510 mL at 100°C and one atmosphere pressure. What is the true (molecular) formula for the compound?

Let us correct the volume of the gaseous compound to STP,

$$\underline{?}\ \text{mL} = 510\ \text{mL} \times \frac{273\ \text{K}}{373\ \text{K}} = 373\ \text{mL (STP)}$$

which allows us to calculate the molecular weight of the compound.

$$\frac{?\ \text{g}}{\text{mol}} = \frac{1.00\ \text{g}}{373\ \text{mL}_{\text{STP}}} \times \frac{22{,}400\ \text{mL}_{\text{STP}}}{1\ \text{mol}} = \underline{60.1\ \text{g/mol}}$$

Alternatively, we could have used the ideal gas law to calculate the molecular weight of this compound as we did in Example 8-12. Now that we know the molecular weight, let us calculate the simplest (empirical) formula from the elemental analysis. As usual we shall use 100 grams of the compound as the basis for our calculations.

$$\underline{?}\ \text{mol C atoms} = 40.0\ \text{g C} \times \frac{1\ \text{mol C atoms}}{12.0\ \text{g C}} = 3.33\ \text{mol C atoms}$$

$$\underline{?}\ \text{mol H atoms} = 6.67\ \text{g H} \times \frac{1\ \text{mol H atoms}}{1.0\ \text{g H}} = 6.67\ \text{mol H atoms}$$

$$\underline{?}\ \text{mol O atoms} = 53.3\ \text{g O} \times \frac{1\ \text{mol O atoms}}{16.0\ \text{g O}} = 3.33\ \text{mol O atoms}$$

From these calculations, we see that the ratio of atoms is a 1:2:1 ratio, and so the simplest formula is CH_2O which has a formula weight of 30.0 amu. Division of the true formula weight by the simplest formula weight gives:

$$\frac{\text{true formula weight}}{\text{simplest formula weight}} = \frac{60.1\ \text{amu}}{30.0\ \text{amu}} = 2$$

which tells us that the true formula is twice the simplest formula, i.e., $(CH_2O)_2$ is $\underline{\underline{C_2H_4O_2}}$. Methyl formate is one compound that has this formula.

8-7 Dalton's Law of Partial Pressures

In 1807 John Dalton's work with gases led him to postulate his Law of Partial Pressures which states that in a mixture of gases, each gas exerts the pressure it would exert if it occupied the volume alone. Stated in another way, the total pressure of a mixture of gases is the sum of the partial pressures of the component gases. If A, B, C, . . . represent different gases in a mixture:

$$P_{\text{Total}} = P_A + P_B + P_C + \ldots$$

In many instances gaseous products of chemical reactions are collected over water. Since water (as well as all other liquids and solids) has a definite vapor pressure at given temperature, the collected gases also include water vapor. Dalton's Law is applied in such cases, to relate the total pressure of the system to the partial pressures of the gases of interest and therefore to the amounts of gaseous products.

8-8 Vapor Pressure

The vapor pressure of a liquid is the pressure molecules of its vapor exert over the surface of the liquid. Solid substances also exhibit definite vapor pressures which are generally much lower than those of liquids. Vapor pressures increase with increasing temperature. Table 8-1 shows the vapor pressures of water near room temperature.

TABLE 8-1

VAPOR PRESSURE OF WATER NEAR ROOM TEMPERATURE

Temperature (°C)	Vapor Pressure of Water (torr)	Temperature (°C)	Vapor Pressure of Water (torr)
15	12.79	24	22.38
16	13.63	25	23.76
17	14.53	26	25.21
18	15.48	27	26.74
19	16.48	28	28.35
20	17.54	29	30.04
21	18.65	30	31.82
22	19.83		
23	21.07		

Example 8-14: Metallic zinc dissolves in hydrochloric acid solution to generate hydrogen. A 500 mL sample of H_2 was collected over water at 30°C at a barometric pressure of 742 torr. What was the mass of the sample of H_2?

$$Zn\ (s) \quad + \quad 2\ HCl\ (aq) \quad \longrightarrow \quad ZnCl_2\ (aq) \ + \ H_2\ (g)$$

We calculate the volume occupied by the H_2 at STP.

$V_1 = 500$ mL $\qquad V_2 = ?$

$T_1 = 303$ K $\qquad T_2 = 273$ K

$P_1 = P_{H_2}$ $\qquad P_2 = 760$ torr

$$P_T = P_{H_2O} + P_{H_2}, \qquad P_{H_2} = P_T - P_{H_2O} = (742 - 32)\text{ torr} = 710\text{ torr}$$

$$\underline{?}\text{ mL} = 500\text{ mL} \times \frac{273\text{ K}}{303\text{ K}} \times \frac{710\text{ torr}}{760\text{ torr}} = 421\text{ mL } H_2\text{ (STP)}$$

V decreases as T decreases; V decreases as P increases

$$\underline{?}\text{ g } H_2 = 421\text{ mL} \times \frac{2.016\text{ g } H_2}{22{,}400\text{ mL}} = \underline{\underline{0.0379\text{ g dry } H_2}}$$

Example 8-15: What is the total pressure exerted by a mixture of 1.00 gram of N_2, 1.00 gram of O_2, and 1.00 gram of H_2 in a 1.00 liter container at STP?

This problem is solved by application of Dalton's Law of Partial Pressures and the Ideal Gas Law.

$$P_{Total} = P_{N_2} + P_{O_2} + P_{H_2} \qquad PV = nRT$$

$$P_{N_2} = \frac{n_{N_2}RT}{V} = \frac{(\frac{1.00\text{ g}}{28.0\text{ g/mol}})(0.0821\ \frac{\text{L}\cdot\text{atm}}{\text{mol}\cdot\text{K}})(273\text{ K})}{1.00\text{ L}} = 0.800\text{ atm}$$

$$P_{O_2} = \frac{n_{O_2}RT}{V} = \frac{(\frac{1.00\text{ g}}{32.0\text{ g/mol}})(0.0821\frac{\text{L}\cdot\text{atm}}{\text{mol}\cdot\text{K}})(273\text{ K})}{1.00\text{ L}} = 0.700\text{ atm}$$

$$P_{H_2} = \frac{n_{H_2}RT}{V} = \frac{(\frac{1.00\text{ g}}{2.02\text{ g/mol}})(0.0821\ \frac{\text{L}\cdot\text{atm}}{\text{mol}\cdot\text{K}})(273\text{ K})}{1.00\text{ L}} = 11.1\text{ atm}$$

$$P_{Total} = 0.800\text{ atm} + 0.700\text{ atm} + 11.1\text{ atm} = \underline{\underline{12.6\text{ atm}}}$$

An alternate solution which does not involve the direct calculation of the partial pressures of the gases makes use of the relationship:

$$n_{Total} = n_{N_2} + n_{O_2} + n_{H_2}$$

$$n_{Total} = \frac{1.00\text{ g }N_2}{28.0\text{ g/mol}} + \frac{1.00\text{ g }O_2}{32.0\text{ g/mol}} + \frac{1.00\text{ g }H_2}{2.02\text{ g/mol}}$$

$$= 0.0357\text{ mol} + 0.0313\text{ mol} + 0.495\text{ mol} = 0.562\text{ mol gas}$$

$$P_{Total} = \frac{(n_{Total})RT}{V} = \frac{(0.562\text{ mol})(0.0821\ \frac{\text{L}\cdot\text{atm}}{\text{mol}\cdot\text{K}})(273\text{ K})}{1.00\text{ L}} = \underline{\underline{12.6\text{ atm}}}$$

8-9 Graham's Law

Graham's Law states that the rate of diffusion of a gas is inversely proportional to the square root of its density (or molecular weight). Thus the ratio of the rates of diffusion of two gases can be expressed as:

$$\frac{R_1}{R_2} = \sqrt{\frac{D_2}{D_1}} \quad \text{or} \quad \frac{R_1}{R_2} = \sqrt{\frac{M_2}{M_1}}$$

Example 8-16: Hydrogen chloride and hydrogen bromide gases are both allowed to diffuse out of a container with a small orifice (at the same temperature and pressure). Which diffuses at a faster rate and how much faster does it diffuse than the other gas?

$$\frac{R_{HCl}}{R_{HBr}} = \sqrt{\frac{M_{HBr}}{M_{HCl}}} = \sqrt{\frac{80.9 \text{ amu}}{36.5 \text{ amu}}} = \sqrt{2.22} = \underline{\underline{1.49}}$$

The HCl diffuses 1.49 times faster than HBr.

Example 8-17: The ratio of the rate of diffusion of an unknown gas to that of carbon dioxide is found experimentally to be 1.14. What is the molecular weight of the unknown gas?

$$\frac{R_X}{R_{CO_2}} = \sqrt{\frac{M_{CO_2}}{M_X}} = 1.14 = \sqrt{\frac{44.0 \text{ amu}}{M_X}}$$

Squaring both sides of the equation gives:

$$1.30 = \frac{44.0 \text{ amu}}{M_X}$$

$$M_X = \frac{44.0 \text{ amu}}{1.30} = \underline{\underline{33.8 \text{ amu}}}$$

8-10 Real Gases-Deviations from Ideality

The van der Waals equation is useful when we deal with real gases at high pressures and low temperatures, i.e., temperatures near the liquefaction point of the gas. It corrects for the attractive forces among gaseous molecules (the "a" term) as well as the volume occupied by the gaseous molecules themselves (the "b" term). The equation is shown in Example 8-18.

Example 8-18: Calculate the pressure exerted by 130 grams of carbon dioxide in a 600 mL container at 30.0°C. For CO_2, $a = 3.59\ L^2 \cdot atm/mol^2$, and $b = 0.0427$ L/mol.

Since P is the only unknown term in the van der Waals equation, we solve for P. (The units are omitted from the solution so that attention can be focused on the numbers. 130 grams of CO_2 is 2.95 moles.)

$$\left(P + \frac{n^2 a}{V^2}\right)(V - nb) = nRT$$

$$\left(P + \frac{(2.95)^2(3.59)}{(0.600)^2}\right)[0.600 - (2.95)(0.0427)] = (2.95)(0.0821)(303)$$

$$(P + 86.8)(0.474) = 73.4$$

$$0.474\ P + 41.1 = 73.4 \qquad 0.474\ P = 32.3$$

$\underline{P = 68.1 \text{ atm}}$, a very high pressure

As a matter of interest, let us calculate the pressure assuming ideal gas behavior.

$$P = \frac{nRT}{V} = \frac{(2.95)(0.0821)(303)}{0.600} = \underline{122 \text{ atm}}$$

Note that this value is nearly twice as large as it should be!! The behavior of CO_2 does not approach ideality under these conditions.

8-11 Gay-Lussac's Law

Gay-Lussac's Law states that at constant temperature and pressure, the volumes of reacting gases, and any gaseous products, can be expressed as a ratio of simple whole numbers. This is consistent with the findings of Amedeo Avogadro who postulated in 1811 that equal volumes of all gases, at the same pressure and temperature, contain the same number of molecules. This postulate has been verified and is now known as Avogadro's Law. It also has been determined that at one atmosphere pressure and 273 K, 22.4 liters of any (ideal) gas contains 6.02×10^{23} molecules, or Avogadro's number of molecules, which is also defined to be one mole of molecules.

Gay-Lussac's Law may be applied only to the gases involved in a chemical reaction, even though solids and/or liquids may also be involved, because the volumes occupied by liquids and solids are not described by the same laws that describe the volumes occupied by gases. Consider the following reaction that occurs in the presence of light:

H_2 (g)	+	Cl_2 (g)	$\xrightarrow{h\nu}$	2 HCl (g)
1 mol		1 mol		2 mol
22.4 L_{STP}		22.4 L_{STP}		2(22.4 L_{STP})
1 volume		1 volume		2 volumes

The volumes of reactants and products, at the same temperature and pressure (not necessarily STP), are always in the same proportions as the coefficients of the balanced equation, 1:1:2.

Consider the following reaction that involves two gases and two liquids.

3 NO_2 (g)	+	H_2O (ℓ)	$\longrightarrow$	2 HNO_3 (ℓ)	+	NO (g)
3 volumes						1 volume

Three "volumes" of NO_2 produce one "volume" of NO, but Gay-Lussac's Law says nothing about the relative volumes of water and nitric acid involved because these compounds are liquids.

<u>Example 8-19:</u> At STP 62.4 mL of N_2 and 34.2 mL of O_2 are mixed and an electrical spark causes the following reaction to occur. After the reaction, 11.2 mL of O_2 remain unreacted (after the mixture is cooled to the original temperature and pressure). How many mL and what weight of NO were produced?

N_2 (g)	+	O_2 (g)	$\longrightarrow$	2 NO (g)
1 volume		1 volume		2 volumes

$$\underline{?}\ mL_{STP}\ O_2 \text{ reacted} = 34.2\ mL - 11.2\ mL = 23.0\ mL_{STP}\ O_2 \text{ reacted}$$

$$\underline{?}\ mL_{STP}\ NO \text{ produced} = 23.0\ mL\ O_2 \times \frac{2 \text{ volumes NO}}{1 \text{ volume } O_2} = \underline{\underline{46.0\ mL_{STP}\ NO}}$$

$$\underline{?}\ g\ NO = 46.0\ mL_{STP} \times \frac{30.0\ g\ NO}{22{,}400\ mL_{STP}} = \underline{\underline{0.0616\ g\ NO}}$$

<u>Example 8-20:</u> At 20.0°C and 2.50 atm pressure, 5.00 L of sulfur dioxide are consumed in the following reaction. How many grams of oxygen were consumed and how many grams of sulfur trioxide were produced?

2 SO_2 (g)	+	O_2 (g)	$\longrightarrow$	2 SO_3 (g)
2 volumes		1 volume		2 volumes

<u>at 20.0°C and 2.50 atm:</u>

$$\underline{?}\ L\ O_2 = 5.00\ L\ SO_2 \times \frac{1 \text{ volume } O_2}{2 \text{ volumes } SO_2} = 2.50\ L\ O_2$$

$$\underline{?}\ L\ SO_3 = 5.00\ L\ SO_2 \times \frac{2 \text{ volumes } SO_3}{2 \text{ volumes } SO_2} = 5.00\ L\ SO_3$$

<u>at STP:</u>

$$\underline{?}\ L\ O_2 = 2.50\ L\ O_2 \times \frac{273\ K}{293\ K} \times \frac{2.50\ atm}{1.00\ atm} = 5.82\ L_{STP}\ O_2 \text{ reacted}$$

$$\underline{?}\ g\ O_2 = 5.82\ L_{STP}\ O_2 \times \frac{32.0\ g}{22.4\ L_{STP}} = \underline{\underline{8.31\ g\ O_2 \text{ reacted}}}$$

$$\underline{?}\ L\ SO_3 = 5.00\ L\ SO_3 \times \frac{273\ K}{293\ K} \times \frac{2.50\ atm}{1.00\ atm} = 11.6\ L_{STP}\ SO_3 \text{ produced}$$

$$\underline{?}\ g\ SO_3 = 11.6\ L_{STP}\ SO_3 \times \frac{80.1\ g}{22.4\ L_{STP}} = \underline{\underline{41.5\ g\ SO_3}} \text{ produced}$$

8-12 Mass-Volume Relationships in Reactions Involving Gases

The following problems illustrate the utilization of several of the concepts discussed in this chapter and in Chapter 2 on Stoichiometry.

Example 8-21: Oxygen can be prepared by heating solid potassium chlorate, $KClO_3$. An impure 58.2 gram sample of potassium chlorate was heated until it yielded no more oxygen. The total volume of dry oxygen collected at 27.0°C and 741 torr was 15.0 liters. What was the percentage of potassium chlorate in the original sample?

$2\ KClO_3$ (s)	$\longrightarrow$	$2\ KCl$ (s)	+	$3\ O_2$ (g)
2 mol 2(122.6 g) 245 g		2 mol		3 mol 3(22.4 L) at STP 67.2 L at STP

We correct the volume of O_2 to STP.

$$\underline{?}\ L\ O_2 = 15.0\ L \times \frac{741\ torr}{760\ torr} \times \frac{273\ K}{300\ K} = 13.3\ L\ O_2\ (STP)$$

$$\underline{?}\ g\ KClO_3 = 13.3\ L\ O_2 \times \frac{245\ g\ KClO_3}{67.2\ L\ O_2} = 48.5\ g\ KClO_3$$

$$\%\ KClO_3 = \frac{\text{mass } KClO_3}{\text{mass sample}} \times 100\% = \frac{48.5\ g\ KClO_3}{58.2\ g\ \text{sample}} \times 100\% = \underline{\underline{83.3\%\ KClO_3}}$$

Example 8-22: Metallic magnesium dissolves in hydrochloric acid according to the equation. If 3.68 grams of magnesium are dissolved in hydrochloric acid, what volume and what weight of hydrogen would be collected over water at 23°C and 780 torr? What volume would the dried hydrogen occupy at STP?

Mg (s)	+	2 HCl (aq)	⟶	$MgCl_2$ (aq)	+	H_2 (g)
1 mol		2 mol		1 mol		1 mol
24.3 g						2.02 g
						22.4 L_{STP}

$$? \text{ g } H_2 = 3.68 \text{ g Mg} \times \frac{2.02 \text{ g } H_2}{24.3 \text{ g Mg}} = \underline{\underline{0.306 \text{ g } H_2 \text{ produced}}}$$

Let us now use the ideal gas law to calculate the volume occupied by the dry hydrogen at 23°C and 780 torr.

$$PV = nRT \qquad V = \frac{nRT}{P}$$

$$n = \frac{0.306 \text{ g } H_2}{2.02 \text{ g/mol}} = 0.151 \text{ mol } H_2$$

$$P_{H_2} = P_{Total} - P_{H_2O} = (780 - 21) \text{ torr} = 759 \text{ torr}$$

$$V = \frac{(0.151 \text{ mol})(0.0821 \frac{\text{L}\cdot\text{atm}}{\text{mol}\cdot\text{K}})(296 \text{ K})}{(\frac{759}{760} \text{ atm})} = \underline{\underline{3.67 \text{ L } H_2}}$$

Thus, the volume occupied by the dry H_2 under the conditions of the experiment is 3.67 L. The dry gas would occupy a volume of 3.39 L at STP.

$$? \text{ L } H_2 = 0.306 \text{ g } H_2 \times \frac{22.4 \text{ L}}{2.02 \text{ g}} = \underline{\underline{3.39 \text{ L } H_2}} \quad \text{(STP)}$$

Example 8-23: Both sodium nitrate and potassium nitrate decompose thermally according to the equations:

$$2 \; NaNO_3 \text{ (s)} \xrightarrow{\Delta} 2 \; NaNO_2 \text{ (s)} + O_2 \text{ (g)}$$

$$2 \; KNO_3 \text{ (s)} \xrightarrow{\Delta} 2 \; KNO_2 \text{ (s)} + O_2 \text{ (g)}$$

A 6.84 gram sample consisting of only $NaNO_3$ and KNO_3 is heated until it is completely decomposed. The total volume of oxygen produced at STP is 0.830 liter. Calculate the masses and percentages of the two salts in the original sample.

$$2\ NaNO_3\ (s) \xrightarrow{\Delta} 2\ NaNO_2\ (s) + O_2\ (g)$$

2 mol	2 mol	1 mol
2(85.0 g)		22.4 L_{STP}
170 g		

$$2\ KNO_3\ (s) \xrightarrow{\Delta} 2\ KNO_2\ (s) + O_2\ (g)$$

2 mol	2 mol	1 mol
2(101 g)		22.4 L_{STP}
202 g		

Let x = g $NaNO_3$ decomposed, (6.84 g - x) = g KNO_3 decomposed

The total volume of O_2 produced may be related to the amounts of the two salts involved.

$$\underbrace{(x\ g\ NaNO_3)\left(\frac{22.4\ L\ O_2}{170\ g\ NaNO_3}\right)}_{L\ O_2\ \text{from}\ NaNO_3} + \underbrace{[(6.84 - x)\ g\ KNO_3]\left(\frac{22.4\ L\ O_2}{202\ g\ KNO_3}\right)}_{L\ O_2\ \text{from}\ KNO_3} = \underbrace{0.830\ L\ O_2}_{\text{Total}\ L\ O_2}$$

$$0.131\bar{8}^{*}\ x + 0.758\bar{5} - 0.110\bar{9}\ x = 0.830$$

$$0.0209\ x = 0.0715$$

$$x = \frac{0.0715}{0.0209} = \underline{\underline{3.42\ g\ NaNO_3}}$$

$$?\ g\ KNO_3 = 6.84\ g - 3.42\ g = \underline{\underline{3.42\ g\ KNO_3}}$$

[masses in original sample]

$$\%\ NaNO_3 = \frac{g\ NaNO_3}{g\ sample} \times 100\%$$

$$= \frac{3.42\ g\ NaNO_3}{6.84\ g\ sample} \times 100\% = \underline{\underline{50.0\%\ NaNO_3}}$$

$$\%\ KNO_3 = 100.0\% - 50.0\% = \underline{\underline{50.0\%\ KNO_3}}$$

*A line over a number means that an extra digit, that is not a significant figure, is being kept.

EXERCISES

1. A sample of carbon monoxide, CO, occupies a volume of 350 mL and exerts a pressure of 1020 torr at 25°C. If the volume expands to 500 mL with no temperature change what pressure will the gas exert?

2. A gas occupies 1.25 liters and exerts a pressure of 718 torr at a certain temperature. If the temperature remains constant what volume must it occupy to exert a pressure of 850 torr?

3. A sample of gas occupies 650 mL at STP. If the pressure remains constant what volume will it occupy at 65.0°C.

4. A sample of nitrogen, N_2, occupies 2500 mL at STP. At constant pressure, what temperature is necessary to decrease the volume to 1000 mL?

5. A gaseous sample occupies 1.00 liter at 120°C and 1.00 atmosphere. What volume will it occupy at STP?

6. A sample of gas occupies 22,600 mL at -40.0°C and 374 torr. What volume will it occupy at 50.0°C and 800 torr?

7. A sample of gas is enclosed in a 5.00 liter vessel at 50.0°C and 480 torr. What pressure will it exert in a 10.5 liter vessel at 35.0°C?

8. A gas occupies 30.0 liters at STP. What will be its temperature if the volume is reduced to 22.5 liters under 3.50 atmospheres pressure?

9. What volume will 12.4 g of CO_2 occupy at STP?

10. A sample of nitrous oxide, N_2O, occupies 16,500 mL at STP. What is the mass of the sample?

11. What volume will 8.41 g of SO_2 occupy at 500°C and 5.40 atmospheres pressure?

12. A 2.50 g sample of a gas occupies 1840 mL at STP. What is the molecular weight of the gas?

13. A sample of hydrogen, H_2, was collected over water at 27.0°C when the barometric pressure was 746 torr. What would be the pressure exerted by the dry hydrogen at 27.0°C?

14. A 10.0 mL sample of oxygen, O_2, was collected over water at 31.0°C when the barometric pressure was 738 torr. What volume would the dry O_2 occupy at STP? How many grams of oxygen were present?

15. What are the partial pressures and the total pressure exerted by a mixture of 14.0 g of N_2, 71.0 g of Cl_2, and 16.0 g of He in a 50.0 liter container at 0°C?

16. A 5.16 g sample of a gas exerts a pressure of 1174 mm of mercury in a 2.50 liter vessel at 40.0°C. What is its molecular weight?

17. A 12.00 liter sample of fluorine, F_2, exerts a pressure of 911 torr at $-20.0^{\circ}C$. What mass of F_2 is present?

18. What is the volume occupied by a mixture of 11.4 g of O_2, 48.2 g of CO_2 and 36.0 g of Ne at $25.0^{\circ}C$ and 2.67 atmospheres pressure?

19. At what temperature will 16.7 g of SO_2 occupy at volume of 8.00 liters and exert a pressure of 806 torr?

20. A gaseous compound is 69.6% oxygen and 30.4% nitrogen. A 4.99 g sample of the gas occupies a volume of 1.00 liter and exerts a pressure of 1.26 atmospheres at $10.0^{\circ}C$. What is its true (molecular) formula?

21. A sample of a gas contains 0.305 g of carbon, 0.407 g of oxygen, and 1.805 g of chlorine. A different sample of the same gas weighing 1.72 g occupies 2.00 liters and exerts a pressure of 992 torr at $1560^{\circ}C$. What is the true (molecular) formula of the gas?

22. How much faster will neon diffuse through an opening of a given size than sulfur dioxide, SO_2, will?

23. What is the molecular weight of a gas that diffuses 0.707 times as fast as nitrogen, N_2?

24. What is the density of gaseous nitrogen, N_2, at STP?

25. What is the density of a gas at STP that diffuses 0.741 times as fast as N_2? Refer to problem 24.

26. What volume of hydrogen reacts with excess chlorine to produce 30.0 liters of hydrogen chloride at constant temperature and pressure assuming 100% yield?

$$H_2 + Cl_2 \longrightarrow 2\,HCl$$

27. What volume of hydrogen reacts with excess nitrogen to produce 20.0 liters of ammonia, NH_3, all volumes measured at $30.0^{\circ}C$ and 1.50 atmospheres pressure? Assume 100% yield.

$$N_2 + 3\,H_2 \longrightarrow 2\,NH_3$$

28. Hypochlorous acid solution, HClO, is decomposed into hydrochloric acid, HCl, and oxygen, O_2, by sunlight.

$$2\,HClO \xrightarrow{\text{sunlight}} 2\,HCl + O_2$$

(a) What volume of dry O_2 is produced at STP by the decomposition of 10.0 g of HClO?

(b) How many g of HCl are produced when 11,200 mL of O_2 at STP are generated by this reaction?

29. What volume of SO_2 (STP) can be produced by complete thermal decomposition of 0.500 mole of barium sulfite, $BaSO_3$?

$$BaSO_3 \xrightarrow{\Delta} BaO + SO_2$$

30. What volume of dry gaseous hydrogen bromide, HBr, can be produced at STP from the reaction of 0.300 mole of phosphorus tribromide, PBr_3, with excess water?

$$PBr_3\ (\ell) \quad + \quad 3\ H_2O\ (\ell) \longrightarrow H_3PO_3\ (\ell) \quad + \quad 3\ HBr\ (g)$$

31. Mangesium reacts with ammonia, NH_3, at high temperature to produce solid magnesium nitride, Mg_3N_2, and hydrogen.

$$3\ Mg\ (s) \quad + \quad 2\ NH_3\ (g) \xrightarrow{\Delta} Mg_3N_2\ (s) \quad + \quad 3\ H_2\ (g)$$

(a) How many g of magnesium must react with 16,400 mL (STP) of ammonia, assuming 85.6% of the ammonia reacts as indicated?
(b) Consider (a). What volume of hydrogen would be produced?
(c) Consider (a). What weight of magnesium nitride would be prepared?

32. How many g of methanol, CH_3OH, could be produced from the reaction of 16.6 liters of carbon monoxide, CO, with 38.5 liters of hydrogen at STP? Assume 77.6% percent yield with respect to the limiting reagent.

$$CO\ (g) \quad + \quad 2\ H_2\ (g) \longrightarrow CH_3OH\ (\ell)$$

33. Ammonia reacts with hot copper(II) oxide to form metallic copper, nitrogen and steam.

$$2\ NH_3\ (g) + 3\ CuO\ (s) \longrightarrow N_2\ (g) + 3\ Cu\ (s) + 3\ H_2O\ (g)$$

Assuming 93.0% yield of metallic copper with respect to copper(II) oxide, what volume of gaseous ammonia reacts with 58.6 g of copper(II) oxide at 1000°C and 5.00 atmospheres pressure?

34. A 4.86 g sample containing only nickel(II) carbonate, $NiCO_3$, and calcium carbonate, $CaCO_3$, was heated to complete decomposition. The liberated carbon dixoide occupied 0.986 liters at STP. (a) What were the masses and percentages of the two carbonates in the original sample? (b) What is the mass of the solid residue?

$$NiCO_3\ (s) \xrightarrow{\Delta} NiO\ (s) \quad + \quad CO_2\ (g)$$
$$CaCO_3\ (s) \xrightarrow{\Delta} CaO\ (s) \quad + \quad CO_2\ (g)$$

ANSWERS TO EXERCISES

1. 714 torr 2. 1.06 L 3. 805 mL 4. -164°C 5. 0.695 L 6. 14.6 L 7. 218 torr 8. 444°C 9. 6.31 L 10. 32.4 g N_2O 11. 1.54 L 12. 30.5 g/mol 13. 719 torr 14. 8.32 mL, 0.0119 g 15. P_{N_2} = 0.22 atm, P_{Cl_2} = 0.45 atm, P_{He} = 1.79 atm, P_{tot} = 2.46 atm 16. 34.4 g/mol 17. 26.3 g 18. 29.7 L 19. 124°C 20. N_2O_4 21. $COCl_2$ 22. 1.78 23. 56.0 g/mol 24. 1.25 g/L 25. 2.28 g/L 26. 15.0 L 27. 30.0 L 28. (a) 2.13 L (b) 36.5 g 29. 11.2 L 30. 20.2 L 31. (a) 22.8 g, (b) 21.1 L (c) 31.6 g 32. CO is limiting reagent, 18.4 g CH_3OH 33. direct approach gives 9.54 L 34. (a) $2.9\bar{7}$ g $NiCO_3$, $1.8\bar{9}$ $CaCO_3$, 61% $NiCO_3$, 39% $CaCO_3$, (b) 2.92 g residue

Chapter Nine

LIQUIDS AND SOLIDS

D, G, W: Chap. 12

HEAT TRANSFER

9-1 Introduction

Heat is absorbed by an object or substance when its temperature (a measure of heat intensity) increases or when it is converted from solid to liquid or liquid to gas. Likewise, the reverse processes are accompanied by the release of heat. In each of these cases heat is transferred between the substance of interest and its surroundings.

Heat can be measured in calories. One calorie is operationally defined as the amount of heat necessary to raise the temperature of one gram of water one Celsius degree. One calorie is 4.184 joules (exactly).

9-2 Specific Heat (Heat Capacity)

The specific heat of a substance is the amount of heat required to raise the temperature of one gram of the substance one Celsius degree; its units are J/g°C or cal/g°C. Thus the specific heat of liquid water is 4.184 J/g°C or 1.000 cal/g°C (Section 1-13 in the text). As the table on page A. 13 in the text shows, specific heats of the same substance in different phases are different. When specific heat is expressed in J/mol °C or cal/mol°C it is called the heat capacity. Heat capacities may also be expressed in kJ/mol °C or kcal/mol °C.

Example 9-1: Calculate the amount of heat required to raise the temperature of 50.0 g of water from 35.0° to 75.0° C.

$$\underline{?}\ J = 50.0\ g \times \frac{4.18\ J}{g\ ^{\circ}C} \times (75.0^{\circ}C - 35.0^{\circ}C)$$

$$= (50.0 \times 4.18 \times 40.0)\ J$$

$$= \underline{\underline{8.36 \times 10^3\ J}}$$

Example 9-2: Calculate the amount of heat required to raise the temperature of 50.0 g of mercury from 35.0°C to 75.0°C. The specific heat of mercury is 0.138 J/g°C.

$$\underline{?}\ \text{J} = 50.0\ \text{g} \times \frac{0.138\ \text{J}}{\text{g}\,^{\circ}\text{C}} \times (75.0^{\circ}\text{C} - 35.0^{\circ}\text{C})$$

$$= (50.0 \times 0.138 \times 40.0)\ \text{J}$$

$$= \underline{\underline{276\ \text{J}}}$$

Since the specific heat of mercury is much less than that of water, much less heat is required to raise the temperature of 50.0 g of mercury by 40.0°C than to cause the same change in temperature for 50.0 g of water.

<u>Example 9-3:</u> If 5000 joules of heat are removed from 500 g of liquid ethanol at 62.0°C what will be its final temperature? The specific heat of ethanol is 2.46 J/g °C.

Let the final temperature = t, a lower temperature than 62.0°C.

$$5000\ \text{J} = 500\ \text{g} \times \frac{2.46\ \text{J}}{\text{g}\,^{\circ}\text{C}} \times (62.0 - t)^{\circ}\text{C}$$

$$5000 = 500 \times 2.46 \times (62.0 - t)$$

$$5000 = 762\overline{60} - 123\overline{0}\,t$$

$$123\overline{4}\,t = 712\overline{60}$$

$$t = \underline{\underline{57.7\,^{\circ}\text{C}}}$$

<u>Example 9-4:</u> If 5000 joules of heat are removed from 500 g of liquid water at 62.0°C what will be its final temperature?

$$5000\ \text{J} = 500\ \text{g} \times \frac{4.18\ \text{J}}{\text{g}\,^{\circ}\text{C}} \times (62.0 - t)^{\circ}\text{C}$$

$$5000\ \text{J} = 500 \times 4.18 \times (62.0 - t)$$

$$5000\ \text{J} = 129\overline{580} - 209\overline{0}\,t$$

$$t = \underline{\underline{59.6\,^{\circ}\text{C}}}$$

<u>Example 9-5:</u> Solve Example 9-4 using calories rather than joules. Is the result the same? (5000 joules = 1195 calories)

$$1195 \text{ cal} = 500 \text{ g} \times \frac{1.000 \text{ cal}}{\text{g}^{\circ}\text{C}} \times (62.0 - t)^{\circ}\text{C}$$

$$1195 = 500 \times 1.000 \times (62.0 - t)$$

$$1195 = 310\overline{00} - 500\,t$$

$$t = \underline{\underline{59.6\ ^{\circ}\text{C}}} \qquad \text{(Of course, this is the same answer.)}$$

<u>Example 9-6</u>: If 5000 joules of heat are added to 500 g of steam at 105°C what will be its final temperature?

Let t = the final temperature, higher than 105°C.

$$5000 \text{ J} = 500 \text{ g} \times \frac{2.03 \text{ J}}{\text{g}^{\circ}\text{C}} \times (t - 105)^{\circ}\text{C}$$

$$5000 = 1.02 \times 10^3\,t - 1.07 \times 10^5$$

$$1.12 \times 10^5 = 1.02 \times 10^3\,t$$

$$t = \underline{\underline{110^{\circ}\text{C}}}$$

When substances at different temperatures are mixed, heat is transferred from the warmer substance to the cooler substance and the final temperature of the mixture is intermediate between the original temperatures of the substances.

<u>Heat</u> <u>gained</u> by cool substance = <u>Heat</u> <u>lost</u> by warm substance

<u>Example 9-7</u>: What will be the final temperature of 100 grams of water at 65.0°C if a 50.0 gram ball of iron at 212°C is placed into the water in an insulated container? The specific heat of iron is 0.444 J/g °C.

Let t = final temperature, higher than 65.0° C and lower than 212 °C.

Heat gained by water (joules) = Heat lost by iron (joules)

$$100 \text{ g} \times \frac{4.18 \text{ J}}{\text{g}\ ^{\circ}\text{C}} \times (t - 65.0)^{\circ}\text{C} = 50.0 \text{ g} \times \frac{0.444 \text{ J}}{\text{g}^{\circ}\text{C}} \times (212 - t)^{\circ}\text{C}$$

$$418t - 271\overline{70} = 470\bar{6} - 22.2\,t$$

$$440.\bar{2}t = 31876$$

$$t = \underline{\underline{72.4^{\circ}\text{C}}}$$

Because the specific heat of iron is so much less than that of liquid water its temperature falls from 212°C to 72.4°C while the temperature of water increases only from 65.0°C to 72.4°C. (The mass of water is twice the mass of iron.)

HEATS OF TRANSFORMATION - PHASE CHANGES

When substances undergo phase changes (changes in state) with no concurrent temperature changes the heat transferred is called a heat of transformation. Various kinds of transformations are described below.

9-3 Transformations Between the Solid and Liquid States

The amount of heat that must be absorbed by one gram of a solid at its melting point to convert it to liquid with no change in temperature is the heat of fusion; its units are usually J/g or cal/g. The molar heat of fusion, ΔH_{fus}, is usually expressed in units of kJ/mol or kcal/mol. The heat of solidification (J/g or cal/g) and molar heat of solidification, ΔH_{sol} (kJ/mol or kcal/mol), apply to the reverse processes (liquid ⟶ solid, or freezing). They are equal in magnitude but opposite in sign to the corresponding heats of fusion.

One gram of ice absorbs 334 J or 79.8 cal as it melts at 0°C. One gram of liquid water releases 334 J or 79.8 cal as it freezes at 0°C. This amount of heat is a measure of the strength of interactions holding particles together in the solid state relative to those binding the particles in the liquid state.

$$H_2O\,(s) \underset{\substack{-334\ J/g \\ (-79.8\ cal/g)}}{\overset{\substack{+334\ J/g \\ (+78.8\ cal/g)}}{\rightleftharpoons}} H_2O\,(\ell)$$

This corresponds to 6.02 kJ/mol or 1.44 kcal/mol for ΔH_{sol}.

$$\frac{?\ kcal}{mol} = \underbrace{\frac{334\ J}{g}}_{\text{ht. of fusion}} \times \frac{18.0\ g}{mol} \times \frac{1\ kJ}{1000\ J} = \underbrace{6.02\ kJ/mol}_{\Delta H_{fus}\text{ for } H_2O}$$

Heats of transformation and transformation temperatures of several substances are given in a table in Appendix A in the text.

9-4 Transformations Between Liquids and Gases

The heats of vaporization and condensation, and molar heats of vaporization and condensation are defined similarly for transformations between one gram or one mole of liquid and gas at the boiling point of the liquid. The heat of vaporization and molar heat of vaporization of liquid water at 100°C are 2260 J/g (540 cal/g) and 40.7 kJ/mol (9.72 kcal/mol), respectively.

$$H_2O\ (\ell) \underset{\substack{-2260\ \text{J/g} \\ (-540\ \text{cal/g})}}{\overset{\substack{(+540\ \text{cal/g}) \\ 2260\ \text{J/g}}}{\rightleftharpoons}} H_2O\ (g)$$

Condensation of a gas is just the reverse of vaporization of a liquid. Heats of condensation are equal in magnitude, but opposite in sign, to heats of vaporization.

Heats of vaporization are related to the magnitude of the forces of attraction among liquid particles relative to the much smaller attractive forces among gaseous particles. In other words, they are a measure of the forces that must be overcome to vaporize a liquid.

9-5 Transformations Between Solids and Gases

Under the appropriate conditions of pressure and temperature, some solids can be converted directly to gases, and vice-versa, without passing through the liquid state. This process is called sublimation and the reverse process is called deposition.

$$\text{Solid} \underset{\text{deposition}}{\overset{\text{sublimation}}{\rightleftharpoons}} \text{Gas}$$

A familiar example is carbon dioxide, CO_2, which can be converted directly from its solid form (called "Dry Ice") to vapor at atmospheric pressure. The heats of sublimation and deposition (J/g or cal/g) and molar heats of sublimation and deposition, ΔH_{subl} and ΔH_{dep}, (kJ/mol or kcal/mol) are associated with these changes.

Example 9-8: Calculate the amount of heat in joules released by 100 g of steam at 100°C when it is condensed to 100 grams of liquid water at 100°C.

There is no change in temperature so the heat release is related only to the heat of condensation for steam.

$$?\ \text{J} = 100\ \text{g} \times \frac{2260\ \text{J}}{\text{g}} = 2.26 \times 10^5\ \text{J}\ \text{(heat released)}$$

Example 9-9: Calculate the amount of heat absorbed when 10.0 moles of liquid water at 100°C are converted to 10.0 moles of steam at 100°C.

It is appropriate to use the molar heat of vaporization of water in this case, 40.7 kJ/mol.

$$\underline{?}\ \text{cal} = 10.0\ \text{mol} \times \frac{40.7\ \text{kJ}}{\text{mol}} = \underline{\underline{407\ \text{kJ}}} \quad \text{(heat absorbed)}$$

Example 9-10: Calculate the amount of heat in calories absorbed by 10.0 g of ice at -15.0°C in converting it to liquid water at 50.0°C.

There are three steps in the process: (1) heating the ice from -15.0°C to 0°C which involves the specific heat of ice, (2) melting the ice at 0°C which involves the heat of fusion of ice, and (3) raising the temperature of the resulting 10.0 g of liquid water to 50.0°C which involves the specific heat of water.

$$(1)\ 10.0\ \text{g} \times \frac{2.09\ \text{J}}{\text{g}\,^\circ\text{C}} \times [0-(-15.0)]\,^\circ\text{C} = 314\ \text{J}$$

$$(2)\ 10.0\ \text{g} \times \frac{334\ \text{J}}{\text{g}} = 334\bar{0}\ \text{J}$$

$$(3)\ 10.0\ \text{g} \times \frac{4.18\ \text{J}}{\text{g}\,^\circ\text{C}} \times (50.0-0)\,^\circ\text{C} = 209\bar{0}\ \text{J}$$

$$\text{Total heat absorbed} = \underline{\underline{5.74 \times 10^3\ \text{J}}}$$

Note that the reverse process, converting 10.0 grams of water at 50.0°C to 10.0 grams of ice at -15.0°C would involve the release of 5.74 kJ of heat.

Example 9-11: Calculate the amount of heat in joules released when 20.0 g of steam at 110.0°C are converted to 20.0 g of ice at -20.0°C.

There are five steps involved in this process: (1) cooling 20.0 g of steam from 110°C to 100.0°C which involves the specific heat of steam, (2) condensing 20.0 g of steam at 100.0 °C to 20.0 g of liquid water at 100.0°C which involves the heat of condensation of steam, (3) cooling 20.0 g of water from 100.0°C to 0.0°C which involves the specific heat of liquid water, (4) freezing 20.0 g of liquid water at 0.0°C which involves the heat of solidification of liquid water, and (5) cooling 20.0 g of ice from 0.0°C to -20.0°C which involves the specific heat of ice.

$$(1)\ 20.0\ g \times \frac{2.03\ J}{g\ ^oC} \times (110.0 - 100.0)^oC = 406\ J$$

$$(2)\ 20.0\ g \times \frac{2260\ J}{g} = 4.52 \times 10^4\ J$$

$$(3)\ 20.0\ g \times \frac{4.18\ J}{g\ ^oC} \times (100.0 - 0.0)^oC = 8.36 \times 10^3\ J$$

$$(4)\ 20.0\ g \times \frac{333\ J}{g} = 6.66 \times 10^3\ J$$

$$(5)\ 20.0\ g \times \frac{2.09\ J}{g\ ^oC} \times [0.0-(-20.0)]\ ^oC = 836\ J$$

Total heat released $= 6.15 \times 10^4$ J

Example 9-12: If 100 g of ice at -20^oC are dropped into 5.0 g of water at 5.0^oC in an insulated container will the water freeze by the time equilibrium is established?

We must calculate the amount of heat absorbed in the warming of 100 g of ice from -20.0^oC to 0.0^oC. This represents the maximum amount of heat that can be absorbed by the ice without melting any of it. This amount of heat must be compared with the amount that would be released in freezing the water. We first calculate the amount of heat absorbed in warming the ice to 0.0^oC.

$$\underline{?}\ cal = 100\ g \times \frac{2.09\ J}{g\ ^oC} \times [0.0-(-20.0^o)]\ ^oC = \underline{\underline{4.18 \times 10^3\ J\ \ (absorbed)}}$$

The amount of heat released in freezing the water is 1.8×10^3 joules as shown below.

$$5.0\ g \times \frac{4.18\ J}{g\ ^oC} \times (5.0 - 0.0)^oC = 1.0 \times 10^2\ J$$

$$5.0\ g \times \frac{334\ J}{g} = 1.7 \times 10^3\ J$$

Total heat released $= 1.8 \times 10^3$ J

Since the ice can absorb all the heat the water must release to freeze and still not melt, the answer is yes.

Example 9-13: What will be the final temperature of the water resulting from the mixing of 10.0 g of steam at 130^oC with 40.0 g of ice at -10.0^oC?

Heat gained by ice = Heat lost by steam. Let t = the final temperature of the water.

$$40.0\ g \times \frac{2.09\ J}{g} \times (0.0-[-10.0])^{o}C + 40.0\ g \times \frac{334\ J}{g} + 40.0\ g \times \frac{4.18\ J}{g\ ^{o}C} \times (t-0.0)^{o}C =$$

$$10.0\ g \times \frac{2.03\ J}{^{o}C} \times (130.0 - 100.0)^{o}C + 10.0\ g \times \frac{2260\ J}{g} + 10.0\ g \times \frac{4.18\ J}{g\ ^{o}C} \times (100.0 - t)^{o}C$$

$$(836 + 133\overline{6}0 + 167\ t)J = (609 + 226\overline{0}0 + 418\overline{0} - 41.8t)J$$

$$209t = 131\overline{93}$$

$$t = 63.1\ ^{o}C$$

THE STRUCTURES OF CRYSTALS

The structures of many solids have been determined very accurately by means of x-ray diffraction (Section 9-14 in the text). Each crystal structure is described by a unit cell, the smallest repeating unit in the crystal. Each unit cell is characterized by its edge lengths, a, b, and c, and the angles between edges α, β, and γ (Fig. 9-19 in the text). Differences and similarities in these six parameters allow each unit cell to be placed into one of seven fundamental crystal systems (Table 9-8 in the text).

In addition, crystals also can be classified in another way based on the kinds of particles (atoms, molecules, ions) that occupy the lattice points that describe the unit cell, and the interactions among these particles. There are four such categories: (1) molecular solids, (2) covalent solids, (3) ionic solids, and (4) metallic solids in which molecules, individual (covalently bonded) atoms, negative and positive ions, and metal cations, respectively, occupy the lattice points. The general physical properties of each kind (Table 9-9 in the text) are consistent with the structures and interparticle interactions.

Within the cubic system there are three subclasses: simple cubic, body-centered cubic, and face-centered cubic. These are shown in Figure 9-1. We shall restrict our attention to crystals with these structures and the hexagonal close-packed structure, which is described Section 9-16.1 in the text.

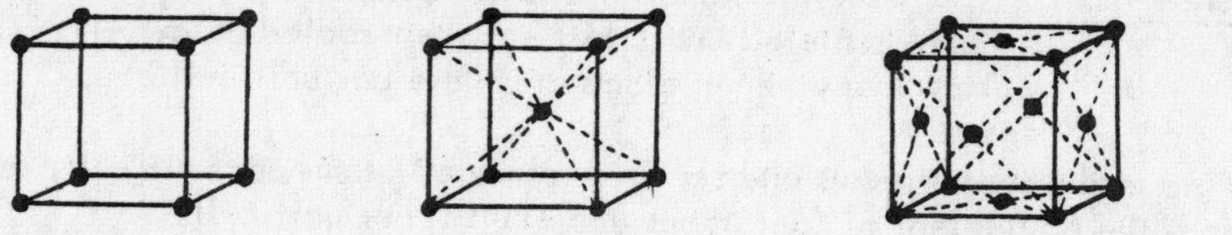

Figure 9-1. The Cubic Subclasses

The hexagonal close-packed and face-centered cubic (also called cubic close-packed) structures represent maximum packing efficiency when non-ionic solids are considered. In ionic solids the unit cells are defined by positions of anions or cations. In "close-packed" ionic solids, such ions must be at least slightly separated to make room for ions of opposite charge. Ideally, in non-ionic close-packed solids 76% of the total volume is due to particles, only 24% is empty space, and each particle is in contact twelve nearest neighbors. Thus, the particles are said to have a coordination number of twelve. These close-packed structures are common for many metals.

The representations of the cubic structures shown in Figure 9-1 depict the particles that define the unit cells as very small spheres relative to the size of the unit cell. In reality, the particles are greatly expanded, to the point that they are in contact with nearest neighbors, or very nearly so. It should be noted also that many nearest neighbors of a particular particle on a corner, edge, or face of a given unit cell are located in adjacent unit cells.

The lines connecting the particles in Figure 9-1 serve as only hypothetical boundaries of a unit cell, and except for particles in the centers of body-centered lattices, only fractional parts of given particles are contained within one unit cell. This situation is summarized below.

Location of Particle	Contribution to One (Cubic) Unit Cell
Corner	1/8
Edge	1/4
Face-center	1/2
Body-center	1

Example 9-14: Oxygen crystallizes as a molecular solid (below -219°C) with O_2 molecules occupying the corners of a simple cubic lattice. How many O_2 molecules are there per unit cell?

Even though parts of eight molecules describe the unit cell, all molecules are located at the corners of the cell, so there is only the equivalent of one molecule in one unit cell.

$$8\ O_2 \text{ molecules} \times 1/8 = 1\ O_2 \text{ molecule in a unit cell}$$

Example 9-15: The noble gas xenon (monatomic molecules) crystallizes in a face-centered cubic lattice when cooled below -112°C. How many xenon atoms are there per unit cell?

There are eight corner atoms and six face-centered atoms per unit cell, and therefore the equivalent of four atoms (molecules) per unit cell.

$$8\ (1/8)\ +\ 6\ (1/2)\ =\ 4 \text{ atoms/unit cell}$$

Example 9-16: Chromium crystallizes in a metallic body-centered cubic lattice. How many chromium atoms are there in one unit cell?

In the body-centered cube there are eight corner atoms and one body-centered atom, and therefore the equivalent of two atoms per unit cell.

$$8\ (1/8)\ +\ 1\ (1)\ =\ 2 \text{ atoms/unit cell}$$

Example 9-17: Iron crystallizes in a cubic lattice (α-Fe) with a unit cell edge length of a = 2.8665 Å. The density of iron is 7.86 g/cm³. Determine whether iron crystallizes in a simple cubic, body-centered cubic, or face-centered cubic lattice.

Since simple, body-centered, and face-centered cubic crystals have 1, 2, and 4 particles per unit cell, we must determine whether there are 1, 2, or 4 Fe atoms per unit cell. First, we find the mass of one unit cell.

$$\frac{?\text{ g Fe}}{\text{unit cell}} = \frac{7.86\text{ g Fe}}{1.00\text{ cm}^3} \times \frac{(2.8665\text{ Å})^3 \times \left(\frac{1.00 \times 10^{-8}\text{ cm}}{1.00\text{ Å}}\right)^3}{1\text{ unit cell}} = 1.85 \times 10^{-22}\text{ g Fe/unit cell}$$

↑ from density ↑ vol. of cubic unit cell = a^3

Now we convert this mass of Fe to the number of Fe atoms per unit cell.

$$\frac{?\text{ Fe atoms}}{\text{unit cell}} = \frac{1.85 \times 10^{-22}\text{ g Fe}}{\text{unit cell}} \times \frac{6.02 \times 10^{23}\text{ Fe atoms}}{55.85\text{ g Fe}} = 1.99\text{ Fe atoms/unit cell}$$

There must be a whole number of atoms per unit cell (two) so there must be two atoms per unit cell and the lattice must be a body-centered cube.

In this problem the three variables were unit cell edge length, density, and type of cubic lattice. We were given the first two and we determined the third. If we are given any two of these three pieces of information, for any cubic crystal, we can determine the third by manipulating the relationships we have just used.

Example 9-18: Beryllium crystallizes in an hexagonal close-packed lattice and has a specific gravity of 1.85. Assume that this is an ideal close-packed lattice, i.e., that 24% of the total volume is empty space. Determine the volume and radius of a beryllium atom in this lattice.

If 24% of the total volume of a unit cell is empty space then 76% must be due to Be atoms. We can first determine the total volume occupied by one mole of Be.

$$\frac{?\ \text{cm}^3}{\text{mol Be atoms}} = \frac{9.012\ \text{g}}{1\ \text{mol Be atoms}} \times \frac{1.00\ \text{cm}^3}{1.85\ \text{g}} = 4.87\ \text{cm}^3/\text{mol Be atoms}$$

(V_{molar}; from atomic wt.; from sp. gr.)

Now we calculate the part of this volume due to the atoms themselves (no empty space).

$$\frac{?\ \text{cm}^3}{\text{mol Be atoms}} = \frac{4.87\ \text{cm}^3}{\text{mol Be atoms}} \times 0.76 = 3.7\ \text{cm}^3/\text{mol Be atoms}$$

From this we calculate the volume of a single Be atom.

$$\frac{?\ \text{cm}^3}{\text{Be atom}} = \frac{3.7\ \text{cm}^3}{\text{mol Be atoms}} \times \frac{1\ \text{mol Be atoms}}{6.02 \times 10^{23}\ \text{Be atoms}} = \underline{\underline{6.1 \times 10^{-24}\ \text{cm}^3/\text{Be atom}}}$$

(V_{atomic})

The volume of a sphere is given by $V = (4/3)\pi r^3$. Thus, the calculated atomic radius of a Be atom is 1.1 Å.

$$r_{atomic} = \sqrt[3]{\frac{V_{atomic}}{(4/3)\pi}} = \sqrt[3]{\frac{6.1 \times 10^{-24}\ \text{cm}^3}{(4/3)\pi}} = 1.1 \times 10^{-8}\ \text{cm} = \underline{\underline{1.1\ \text{Å}}}$$

The atomic radius of Be is listed as 1.11 Å in Figure 4-2.

Example 9-19: Nickel (the γ-form) crystallizes in a face-centered cubic lattice with a = 3.525 Å. Calculate the radius of a Ni atom.

We can visualize one face of the face-centered cube as shown at the right, assuming contact between the corner atoms and the face-centered atom. The length of the face diagonal, h, is related to the edge length, a, by the Pythagorean theorem.

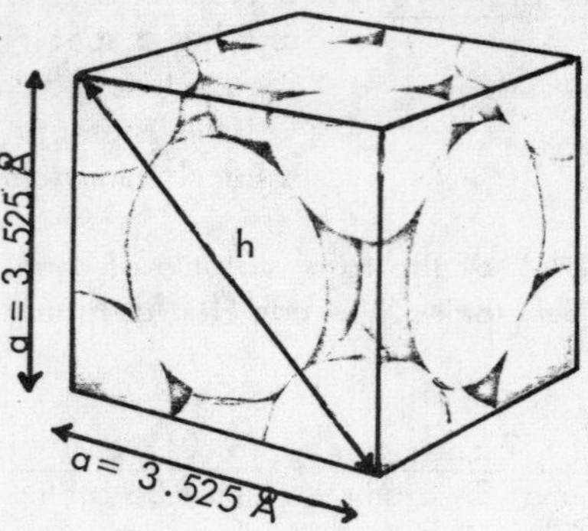

$$h^2 = a^2 + a^2$$

$$h = \sqrt{a^2 + a^2} = \sqrt{2\,a^2} = \sqrt{2}\,a = \sqrt{2}(3.525\ \text{Å}) = 4.985\ \text{Å}$$

We can also see from the diagram that h = four times the radius of a Ni atom and therefore $r_{Ni} = 1.246$ Å.

$$r_{Ni} = \frac{4.985\ \text{Å}}{4} = \underline{\underline{1.246\ \text{Å}}} \qquad \text{(tabulated value = 1.24 Å)}$$

Example 9-20: Rubidium crystallizes in a body-centered cubic lattice with a = 5.63 Å. Calculate the radius of a rubidium atom.

The body-centered cubic lattice is shown. We have already seen (Example 9-19) that the face diagonal, h, of a cube is $\sqrt{2}a$. Thus, by the Pythagorean theorem, the body diagonal, j, of a cube is related to a, the edge length, by the following equation.

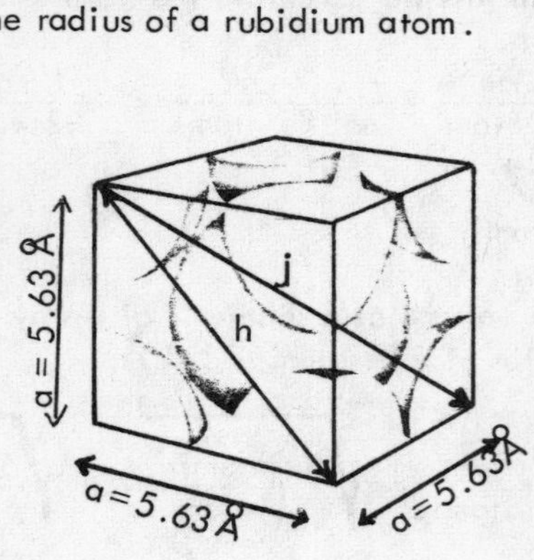

$$j^2 = a^2 + h^2$$

$$j = \sqrt{a^2 + h^2} = \sqrt{a^2 + 2\,a^2} = \sqrt{3\,a^2} = \sqrt{3}\,a$$

In this case a = 5.63 Å and j = 9.75 Å. Assuming atom-atom contact along the body diagonal j gives: $j = 4r_{Rb}$ and $r_{Rb} = 2.44$ Å.

$$j = \sqrt{3}\,a = \sqrt{3}\,(5.63\ \text{Å}) = 9.75\ \text{Å}$$

$$r_{Rb} = \frac{9.75\ \text{Å}}{4} = \underline{\underline{2.44\ \text{Å}}} \qquad \text{(tabulated value = 2.44 Å)}$$

Ionic lattices are not quite as simple as metallic lattices in that a unit cell is defined by one kind of ion and the other kinds occupy spaces among them. This is illustrated in the next example.

Example 9-21: Barium titanate, $BaTiO_3$ crystallizes in the perovskite structure in which the titanium(IV) atoms lie at the corners of a simple cube, the oxygen atoms are in the middle of each edge and the barium ion is the "body-centered" ion. Calculate the number of formula units of $BaTiO_3$ in one unit cell.

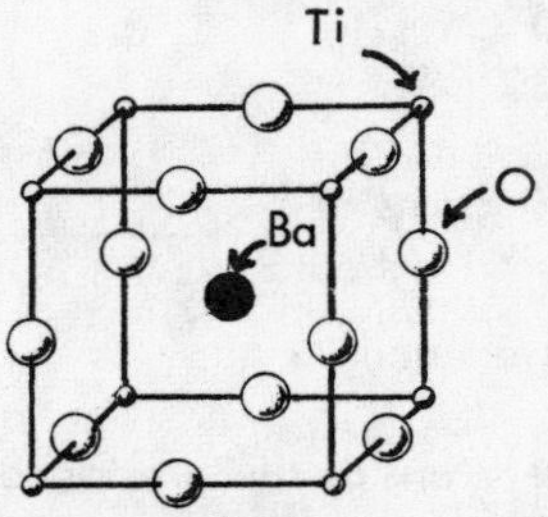

To give a better perspective, the atoms (ions) are shown as smaller spheres relative to the unit cell than in reality. $BaTiO_3$ consists of Ba^{2+} ions and TiO_3^{2-} ions.

We calculate the number of atoms (ions) of each kind in one unit cell just as we did in Examples 9-14 to 9-16.

Ba^{2+}	1 ion x 1	=	1 ion	
Ti	8 atoms x 1/8	=	1 atom	1 formula unit of $BaTiO_3$ per unit cell
O	12 atoms x 1/4	=	3 atoms	

Example 9-22: Cesium chloride, CsCl, crystallizes in a lattice in which there are chloride ions at each corner of a cube and a cesium ion in a body-center position. The edge length, a, is 4.121 Å and the radius of a chloride ion is 1.81 Å. Assuming anion-cation contact along the body diagonal, determine the ionic radius of Cs^+.

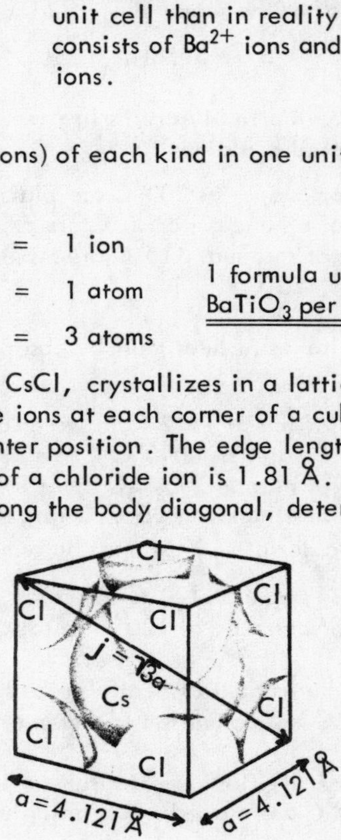

Since this is a cubic lattice we know from Example 9-20 that the length of the body diagonal, j, is $\sqrt{3}\,a = 7.138$ Å, and this (see figure) is equal to $2\,r_{Cs^+} + 2\,r_{Cl^-}$.

$$j = \sqrt{3}\,a = \sqrt{3}\,(4.121\text{ Å}) = 7.138\text{ Å} = 2\,r_{Cs^+} + 2\,r_{Cl^-}$$

$$r_{Cs^+} = \frac{7.138\text{ Å} - 2(1.81\text{ Å})}{2} = \underline{\underline{1.76\text{ Å}}} \quad \text{(tabulated value is 1.69 Å)}$$

EXERCISES

Tables 9-3 and 9-6 and Appendix E in the text may be used in solving the following problems.

1. How much heat must be absorbed to raise the temperature of:

   (a) 35.0 g of liquid benzene from 46.0°C to 72.0°C?
   (b) 35.0 g of water from 46.0°C to 72.0°C?
   (c) 35.0 g of copper from 46.0°C to 72.0°C?

2. What will be the new temperature of 200 grams of __________ initially at 70.0°C after the removal of 3138 joules of heat?

   (a) water, (b) ethanol, (c) aluminum

3. What will be the final equilibrium temperature of the substances present in an insulated container if:

   (a) 500 g of copper at 200°C are placed in 100 g of water at 30.0°C?
   (b) 200 g of ethanol at 65.0°C are mixed with 200 g of water at 10.0°C?
   (c) 50.0 g of steam at 115°C are mixed with 100 grams of diethyl ether (gas) at 150°C?

4. How many joules of heat must be absorbed, or released, to change the temperature of:

   (a) 25.0 g of ice at -112°C to water at 60.0°C?
   (b) 100 g of water at 40.0°C to ice at -7.0°C?
   (c) 65.0 g of ethanol, C_2H_5OH, at -15.0°C to 60.0°C?
   (d) 35.0 g of ethanol at 25.0°C to gaseous ethanol at 95.0°C?

5. How many calories of heat must be removed in cooling:

   (a) 150 g of steam at 120°C to ice at -5.0°C?
   (b) 150 g of benzene at 120°C to solid benzene at 5.48°C?

6. Calculate the amount of heat absorbed in kilojoules when 7.50 moles of ice at 0.0°C are converted to steam at 110°C.

7. What will be the final temperature of the substances present if 100 g of iron at 400°C are placed in an insulated container with 50 g of ice at -6.0°C?

8. Will 10.0 g of water at 12.0°C freeze if mixed with one kilogram of ice at -30.0°C in an insulated container?

9. Will 55.0 g of steam at 100°C condense if mixed with 100 g of liquid water at 0.0°C in an insulated container?

10. The unit cell of the cesium chloride structure is shown at the right. Calculate the number of formula units of CsCl in one unit cell.

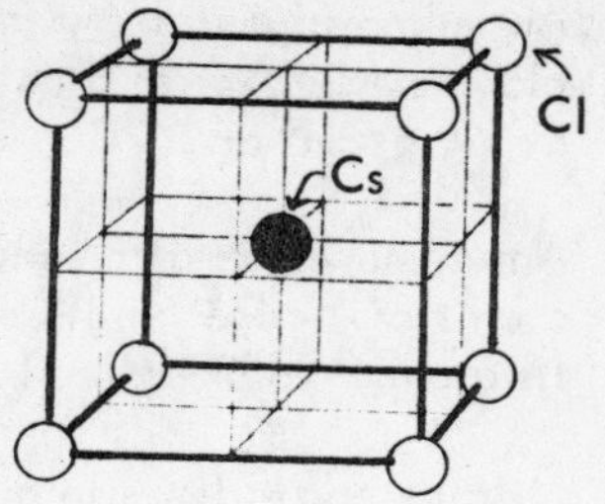

11. The cubic unit cell of the zinc blende, ZnS, structure is shown at the right. Calculate the number of formula units of ZnS in one unit cell.

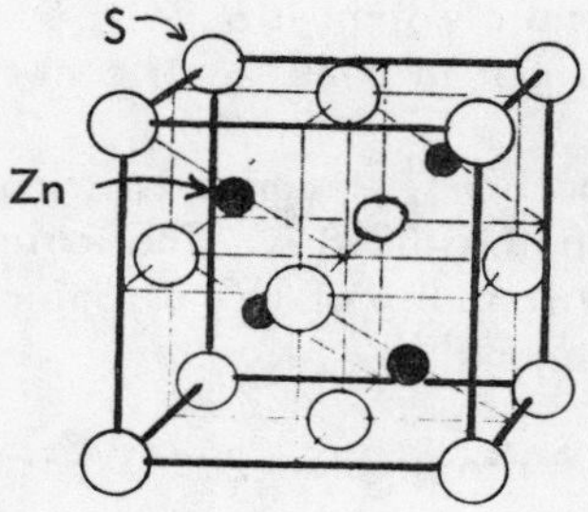

12. The cubic unit cell of the fluorite, CaF_2, structure is shown at the right. Calculate the number of formula units of CaF_2 in one unit cell.

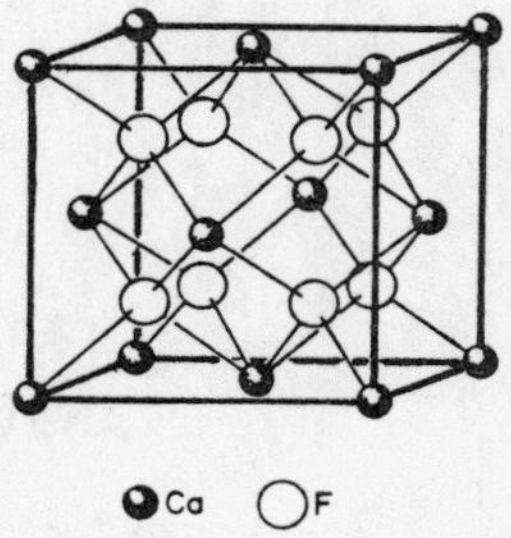

13. Iridium crystallizes in a face-centered cubic lattice with an edge length of 3.839 Å. Calculate the atomic radius and the density of iridium.

14. Vanadium crystallizes in a body-centered cubic lattice. Its specific gravity is 6.11. Calculate the edge length, a, of the unit cell.

15. Potassium bromide crystallizes in the sodium chloride type structure (see Figure 9-21 in the text). The ionic radii of K^+ and Br^- are 1.33 Å and 1.95 Å, respectively. Calculate the edge length of the unit cell assuming anion-cation contact along the edge.

16. Refer to question 15. Calculate the density of KBr.

17. An unknown metal crystallizes in a body-centered cubic lattice (cubic close-packed) with a unit cell edge length of 3.301 Å. Its density is 8.569 g/cm^3 at 20°C. Identify the metal.

18. An unknown metal crystallizes in a hexagonal close-packed lattice with a unit cell edge length of 5.344 Å. Its density is 0.862 g/cm^3 at 20°C. Its atomic volume is 51.63 $Å^3$. Identify the metal.

19. Lithium crystallizes in a body-centered cubic lattice with a unit cell edge length of 3.5087 Å. The density of lithium is 0.534 g/cm^3 and its atomic weight is 6.941 g/mol. From this information calculate Avogadro's number to three significant figures.

20. Thallium(I) bromide is isomorphous with CsCl. The unit cell edge length is 3.98 Å. The ionic radius of Br^- is 1.95 Å. Calculate the ionic radius of Tl^+ assuming anion-cation contact along the body diagonal.

21. Refer to problem 20. What is the density of TlBr?

ANSWERS FOR EXERCISES

1. (a) 378 cal (b) 910 cal (c) 84 cal 2. (a) 66.3°C (b) 63.6°C (c) 52.6°C 3. (a) 83.6°C (b) 30.4°C (c) 139°C 4. (a) 2.04×10^4 J absorbed (b) 5.15×10^4 J released (c) 1.20×10^4 J absorbed. (d) 3.50×10^4 J absorbed 5. (a) 1.10×10^5 cal (b) 2.48×10^4 cal 6. 409 kJ 7. 1.6 °C 8. Yes, the ice can absorb more than enough heat to freeze the water without any ice melting. 9. No, the cold water will reach its boiling point before all the steam condenses. 10. one 11. four 12. four 13. 1.37 Å, 22.6 g/cm^3 14. 3.03 Å 15. 6.56 Å 16. 2.80 g/cm^3 17. Nb 18. K 19. 6.02×10^{23} 20. 1.50 Å 21. 7.49 g/cm^3

Chapter Ten

SOLUTIONS

D, G, W: Chap. 13

Simple solutions are homogeneous mixtures that consist of one or more solutes and a solvent. The solvent is the dissolving or dispersing medium. A solute is the dissolved or dispersed substance. Although there are many kinds of solutions, the most common ones are those in which the solvent is a liquid.

CONCENTRATIONS OF SOLUTIONS

Concentrations of solutes are often expressed by one of four methods. They are: (1) percent by mass (often called simply mass percent), (2) molarity, (3) normality, and (4) molality. Normality will be discussed in Chapters 11 and 12.

10-1 Percent by Mass

Percent by mass or mass percent (mass %) of solute is the percentage of the total mass of a sample of solution which is due to a solute alone.

$$\text{mass \%} = \frac{\text{mass of solute}}{\text{mass of solution}} \times 100\% = \frac{\text{mass of solute}}{\text{mass of solute + mass of solvent}} \times 100\%$$

Assume the solvent is water in the following examples unless otherwise indicated.

Example 10-1: What mass of a solution that is 20.0% KCN by mass contains 35.0 grams of KCN?

Since one hundred grams of 20.0% KCN solution contains 20.0 grams of KCN and 80.0 grams of H_2O, we use this information to construct the appropriate unit factor.

$$\underline{?}\ \text{g sol'n} = 35.0\ \text{g KCN} \times \frac{100\ \text{g sol'n}}{20.0\ \text{g KCN}} = \underline{\underline{175\ \text{g sol'n}}}$$

Example 10-2: What mass of 20.0% KCN solution contains 40.0 grams of H_2O?

We construct the appropriate unit factor from the information given in the previous example.

$$\underline{?}\ \text{g sol'n} = 40.0\ \text{g } H_2O \times \frac{100\ \text{g sol'n}}{80.0\ \text{g } H_2O} = \underline{\underline{50.0\ \text{g sol'n}}}$$

Example 10-3: How many grams of KCN are contained in 65.0 grams of 20.0% KCN solution?

The unit factor used here is the reciprocal of that used in Example 10-1.

$$\underline{?}\ \text{g KCN} = 65.0\ \text{g sol'n} \times \frac{20.0\ \text{g KCN}}{100\ \text{g sol'n}} = \underline{\underline{13.0\ \text{g KCN}}}$$

Example 10-4: How many grams of H_2O are contained in a 20.0% KCN solution that contains 55.0 grams of KCN?

The desired unit factor must relate grams KCN to grams H_2O.

$$\underline{?}\ \text{g H}_2\text{O} = 55.0\ \text{g KCN} \times \frac{80.0\ \text{g H}_2\text{O}}{20.0\ \text{g KCN}} = \underline{220\ \text{g H}_2\text{O}}$$

Example 10-5: What volume of 15.0% NaBr solution contains 32.0 grams of NaBr? The density of 15.0% NaBr solution is 1.13 g/mL.

We must use an additional factor that relates mass to volume of the solution.

$$\underline{?}\ \text{mL sol'n} = 32.0\ \text{g NaBr} \times \frac{100\ \text{g sol'n}}{15.0\ \text{g NaBr}} \times \frac{1\ \text{mL sol'n}}{1.13\ \text{g sol'n}} = \underline{\underline{189\ \text{mL sol'n}}}$$

Example 10-6: The specific gravity of 26.0% $NaCH_3COO$ (sodium acetate) solution is 1.14. What mass of $NaCH_3COO$ is contained in 45.0 mL of solution?

Since the specific gravity of the solution is 1.14, we know that its density is 1.14 g/mL.

$$\underline{?}\text{g NaCH}_3\text{COO} = 45.0\ \text{mL sol'n} \times \frac{1.14\ \text{g sol'n}}{\text{mL sol'n}} \times \frac{26.0\ \text{g NaCH}_3\text{COO}}{100\ \text{g sol'n}}$$

$$= \underline{\underline{13.3\ \text{g NaCH}_3\text{COO}}}$$

10-2 Molarity

Probably the most commonly used method for expressing concentrations of solutions is molarity ($\underline{M}$). The molarity of a solute in a solution is the number of moles of solute per liter of solution [or, the number of millimoles of solute per milliliter of solution].

$$M = \frac{\text{no of mol solute}}{\text{liter of solution}} \quad \text{or} \quad \underline{M} = \frac{\text{no of mmol solute}}{\text{mL of solution}}$$

Example 10-7: What is the molarity of a solution that contains 16.7 grams of KOH per liter of solution?

We must convert grams of KOH per liter into moles of KOH per liter.

$$\frac{?\ \text{mol KOH}}{\text{L sol'n}} = \frac{16.7\ \text{g KOH}}{\text{L sol'n}} \times \frac{1\ \text{mol KOH}}{56.1\ \text{g KOH}} = \frac{0.298\ \text{mol KOH}}{\text{L sol'n}}$$

$$= \underline{\underline{0.298\ \underline{M}\ \text{KOH}}}$$

Example 10-8: What is the molarity of a solution containing 64.0 grams of H_3PO_4 in 640 mL of solution?

$$\frac{?\ \text{mol } H_3PO_4}{\text{L sol'n}} = \frac{64.0\ \text{g } H_3PO_4}{0.640\ \text{L sol'n}} \times \frac{1\ \text{mol } H_3PO_4}{98.0\ \text{g } H_3PO_4} = \frac{1.02\ \text{mol } H_3PO_4}{\text{L sol'n}}$$

$$= \underline{\underline{1.02\ \underline{M}\ H_3PO_4}}$$

Example 10-9: What volume of 0.100 $\underline{M}$ H_3PO_4 solution contains 16.0 grams of H_3PO_4?

$$?\ \text{L sol'n} = 16.0\ \text{g } H_3PO_4 \times \frac{1\ \text{mol } H_3PO_4}{98.0\ \text{g } H_3PO_4} \times \frac{1\ \text{L sol'n}}{0.100\ \text{mol } H_3PO_4}$$

$$= \underline{\underline{1.63\ \text{L sol'n}}}$$

Example 10-10: How many moles of NH_4NO_3 are contained in 15.0 liters of 0.400 molar NH_4NO_3 solution?

$$?\ \text{mol } NH_4NO_3 = 15.0\ \text{L} \times \frac{0.400\ \text{mol } NH_4NO_3}{1\ \text{L}} = \underline{\underline{6.00\ \text{mol } NH_4NO_3}}$$

Example 10-11: What is the molarity of 18.0% $(NH_4)_2SO_4$ solution? Its specific gravity is 1.11.

$$\frac{?\ \text{mol } (NH_4)_2SO_4}{\text{L sol'n}} = \frac{1.11\ \text{g sol'n}}{\text{mL}} \times \frac{1000\ \text{mL}}{1\ \text{L}} \times \frac{18.0\ \text{g } (NH_4)_2SO_4}{100\ \text{g sol'n}}$$

$$\times \frac{1\ \text{mol } (NH_4)_2SO_4}{132\ \text{g } (NH_4)_2SO_4} = \underline{\underline{1.51\ \underline{M}\ (NH_4)_2SO_4}}$$

Example 10-12: What is the density of 10.0% $(NH_4)_2SO_4$ solution? It is 0.800 $\underline{M}$ in $(NH_4)_2SO_4$.

$$\frac{?\ \text{g sol'n}}{\text{mL sol'n}} = \frac{100\ \text{g sol'n}}{10.0\ \text{g } (NH_4)_2SO_4} \times \frac{132\ \text{g } (NH_4)_2SO_4}{1\ \text{mol } (NH_4)_2SO_4}$$

$$\times \frac{0.800\ \text{mol } (NH_4)_2SO_4}{1000\ \text{mL sol'n}} = \underline{\underline{1.06\ \text{g sol'n/mL}}}$$

10-3 Dilution

Dilution is the process in which a solution is mixed with more solvent. In any dilution, the concentration of solute decreases, but the total number of moles of solute remains constant. The amount of solute (number of moles or millimoles) in a sample of solution can be obtained by multiplying the volume of the solution by its concentration.

volume × concentration = amount

$$\text{liters} \times \frac{\text{moles}}{\text{liter}} = \text{moles} \quad \underline{\text{or}} \quad \text{mL} \times \frac{\text{mmol}}{\text{mL}} = \text{mmol}$$

Since the amount of solute present in a solution before and after dilution is constant, then the above relationship can be written in a useful form as follows. The subscripts 1 and 2 refer to the solution before and after dilution.

$$L_1 \times \frac{\text{mol}}{L_1} = L_2 \times \frac{\text{mol}}{L_2} \quad \text{or} \quad L_1 \times \underline{M}_1 = L_2 \times \underline{M}_2$$

$$mL_1 \times \frac{\text{mmol}}{mL_1} = mL_2 \times \frac{\text{mmol}}{mL_1} \quad \text{or} \quad mL_1 \times \underline{M}_1 = mL_2 \times \underline{M}_2$$

Example 10-13: If 200 mL of 3.00 $\underline{M}$ H_2SO_4 solution are added to 100 mL of water what is the molarity of H_2SO_4 in the new solution?

$$mL_1 \times \underline{M}_1 = mL_2 \times \underline{M}_2$$

The volume of the dilute solution is 100 mL + 200 mL = 300 mL = mL_2.

$$\underline{M}_2 = \frac{mL_1 \times \underline{M}_1}{mL_2}$$

$$\underline{M}_2 = \frac{200 \text{ mL} \times 3.00 \underline{M}}{300 \text{ mL}} = \underline{\underline{2.00 \underline{M} H_2SO_4}}$$

Example 10-14: What volume of 0.500 $\underline{M}$ H_2SO_4 solution must be added to 100 mL of water to give a solution that is 0.150 $\underline{M}$ H_2SO_4?

First we calculate the volume of the dilute solution.

$$mL_1 \times \underline{M}_1 = mL_2 \times \underline{M}_2$$

$$mL_2 = \frac{mL_1 \times \underline{M}_1}{\underline{M}_2} = \frac{100 \text{ mL} \times 0.500 \underline{M}}{0.150 \underline{M}} = 333 \text{ mL}$$

$\underline{?}$ mL sol'n added = 333 mL − 100 mL = $\underline{\underline{233 \text{ mL}}}$

Example 10-15: A volume of 3.00 liters of NaOH solution of unknown concentration is mixed with 4.00 liters of water. The resulting solution is found to be 0.140 $\underline{M}$ in NaOH. What was the molarity of the original solution?

$$L_1 \times \underline{M}_1 = L_2 \times \underline{M}_2$$

$$\underline{M}_1 = \frac{L_2 \times \underline{M}_2}{L_1} = \frac{7.00\ L \times 0.140\ \underline{M}}{3.00\ L} = \underline{\underline{0.327\ \underline{M}}}$$

10-4 Molality

The molality ($\underline{m}$) of a solute is defined as the number of moles of solute per kilogram of solvent (not solution).

$$\text{molality} = \frac{\text{no of mol solute}}{\text{kilogram solvent}}$$

Example 10-16: Calculate the molality of an aqueous solution that has a mass of 500 grams and contains 36.0 grams of NH_4NO_3.

The number of moles of NH_4NO_3 is:

$$\underline{?}\ \text{mol}\ NH_4NO_3 = 36.0\ \text{g}\ NH_4NO_3 \times \frac{1\ \text{mol}}{80.0\ \text{g}} = 0.450\ \text{mol}\ NH_4NO_3$$

The mass of water is 500 g - 36.0 g = 464 g H_2O.

$$\frac{?\ \text{mol}\ NH_4NO_3}{\text{kg}\ H_2O} = \frac{0.450\ \text{mol}\ NHNO_3}{0.464\ \text{kg}\ H_2O} = \underline{\underline{0.970\ \underline{m}\ NH_4NO_3}}$$

COLLIGATIVE PROPERTIES OF SOLUTIONS

Physical properties of solutions that are affected by the number and not the kind of solute particles are called colligative properties. These include (1) vapor pressure, (2) osmotic pressure, (3) boiling point and (4) freezing point. Changes in these properties are directly proportional to the molalities of solutes.

10-5 Vapor Pressure Lowering

The vapor pressure exerted by a solvent at a particular temperature is always lowered by dissolving a nonvolatile solute in it. Raoult's Law states that the lowering of the vapor pressure of a solvent is directly proportional to the number of moles of nonvolatile solute dissolved in a definite mass of solvent. Dissolution of a volatile solute decreases the vapor pressure of the solvent, but since a volatile solute also exhibits a significant vapor pressure, we shall restrict our considerations to nonvolatile solutes.

10-6 Boiling Point Elevation

The boiling point of a solution that contains a nonvolatile solute is always higher than that of the pure solvent. The elevation of the boiling point of a solution is directly proportional to the molality of the solute.

$$\Delta T_b = K_b \underline{m}$$

In this relationship, ΔT_b is the increase in boiling point, K_b is the boiling point elevation constant, which is characteristic of the solvent only, and $\underline{m}$ is the molality.

Example 10-17: Calculate boiling point of a solution that contains 12.2 grams of benzoic acid dissolved in 250 grams of nitrobenzene. The boiling point of pure nitrobenzene is 210.88°C and K_b for nitrobenzene is 5.24°C/$\underline{m}$. Benzoic acid is $C_7H_6O_2$ and nitrobenzene is $C_6H_5NO_2$.

We first calculate the molality of the solute, benzoic acid.

$$\frac{?\ \text{mol}\ C_7H_6O_2}{\text{kg}\ C_6H_5NO_2} = \frac{12.2\ \text{g}\ C_7H_6O_2}{0.250\ \text{kg}\ C_6H_5NO_2} \times \frac{1\ \text{mol}\ C_7H_6O_2}{122\ \text{g}\ C_7H_6O_2} = 0.400\ \underline{m}$$

$$\Delta T_b = K_b \underline{m} = (5.24\,^\circ\text{C}/\underline{m})(0.400\ \underline{m}) = 2.10^\circ\text{C}$$

The boiling point of the solution is 210.88°C + 2.10°C = 212.98°C

10-7 Freezing Point Depression

The freezing point of a solution is always lower than that of the pure solvent. The freezing point depression, ΔT_f, is proportional to K_f, a constant characteristic of the solvent only, and to the molality of the solute.

$$\Delta T_f = K_f \underline{m}$$

Example 10-18: Calculate the freezing point of the solution described in Example 10-17. K_f for nitrobenzene is 7.00°C/$\underline{m}$ and pure nitrobenzene freezes at 5.67°C.

The molality of the solution is 0.400 $\underline{m}$ (from Example 10-17).

$$\Delta T_f = K_f \underline{m} = (7.00\,^\circ\text{C}/\underline{m})(0.400\ \underline{m}) = 2.80\,^\circ\text{C}$$

Thus the solution freezes 2.80°C below 5.67°C, or at 2.87°C.

10-8 Determination of Molecular Weights Using Freezing Point Depressions

The molecular weight of an unknown solid or liquid nonelectrolyte can be determined by dissolving a known mass in a known mass of solvent.

and measuring the resulting freezing point depression. This is illustrated in the following example.

Example 10-19: Elemental analysis indicates the simplest formula of an unknown nonelectrolyte is C_2H_4O. When 0.735 gram of the compound is dissolved in 50.0 grams of benzene, the solution freezes at 5.06°C. What are the molecular weight and true molecular formula of the compound? For benzene, $K_f = 5.12°C/\underline{m}$ and its melting point (freezing point) = 5.48°C.

The freezing point depression is 5.48°C-5.06°C = 0.42°C, and we can represent molality as mol solute/kg solvent.

$$\Delta T_f = K_f \underline{m} = K_f \left(\frac{\text{mol solute}}{\text{kg solvent}}\right)$$

Since moles solute = mass solute/mol wt, then we can also represent molality in the above relationship as follows.

$$\Delta T_f = K_f \left(\frac{\text{mass solute}}{\text{mol wt x kg solvent}}\right)$$

Solving for molecular weight gives:

$$\text{mol wt} = \frac{(K_f)(\text{mass solute})}{(\Delta T_f)(\text{kg solvent})} = \frac{(5.12°C/\underline{m})(0.735\text{ g})}{(0.42°C)(0.0500\text{ kg})}$$

$$\text{mol wt} = \underline{\underline{179\text{ g/mol}}}$$

The formula weight of C_2H_4O is 44 g/mol.

The true formula must be $(C_2H_4O)_4$ since 4 x 44 g/mol = 176 g/mol. This result agrees well with the experimental value of 179 g/mol. Experimental values usually differ slightly from true values due to inherent experimental limitations. Of course, the true molecular weight must be an exact multiple of the simplest formula weight. The compound has the molecular formula $C_8H_{16}O_4$ and a molecular weight of $\underline{\underline{176\text{ g/mol}}}$.

10-9 Percent Ionization of Weak Electrolytes from Freezing Point Depressions

Since colligative properties depend upon the number of solute particles in solution only, greater effects are observed when ionized or partially ionized substances are solutes. This effect can be utilized in determining the percent ionization in solutions of weak electrolytes and the "apparent" percent ionization in solutions of strong electrolytes. The next example demonstrates this.

Example 10-20: A 0.1096 molal aqueous solution of the weak monoprotic (one ionizable H) acid, formic acid, HCOOH freezes at $-0.210^{\circ}C$. Determine the percentage ionization of the acid in this solution. For water $K_f = 1.86^{\circ}C/\underline{m}$.

The acid ionizes partially as shown below and we let x represent the molality of HCOOH that ionizes. Every molecule of HCOOH that ionizes produces one H^+ ion and one $HCOO^-$ ion.

$$\begin{array}{cccc} HCOOH & \rightleftharpoons & H^+ & + & HCOO^- \\ (0.1096-x)\underline{m} & & x\,\underline{m} & & x\,\underline{m} \end{array}$$

Because ΔT_f depends upon the total number of particles (molecules or ions) in solution, it is proportional to the total molality of the solute, $\underline{m}_{total}$, which is the sum of the molalities of HCOOH, H^+, and $HCOO^-$ in this solution.

$$\underline{m}_{total} = (0.1096 - x)\underline{m} + x\,\underline{m} + x\,\underline{m}$$

$$= (0.1096 + x)\underline{m}$$

Since $\Delta T_f = K_f\,\underline{m}_{total}$, $\quad \underline{m}_{total} = \dfrac{\Delta T_f}{K_f} = \dfrac{0.210^{\circ}C}{1.86^{\circ}C/\underline{m}} = 0.113\ \underline{m}$

We have now determined that $\underline{m}_{total}$ is $0.113\underline{m}$ and we also represented it as as $(0.1096 + x)\underline{m}$. Things equal to the same thing are equal to each other, and therefore:

$$(0.1096 + x)\underline{m} = 0.113\ \underline{m}$$

$$x = 0.003\ \underline{m} = \underline{m}_{H^+} = \underline{m}_{HCOO^-} = \underline{m}_{HCOOH\ (ionized)}$$

$$\%\ \text{ionization} = \frac{\underline{m}_{HCOOH\ (ionized)}}{\underline{m}_{HCOOH\ (original)}} \times 100\%$$

$$= \frac{0.003\ \underline{m}}{0.1096\ \underline{m}} \times 100\%$$

$$= \underline{\underline{3\%\ \text{ionized}}}$$

We have shown that HCOOH is 3% ionized and 97% un-ionized in 0.1096 $\underline{m}$ solution.

10-10 Percent Ionization of Strong Electrolytes from Freezing Point Depressions

Thus far we have assumed that strong electrolytes are 100% ionized in dilute aqueous solution. However, solutions of strong electrolytes behave as if the strong electrolytes are less than completely ionized.

At any instant in a solution of a strong electrolyte a statistical percentage of the cations and anions, which are in constant motion, will collide with each other and "stick together". While an ion pair "sticks together" it behaves as an "un-ionized" ion pair. (This statement does not imply that molecules of ionic substances exist in aqueous solution.) A strong electrolyte is assumed to be 100% ionized in an infinitely dilute solution. As the concentration of a solution increases, more of these collisions occur and the "apparent" percentage ionization decreases.

Example 10-21: Determine the "apparent" percentage ionization in 0.0531 $\underline{m}$ aqueous solution of $MgCl_2$, which freezes at -0.255°C. K_f for water is 1.86°C/$\underline{m}$.

Let x be the molality of $MgCl_2$ that is apparently ionized. For each formula unit of $MgCl_2$ that ionizes, one Mg^{2+} ion and two Cl^- ions are produced.

$MgCl_2$	$\longrightarrow$	Mg^{2+}	+	$2\ Cl^-$
$(0.0531 - x)\underline{m}$		$x\ \underline{m}$		$2x\ \underline{m}$

$$\underline{m}_{total} = \underline{m}_{MgCl_2} + \underline{m}_{Mg^{2+}} + \underline{m}_{Cl^-}$$

$$= 0.0531\ \underline{m} - x\ \underline{m} + x\ \underline{m} + 2x\ \underline{m}$$

$$\underline{m}_{total} = (0.0531 + 2x)\underline{m}$$

We shall use a strictly algebraic approach that differs from the approach used in Example 10-20.

ΔT_f is directly proportional to $\underline{m}_{total}$.

$$\Delta T_f = K_f\ \underline{m}_{total}$$

$$0.255^\circ C = (1.86^\circ C/\underline{m})(0.0531 + 2x)\underline{m}$$

Solving for x,

$$0.255 = 0.0988 + 3.72x$$

$$x = \frac{0.156}{3.72} = 0.0419\ \underline{m}$$

$$\underline{m}_{Mg^{2+}} = 1/2\ \underline{m}_{Cl^-} = \underline{m}_{MgCl_2\ (ionized)} = 0.0419\ \underline{m}$$

$$\text{apparent \% ionization} = \frac{\underline{m}_{MgCl_2}\text{ (ionized)}}{\underline{m}_{MgCl_2}\text{ (original)}} \times 100\% = \frac{0.0419\ \underline{m}}{0.0531\ \underline{m}} \times 100\%$$

$$= \underline{\underline{78.9\%\ \text{apparently ionized}}}$$

10-11 Osmotic Pressure

If a solution is separated from pure solvent by a membrane that is permeable only to solvent molecules, the natural tendency is for solvent molecules to migrate more rapidly from pure solvent into the solution than in the reverse direction. This process, which dilutes the solution, is called osmosis. Osmosis can be counterbalanced by an opposing pressure called the osmotic pressure, π. The osmotic pressure for a given solution is given by the following expression.

$$\pi = \underline{M}RT$$

in which $\underline{M}$ is the molarity of the solute, R is the universal gas constant, and T is the absolute temperature. For dilute aqueous solutions molarity is nearly equal to molality and the following approximation can be made.

$$\pi\ \text{(in atm.)} = \underline{m}RT \qquad \text{(for dilute aqueous solutions)}$$

Measurements of osmotic pressures of dilute solutions of large molecules are often used to determine molecular weights of solutes because only a small number of moles gives rise to easily observable osmotic pressures.

Example 10-22: What is the osmotic pressure (in torr) of a 0.00100 $\underline{M}$ solution of a nonelectrolyte in water at 25.0°C?

$$\pi = \underline{M}RT$$

$$= (0.00100\ \underline{M})(0.0821\ \frac{L\cdot atm}{mol\cdot K})(298\ K) = 2.45 \times 10^{-2}\ atm$$

$$?\ torr = 2.45 \times 10^{-2}\ atm \times \frac{760\ torr}{atm} = \underline{\underline{18.6\ torr}}$$

Example 10-23: What is the molecular weight of a biological compound if 4.00 grams of it dissolved in enough benzene to give 2.00 liters of solution has an osmotic pressure of 3.20 torr at 25.0°C?

If we determine the molarity of the solution and then multiply it by 2.00 liters we will know the fraction of a mole that 4.00 grams correspond to.

$$\pi = \underline{M}RT$$

$$\underline{M} = \frac{\pi}{RT} = \frac{3.20 \text{ torr} \times \left(\frac{1.00 \text{ atm}}{760 \text{ torr}}\right)}{\left(0.0821 \frac{\text{L}\cdot\text{atm}}{\text{mol K}}\right)(298 \text{ K})} = 1.72 \times 10^{-4} \underline{M}$$

$$? \text{ mol solute} = 2.00 \text{ L} \times \frac{1.72 \times 10^{-4} \text{ mol solute}}{\text{L}} = 3.44 \times 10^{-4} \text{ mol solute}$$

The molecular weight is simply the number of grams of solute per mole.

$$\frac{? \text{ g}}{\text{mol}} = \frac{4.00 \text{ g}}{3.44 \times 10^{-4} \text{ mol}} = 11{,}600 \text{ g/mol} = \underline{\underline{1.16 \times 10^{4} \text{ g/mol}}}$$

Typical of many important biological molecules, this is a fairly large and complex compound.

EXERCISES

1. If 4.16 g of silver nitrate, $AgNO_3$, are contained in 65.0 g of solution, what is the percent by weight of $AgNO_3$?

2. How many g of ammonium bromide, NH_4Br, are contained in 115 mL of 2.50 NH_4Br solution?

3. How many g of NH_4Cl are contained in 115 mL of 2.50 $\underline{M}$ NH_4Cl solution?

4. How many g of solid zinc sulfate heptahydrate, $ZnSO_4 \cdot 7H_2O$, must be dissolved to give 150 mL of 3.00 $\underline{M}$ zinc sulfate, $ZnSO_4$, solution?

5. The density of 16.0% $C_{12}H_{22}O_{11}$ (sucrose) solution is 1.064 g/mL (a) What volume of solution contains 50.0 g of sucrose? (b) What mass of solution contains 50.0 g of sucrose? (c) What volume of solution contains one mole of sucrose?

6. How many milliliters of 6.00% sodium bicarbonate, $NaHCO_3$, solution contain 25.0 g of $NaHCO_3$? The density of the solution is 1.041 g/mL.

7. What is the molarity of 12.00% sodium nitrate, $NaNO_3$, solution? Its density is 1.082 g/mL.

8. Calculate the molarity of 650 mL of a solution that contains 17.6 g of magnesium chloride, $MgCl_2$.

9. Calculate the molarity of 300 mL of a solution that contains 12.20 mg of lead(II) acetate, $Pb(CH_3COO)_2$.

10. Calculate the molarity of $Na_2Cr_2O_7$ in a solution prepared by dissolving 184 g of sodium dichromate dihydrate, $Na_2Cr_2O_7 \cdot 2H_2O$ in enough water to make 550 mL of solution.

11. What is the percent by mass of sodium bromide, NaBr, in 4.405 $\underline{M}$ NaBr solution which has a specific gravity of 1.333?

12. How many grams of solute are necessary to prepare each of the following?

    (a) 300 mL of 1.25 $\underline{M}$ H_3PO_4
    (b) 1200 mL of 0.600 $\underline{M}$ $NaNO_3$
    (c) 2.67 L of 0.450 $\underline{M}$ $NaCH_3COO$
    (d) 1.56 L of 1.00 $\underline{M}$ $CuSO_4$ from $CuSO_4 \cdot 5H_2O$
    (e) 3.50 mL of 0.600 $\underline{M}$ $AlCl_3$
    (f) 750 mL of 5.0×10^{-4} $\underline{M}$ $MgBr_2$
    (g) 1.28 L of 1.7×10^{-6} $\underline{M}$ $(NH_4)_2SO_4$

13. How many grams of 92.0% pure solid sodium hydroxide, NaOH, must be dissolved to give 1.00 liter of 1.00 $\underline{M}$ NaOH?

14. What is the molarity of the solution resulting from addition of:

   (a) 50.0 mL of 8.00 M HNO_3 to 200 mL of water
   (b) 500 mL of 3.00 $\underline{M}$ H_2SO_4 to 500 mL of water?
   (c) 20.0 mL of 12.0 $\underline{M}$ HCl to 1.20 liters of water?

15. What is the molarity of the solution resulting from the addition of:

   (a) 350 mL of 0.100 $\underline{M}$ KOH to 1.00 L of 5.25 M KOH?
   (b) 1500 mL of 1.00 × 10^{-3} $\underline{M}$ HCl to 500 mL of 3.33 × 10^{-1} $\underline{M}$ HCl?

16. If 4.27 g of sucrose, $C_{12}H_{22}O_{11}$ are dissolved in 15.2 grams of water what will be the freezing point and boiling point of the resulting solution? Sucrose is a nonvolatile covalent compound. $K_b = 0.512°C/m$ and $K_f = 1.86°C/\underline{m}$ for water.

17. When 1.150 g of an unknown nonelectrolyte dissolves in 10.0 g of water the solution freezes at -2.16°C. What is the molecular weight of the compound?

18. What will be the freezing point and boiling point of a solution prepared by dissolving 10.0 g of naphthalene, $C_{10}H_8$, in 30.0 g of benzene, C_6H_6? For benzene, $K_b = 2.53°C/m$, $K_f = 5.12°C/m$ and its freezing and boiling points are 5.48°C and 80.1°C, respectively.

19. The freezing point of a solution of 1.048 g of an unknown nonelectrolyte dissolved in 36.21 g of benzene is 1.39°C. Pure benzene freezes at 5.48°C and its K_f value is 5.12°C/$\underline{m}$. What is the molecular weight of the compound?

20. A 0.294 molal solution of aqueous ammonia, NH_3, freezes at -0.550°C. Calculate the percent ionization of the NH_3 as a weak base.

$$NH_3 \quad + \quad H_2O \quad \rightleftharpoons \quad NH_4^+ \quad + \quad OH^-$$

21. Methylamine is a weak base that ionizes as follows in water.

$$CH_3NH_2 \quad + \quad H_2O \quad \rightleftharpoons \quad CH_3NH_3^+ \quad + \quad OH^-$$

   A 0.0100 molal aqueous solution of methylamine is 18.9% ionized. What is the freezing point of the solution?

22. A 0.0490 molal aqueous NaBr solution freezes at -0.173°C. What is its apparent percent ionization in this solution?

$$NaBr \quad \rightarrow \quad Na^+ \quad + \quad Br^-$$

23. A 0.0620 molal aqueous $ZnSO_4$ solution freezes at -0.150°C. What is its apparent percent ionization in this solution?

$$ZnSO_4 \quad \rightarrow \quad Zn^{2+} \quad + \quad SO_4^-$$

24. A 0.0308 molal aqueous $SrCl_2$ solution freezes at -0.158°C. What is its apparent percent ionization in this solution?

$$SrCl_2 \quad \rightarrow \quad Sr^{2+} \quad + \quad 2\ Cl^-$$

25. A 0.686 g sample of a high molecular weight compound dissolves in benzene to give 80.0 mL of a solution that exhibits an osmotic pressure of 11.39 torr at 25.0°C. Estimate the molecular weight of the compound.

26. (a) What is the osmotic pressure in torr associated with 0.0010 M sucrose in water at 25°C? (b) Compare the magnitude of the osmotic pressure of this solution with the boiling point elevation and freezing point depression of the same solution. Assume that for this dilute solution the molality is equal to the molarity. For water K_b = 0.512°C/m and K_f = 1.86°C/m.

ANSWERS FOR EXERCISES

1. 6.4% $AgNO_3$ 2. 28.1 g NH_4Br 3. 15.4 g NH_4Cl 4. 130 g $ZnSO_4 \cdot 7H_2O$ 5. (a) 294 mL solution (b) 313 g solution (c) 2.01 L solution 6. 400 mL solution 7. 1.53 M $NaNO_3$ 8. 0.284 M $MgCl_2$ 9. 1.25×10^{-4} M $Pb(C_2H_3O_2)_2$ 10. 1.12 M $Na_2Cr_2O_7$ 11. 34.00% NaBr 12. (a) 36.8 g H_3PO_4 (b) 61.2 g $NaNO_3$ (c) 98.5 g $NaC_2H_3O_2$ (d) 390 g $CuSO_4 \cdot 5H_2O$ (e) 0.279 g $AlCl_3$ (f) 0.069 g $MgBr_2$ (g) 2.87×10^{-4} g $(NH_4)_2SO_4$ 13. 43.5 g impure NaOH 14. (a) 1.60 M HNO_3 (b) 1.50 M H_2SO_4 (c) 0.197 M HCl 15. (a) 3.91 M KOH (b) 0.0840 M HCl 16. T_b = 100.421°C, T_f = -1.53°C 17. 99.1 g/mol 18. T_b = 86.7°C, T_f = -7.8°C 19. 36.3 g/mol 20. 0.7% ionized 21. -0.0221°C 22. 89.8% 23. 30.0% (This unexpectedly low percent ionization is due to the high degree of covalent character exhibited by many zinc salts.) 24. 88.0% 25. 1.40×10^4 g/mol 26. (a) 18.6 torr (b) 0.000512°C, 0.00186°C [Note that (b) and (c) are very small values compared to (a).]

Chapter Eleven

ACIDS, BASES, AND SALTS

D, G, W: Sec. 9-4.2, .3, .4 and 9-6.1, .2; Sec. 17-1 through 17-4 and Sec. 17-10

Several theories of acid-base behavior have been developed. Among the more useful ones are (1) the Arrhenius theory, (2) the Bronsted-Lowry theory and (3) the Lewis theory.

11-1 The Arrhenius Theory

In 1884 Arrhenius defined acids and bases as follows:

Acid - any substance containing hydrogen that ionizes in aqueous solutions to produce hydrogen ions, H^+. [We usually represent hydrated hydrogen ions as H_3O^+ or H^+ (aq) in aqueous solution.]

Base - any substance containing OH groups that ionizes in aqueous solutions to produce hydroxide, ions, OH^-.

Some examples of Arrhenius acids and bases are given below.

1. $HCl\,(aq) + H_2O\,(\ell) \rightarrow H_3O^+\,(aq) + Cl^-\,(aq)$

hydrochloric acid

or in simplified form: $HCl\,(aq) \rightarrow H^+\,(aq) + Cl^-\,(aq)$

2. $HNO_3\,(aq) + H_2O\,(\ell) \rightarrow H_3O^+\,(aq) + NO_3^-\,(aq)$

nitric acid

$HNO_3\,(aq) \rightarrow H^+\,(aq) + NO_3^-\,(aq)$

3. $HCN\,(aq) + H_2O\,(\ell) \rightleftharpoons H_3O^+\,(aq) + CN^-\,(aq)$

hydrocyanic acid
(a weak acid)

$HCN\,(aq) \rightleftharpoons H^+\,(aq) + CN^-\,(aq)$

4. $NaOH\,(s) \xrightarrow{H_2O} Na^+\,(aq) + OH^-\,(aq)$

sodium hydroxide
(a base)

5. KOH (s) $\xrightarrow{H_2O}$ K^+ (aq) + OH^- (aq)

potassium hydroxide
(a base)

The strong soluble bases (Table 7-3 in the text) are ionic in the solid state. We indicate dissociation of these compounds by writing H_2O "over the arrow."

Neutralization reactions are those in which an acid and a base react to form a salt and (usually) water. A salt contains the anion of the acid and cation of the base from which it is derived. Examples are given by the following molecular equations.

1. HI (aq) + NaOH (aq) $\rightarrow$ NaI (aq) + H_2O (ℓ)

hydroiodic acid — sodium hydroxide — sodium iodide (a salt)

2. HCN (aq) KOH (aq) $\rightarrow$ KCN (aq) + H_2O (ℓ)

hydrocyanic acid — potassium hydroxide — potassium cyanide (a salt)

3. 2 HNO_3 (aq) + $Ca(OH)_2$ (aq) $\rightarrow$ $Ca(NO_3)_2$ (aq) + 2 H_2O (ℓ)

nitric acid — calcium hydroxide — calcium nitrate (a salt)

11-2 The Bronsted-Lowry Theory

According to this theory, presented in 1923, acids and bases are defined as follows:

Acid - a proton (H^+) donor.

Base - a proton (H^+) acceptor.

This is a broader and more general classification of acids and bases than the Arrhenius classification. All Arrhenius acids and bases are also classified as acids and bases by the Bronsted-Lowry theory. Additionally, some substances that are not classified as acids or bases according to the Arrhenius definitions are classified as such by the Bronsted definitions. An example is aqueous ammonia, a base that does not contain an OH group, but which reacts with water to produce low concentrations of OH^- ions.

NH_3 (aq)	+	H_2O (ℓ)	⇌	NH_4^+ (aq)	+	OH^- (aq)
ammonia (proton acceptor)		water (proton donor)				

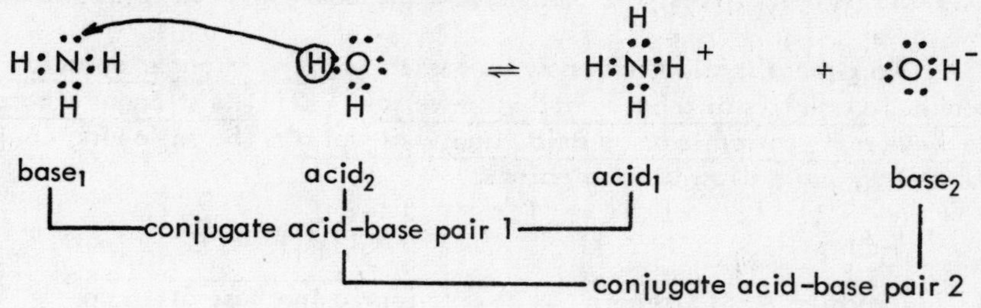

According to this theory, the products as well as the reactants can be considered acids and bases; they are described as conjugate acid-base pairs. In the forward reaction above, ammonia, the proton acceptor, is a base. The product of its protonation, the ammonium ion, is its conjugate acid because it acts as a proton donor in the reverse reaction. Likewise, water is classified as an acid because it is a proton donor, and its conjugate base, the hydroxide ion, acts as a proton acceptor in the reverse reaction.

Water can also act as a base when in contact with a substance more acidic than itself. (Ammonia is more basic than water.) For example, water acts as a base, or proton acceptor, when in contact with hydrofluoric acid, HF, or any other substance more acidic than H_2O.

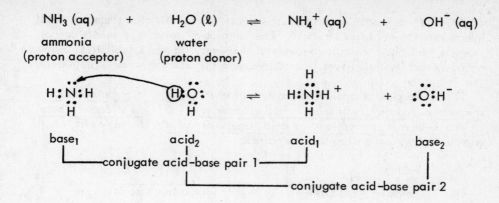

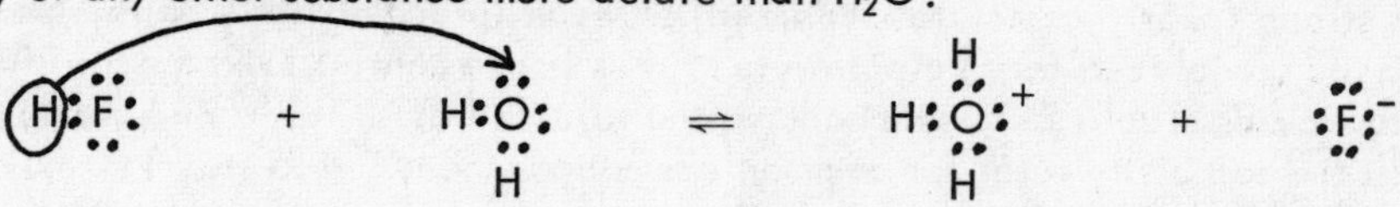

hydrofluoric acid	water	hydronium ion	fluoride ion
$acid_1$	$base_2$	$acid_2$	$base_1$

The last two reactions illustrate that whether a substance acts as an acid or base depends upon its environment. They also illustrate the <u>amphoterism</u> of water, i.e., its ability to act as either an acid or base. Note that according to the Bronsted-Lowry theory <u>the anion of an acid is a base</u> and <u>the cation of a base is an acid</u>.

Recall that the strength of an acid or base refers to its degree of ionization in aqueous solution. Those that are essentially completely ionized in dilute aqueous solution are strong and those that are predominantly un-ionized are weak. Lists of the common strong acids and strong soluble bases were given in Tables 7-1 and 7-3 in the text. These short lists should be learned because you may assume (unless otherwise specified) that most other acids you encounter in general chemistry are weak. All other common metal hydroxides are insoluble in water. The "insoluble" metal hydroxides cannot produce very basic solutions because of their very limited solubilities. Aqueous ammonia and its derivatives, the amines, are the common weak soluble bases.

Two generalizations can now be made: (1) The stronger an acid is, the weaker its anion is as a base, and vice-versa. (2) The stronger a base is, the weaker its cation is as an acid, and vice-versa. The following conjugate acid-base pairs illustrate the point.

$$HCl > HF > HCN \qquad Cl^- < F^- < CN^-$$

<— Increasing Acid Strength —— Increasing Base Strength —>

$$NaOH > CH_3NH_2 > NH_3 \qquad Na^+ < CH_3NH_3^+ < NH_4^+$$

<— Increasing Base Strength —— Increasing Acid Strength —>

11-3 Writing Equations for Acid-Base Reactions

In Section 7-3, we introduced and distinguished among (1) molecular, (2) total ionic and (3) net ionic equations. In order to write correctly the latter two kinds of equations for acid-base reactions in aqueous solution it is necessary to know the following about each acid, base, and salt of interest: (1) whether it is soluble or insoluble in water and (2) if it is soluble, whether it is a strong (nearly completely ionized) or weak (predominantly in the molecular or un-ionized form) electrolyte. Thus it is essential to know the lists of strong acids (Table 7-1, text) and strong soluble bases (Table 7-3, text) as well as the solubility rules for common compounds (Sec. 7-1.5, text). It may be assumed, unless otherwise specified, that water-soluble salts are also predominantly ionized in aqueous solution. There are a few exceptions such as the soluble covalent (un-ionized) salts, lead acetate, $Pb(CH_3COO)_2$ and mercury(II) chloride, $HgCl_2$. Water is predominantly un-ionized, and therefore is written in molecular form.

Example 11-1: Write (a) molecular, (b) total ionic, and (c) net ionic equations for each of the following acid-base reactions in dilute aqueous solution. Assume that only normal salts are formed.

1. $CH_3COOH + Ca(OH)_2 \longrightarrow$

2. $Ba(OH)_2 + H_3PO_4 \longrightarrow$

3. $HCl + Ni(OH)_2 \longrightarrow$

4. NH_3 + H_2S $\longrightarrow$

1. (a) $2\ CH_3COOH\ (aq) + Ca(OH)_2\ (aq) \longrightarrow Ca(CH_3COO)_2\ (aq) + 2\ H_2O\ (\ell)$

weak acid strong soluble base soluble salt

(b) $2\ CH_3COOH\ (aq) + [Ca^{2+}\ (aq) + 2\ OH^-\ (aq)] \longrightarrow$

$[Ca^{2+}\ (aq) + 2\ CH_3COO^-\ (aq)] + 2\ H_2O\ (\ell)$

(c) $2\ CH_3COOH\ (aq) + 2\ OH^-\ (aq) \longrightarrow 2\ CH_3COO^-\ (aq) + 2\ H_2O\ (\ell)$

Division of all coefficients by 2 gives the net ionic equation.

$CH_3COOH\ (aq) + OH^-\ (aq) \longrightarrow CH_3COO^-\ (aq) + H_2O\ (\ell)$

2. (a) $3\ Ba(OH)_2\ (aq) + 2\ H_3PO_4\ (aq) \longrightarrow Ba_3(PO_4)_2\ (s) + 6\ H_2O\ (\ell)$

strong soluble base weak acid insoluble salt

(b) $3[Ba^{2+}\ (aq) + 2\ OH^-\ (aq)] + 2\ H_3PO_4\ (aq) \longrightarrow Ba_3(PO_4)_2\ (s) + 6\ H_2O\ (\ell)$

(c) $3\ Ba^{2+}\ (aq) + 6\ OH^-\ (aq) + 2\ H_3PO_4\ (aq) \longrightarrow Ba_3(PO_4)_2\ (s) + 6\ H_2O\ (\ell)$

3. (a) $2\ HCl\ (aq) + Ni(OH)_2\ (s) \longrightarrow NiCl_2\ (aq) + 2\ H_2O\ (\ell)$

strong acid insoluble base soluble salt

(b) $2[H_3O^+\ (aq) + Cl^-\ (aq)] + Ni(OH)_2\ (s) \longrightarrow$

$[Ni^{2+}\ (aq) + 2\ Cl^-\ (aq)] + 4\ H_2O\ (\ell)$

(c) $2\ H_3O^+\ (aq) + Ni(OH)_2\ (s) \longrightarrow Ni^{2+}\ (aq) + 4\ H_2O\ (\ell)$

4. (a) $2\ NH_3\ (aq) + H_2S\ (aq) \longrightarrow (NH_4)_2S\ (aq)$

weak base weak acid soluble salt

(b) $2\ NH_3\ (aq) + H_2S\ (aq) \longrightarrow [2\ NH_4^+\ (aq) + S^{2-}\ (aq)]$

(c) $2\ NH_3\ (aq) + H_2S\ (aq) \longrightarrow 2\ NH_4^+\ (aq) + S^{2-}\ (aq)$

Example 11-2: Write the same kinds of equations as in Example 11-1 for the acid-base reactions that produce the following normal salts.

1. rubidium nitrate, $RbNO_3$
2. copper(II) sulfate, $CuSO_4$
3. ammonium carbonate, $(NH_4)_2CO_3$
4. magnesium chloride, $MgCl_2$
5. strontium arsenate, $Sr_3(AsO_4)_2$

1. (a) $HNO_3\,(aq) + RbOH\,(aq) \rightarrow RbNO_3\,(aq) + H_2O\,(\ell)$

strong acid — strong soluble base — soluble salt

(b) $[H_3O^+\,(aq) + NO_3^-\,(aq)] + [Rb^+\,(aq) + OH^-\,(aq)] \rightarrow$

$[Rb^+\,(aq) + NO_3^-\,(aq)] + 2\,H_2O\,(\ell)$

(c) $H_3O^+\,(aq) + OH^-\,(aq) \longrightarrow 2\,H_2O\,(\ell)$

2. (a) $Cu(OH)_2\,(s) + H_2SO_4 \rightarrow CuSO_4\,(aq) + 2\,H_2O\,(\ell)$

insoluble base — strong acid — soluble salt

(b) $Cu(OH)_2\,(s) + [2\,H_3O^+\,(aq) + SO_4^{2-}\,(aq)] \rightarrow$

$[Cu^{2+}\,(aq) + SO_4^{2-}\,(aq)] + 4\,H_2O\,(\ell)$

(c) $Cu(OH)_2\,(s) + 2\,H_3O^+\,(aq) \rightarrow Cu^{2+}\,(aq) + 4\,H_2O\,(\ell)$

3. (a) $2\,NH_3\,(aq) + H_2CO_3\,(aq) \rightarrow (NH_4)_2CO_3\,(aq)$

weak base — weak acid — soluble salt

(b) $2\,NH_3\,(aq) + H_2CO_3\,(aq) \rightarrow [2\,NH_4^+\,(aq) + CO_3^{2-}\,(aq)]$

(c) $2\,NH_3\,(aq) + H_2CO_3\,(aq) \rightarrow 2\,NH_4^+\,(aq) + CO_3^{2-}\,(aq)$

4. (a) $2\,HCl\,(aq) + Mg(OH)_2\,(s) \rightarrow MgCl_2\,(aq) + 2\,H_2O\,(\ell)$

strong acid — insoluble base — soluble salt

(b) $2\,[H_3O^+\,(aq) + Cl^-\,(aq)] + Mg(OH)_2\,(s) \rightarrow$

$[Mg^{2+}\,(aq) + 2\,Cl^-\,(aq)] + 4\,H_2O\,(\ell)$

(c) $2\,H_3O^+\,(aq) + Mg(OH)_2\,(s) \rightarrow Mg^{2+}\,(aq) + 4\,H_2O\,(\ell)$

5. (a) $3\ Sr(OH)_2\ (aq) + 2\ H_3AsO_4\ (aq) \longrightarrow Sr_3(AsO_4)_2\ (s) + 6\ H_2O\ (\ell)$

strong soluble base   weak acid   insoluble salt

(b) $3[Sr^{2+}(aq) + 2\ OH^-\ (aq)] + 2\ H_3AsO_4\ (aq) \longrightarrow Sr_3(AsO_4)_2\ (s) + 6\ H_2O\ (\ell)$

(c) $3\ Sr^{2+}\ (aq) + 6\ OH^-\ (aq) + 2\ H_3AsO_4\ (aq) \longrightarrow Sr_3(AsO_4)_2\ (s) + 6\ H_2O\ (\ell)$

11-4 The Lewis Theory

The basic ideas of the most comprehensive and general classification of acids and bases were stated by G. N. Lewis in 1923, but the ideas were not developed and utilized widely until 1938.

Acid - any species that accepts a share in an electron pair.

Base - any species that makes available a share in an electron pair.

Neutralization - the formation of a coordinate covalent bond.

This theory includes all acids and bases defined by the Bronsted-Lowry theory plus many others. Some examples are given below.

1. 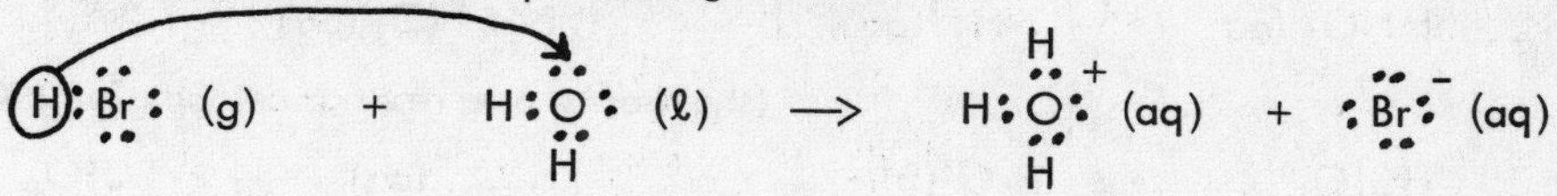

acid, hydrogen bromide, H^+ accepts a share in an electron pair

base, water makes available an electron pair

2. 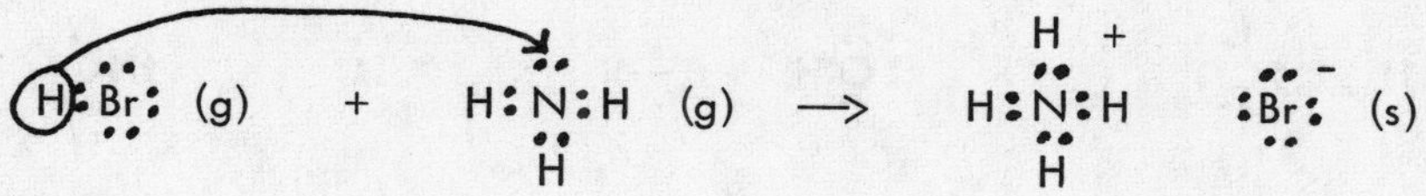

acid   base

3. 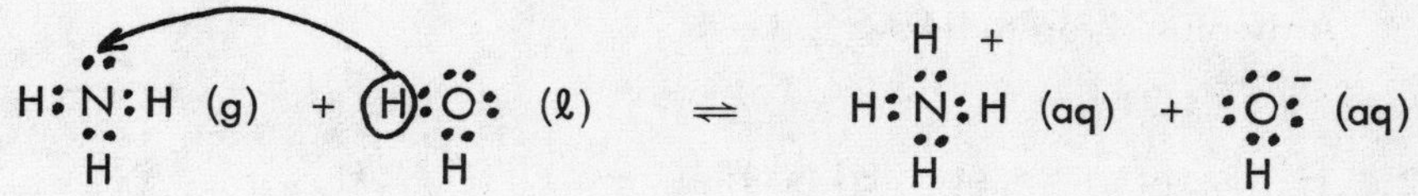

base   acid

4.

$$H:\overset{\cdot\cdot}{N}:H \quad\overset{\curvearrowright}{\longrightarrow}\quad \overset{Cl}{\underset{Cl}{B:Cl}} \quad\longrightarrow\quad \overset{H}{\underset{H}{H:\overset{\cdot\cdot}{N}}}:\overset{Cl}{\underset{Cl}{B}}:\overset{\cdot\cdot}{\underset{\cdot\cdot}{Cl}}$$

      base   acid

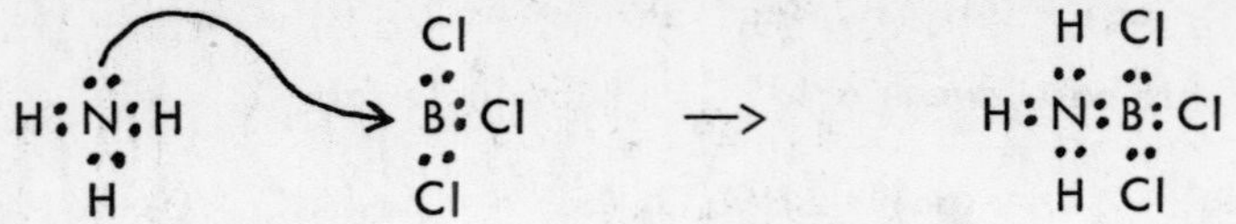

Reaction (4) is not an acid-base reaction according to either the Arrhenius or Bronsted definitions, but it is according to the Lewis definitions. A bond formed between a Lewis acid and base (involving the electron pair made available by the Lewis base) is called a <u>coordinate covalent bond</u>.

<u>Example 11-3:</u>   Write the equations below with dot formulas and classify them as one or more of the following kinds of acid-base reactions: (1) Arrhenius, (2) Bronsted-Lowry, and (3) Lewis. For those in category (2) identify conjugate acid-base pairs.

1. HCl (aq)   + KOH (aq) $\longrightarrow$ KCl (aq)   + H_2O (ℓ)

2. HNO_3 (aq) + NH_3 (aq) $\longrightarrow$ NH_4NO_3 (aq)

                   (show only one resonance form for NO_3^-)

3. $HClO_4$ (ℓ) + H_2O (ℓ) $\longrightarrow$ H_3O^+ (aq) + ClO_4^- (aq)

4. CH_3NH_2 (ℓ) + H_2O (ℓ) $\rightleftharpoons$ $CH_3NH_3^+$ (aq) + OH^- (aq)

   methlyamine   methylammonium ion

5. BCl_3 (ℓ)   + CH_3NH_2 (ℓ) $\longrightarrow$ $CH_3NH_2BCl_3$ (s)

6. SiF_4 (g)   + $2\ F^-$ (aq) $\longrightarrow$ SiF_6^{2-} (aq)

1. H^+, $:\overset{\cdot\cdot}{\underset{\cdot\cdot}{Cl}}:^-$ + K^+, $:\overset{\cdot\cdot}{\underset{\cdot\cdot}{O}}:H^-$ $\longrightarrow$ K^+, $:\overset{\cdot\cdot}{\underset{\cdot\cdot}{Cl}}:^-$ + $\overset{H:\overset{\cdot\cdot}{O}:}{\underset{H}{}}$

   acid$_1$ base$_2$ base$_1$ acid$_2$

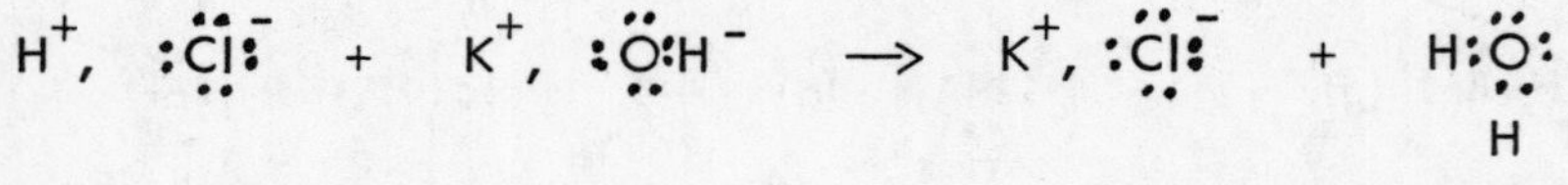

Arrhenius, Bronsted-Lowry, Lewis

2. H^+, $\overset{:\overset{\cdot\cdot}{O}:}{\underset{:O\cdot\ \ \cdot O:}{N}}^-$ + $H:\overset{\cdot\cdot}{\underset{H}{N}}:H$ $\longrightarrow$ $H:\overset{H}{\underset{H}{N}}:H$, $\overset{:\overset{\cdot\cdot}{O}:^-}{\underset{:O:\ \ :O:}{N}}$

   acid$_1$ base$_2$ acid$_1$ base$_2$

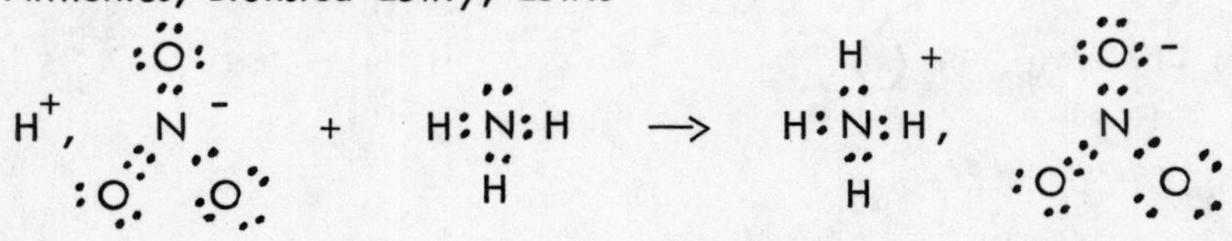

Bronsted-Lowry, Lewis (only one resonance form shown for NO_3^-)

3. 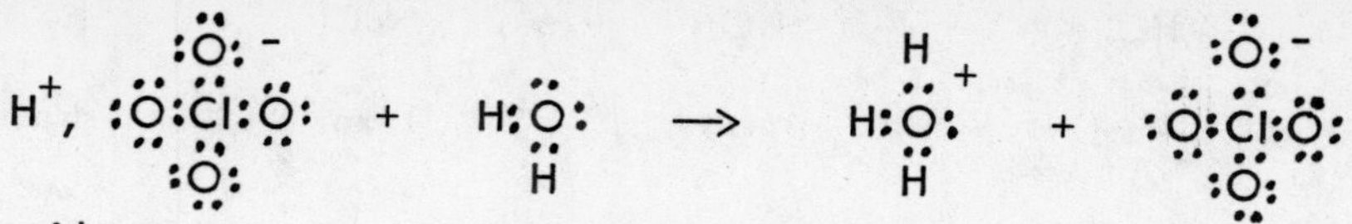

$acid_1$ $base_2$ $acid_2$ $base_1$

Bronsted-Lowry, Lewis

4. 

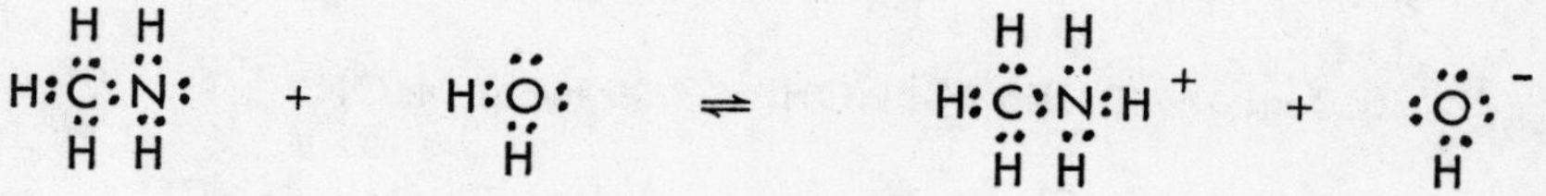

$base_1$ $acid_2$ $acid_1$ $base_2$

Bronsted-Lowry, Lewis

5. 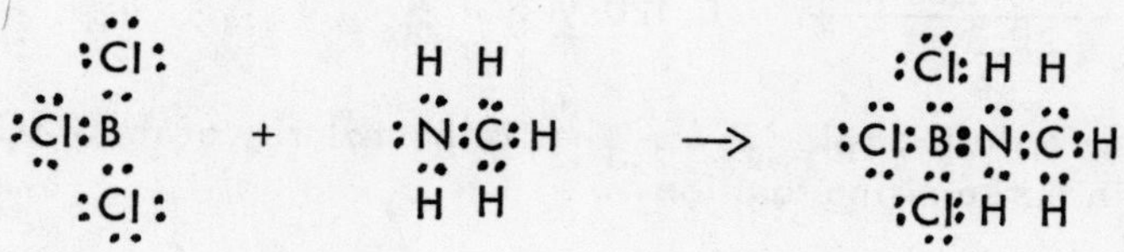

Lewis only

6. 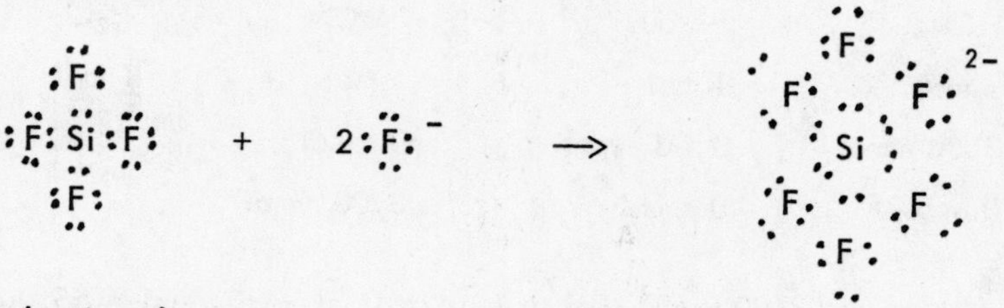

Lewis only

QUANTITATIVE RELATIONSHIPS IN ACID-BASE REACTIONS

11-5 Calculations Involving Molarity (A Review)

Let us now examine some aspects of the quantitative relationships that are useful in describing acid-base reactions in aqueous solutions. We begin with a review of the use of molarity as a unit of concentration.

Example 11-4: A volume of 30.00 mL of 0.100 M NaOH requires 25.00 mL of an HCl solution for complete neutralization. What is the molarity of the HCl solution?

Recall that an acid reacts with a base to produce a salt and water. The equation for this reaction is:

	HCl	+	$NaOH$	$\longrightarrow$	$NaCl$	+	H_2O
rxn ratio:	1 mol		1 mol		1 mol		1 mol

Since 1 mole of HCl reacts with 1 mole of NaOH, 25.0 mL of the HCl solution must contain the same number of moles of HCl as 30.0 mL of the NaOH solution contains moles of NaOH.

$$\text{mL} \times \underline{M} = \text{no mmol solute}$$

$$30.0 \text{ mL} \times 0.100\ \underline{M}\ \text{NaOH} = 3.00 \text{ mmol NaOH}$$

Therefore, 25.0 mL of HCl solution must contain 3.00 mmol of HCl.

$$25.0 \text{ mL} \times \underline{M}_{HCl} = 3.00 \text{ mmol HCl}$$

$$\underline{M}_{HCl} = \frac{3.00 \text{ mmol HCl}}{25.0 \text{ mL}} = 0.120\ \underline{M}\ \text{HCl}$$

<u>Example 11-5:</u> Refer to Example 11-4. What is the molarity of the NaCl in the resulting solution?

Inspection of the following relationships shows that 3.00 mmol of NaCl are produced.

	HCl	+	$NaOH$	$\longrightarrow$	$NaCl$	+	H_2O
rxn ratio:	1 mol		1 mol		1 mol		1 mol
start:	3.00 mmol		3.00 mmol		0 mmol		
after rxn:	0 mmol		0 mmol		3.00 mmol		

After reaction there are 3.00 mmol of NaCl in 25.0 mL + 30.0 mL = 55.0 mL of solution.

$$\underline{M}_{NaCl} = \frac{3.00 \text{ mmol NaCl}}{55.0 \text{ mL}} = \underline{\underline{0.0545\ \underline{M}\ \text{NaCl}}}$$

<u>Example 11-6:</u> What are the molarities of HBr, KOH and KBr after reaction of 20.0 mL of 0.100 $\underline{M}$ HBr with 15.0 mL of 0.100 $\underline{M}$ KOH?

Let us first determine the initial numbers of mmol of HBr and KOH.

$$\text{initial mmol HBr} = 20.0 \text{ mL} \times 0.100\ \underline{M} = 2.00 \text{ mmol HBr}$$

$$\text{initial mmol KOH} = 15.0 \text{ mL} \times 0.100\ \underline{M} = 1.50 \text{ mmol KOH}$$

All of the KOH (limiting reagent) is consumed in the reaction.

	HBr	+	KOH	→	KBr	+	H_2O
rxn ratio:	1 mol		1 mol		1 mol		
start:	2.00 mmol		1.50 mmol		0 mmol		
after rxn:	0.50 mmol		0 mmol		1.50 mmol		

The total volume of the solution after reaction is 20.0 mL + 15.0 mL = 35.0 mL. We can now calculate the final concentrations.

$$\underline{M}_{HBr} = \frac{0.50 \text{ mmol}}{35.0 \text{ mL}} = \underline{\underline{0.014\ \underline{M}\ HBr}}$$

$$\underline{M}_{KOH} = \frac{0 \text{ mmol}}{35.0 \text{ mL}} = \underline{\underline{0\ \underline{M}\ KOH}}$$

$$\underline{M}_{KBr} = \frac{1.50 \text{ mmol}}{35.0 \text{ mL}} = \underline{\underline{0.0429\ \underline{M}\ KBr}}$$

11-6 Standardization and Acid-Base Titrations

Standardization of acids or bases is the process by which the concentration of a solution of an acid or base is determined accurately by allowing a known volume to react with an accurately determined mass of a primary standard base or acid, or with an accurately determined volume of a solution of base or acid of known concentration. The following examples illustrate the process.

1. The Mole Method and Molarity

Example 11-7: Potassium hydrogen phthalate (KHP, 1 mol = 204.2 g), which has one ionizable hydrogen, is often used as a primary standard for the standardization of solutions of bases. A volume of 30.00 mL of $Ba(OH)_2$ solution reacts with 171 mg of KHP. What is the molarity of the NaOH solution?

$Ba(OH)_2$	+	2 KHP	⟶	K_2P	+	BaP	+	2 H_2O
1 mmol		2 mmol 0.2042 g						

We first calculate how many millimoles of $Ba(OH)_2$ react with 0.171 g KHP.

$$\underline{?} \text{ mol } Ba(OH)_2 = 0.171 \text{ g KHP} \times \frac{1 \text{ mmol KHP}}{0.2042 \text{ g KHP}} \times \frac{1 \text{ mmol } Ba(OH)_2}{2 \text{ mmol KHP}}$$

$$= 0.419 \text{ mmol } Ba(OH)_2$$

Now we calculate the molarity of the $Ba(OH)_2$ solution.

$$\frac{?\ \text{mmol Ba(OH)}_2}{\text{mL}} = \frac{0.419\ \text{mmol Ba(OH)}_2}{30.00\ \text{mL}} = \underline{0.0140\ \underline{M}\ \text{Ba(OH)}_2}$$

Example 11-8: What volume of 0.0210 M H_2SO_4 is required to neutralize 25.0 mL of 0.0176 M NaOH solution?

H_2SO_4	+	2 NaOH	$\longrightarrow$	Na_2SO_4	+	2 H_2O
1 mmol		2 mmol				

First we calculate the number of millimoles (or moles) of NaOH involved.

$$?\ \text{mmol NaOH} = 25.0\ \text{mL} \times \frac{0.0176\ \text{mmol NaOH}}{\text{mL}} = 0.440\ \text{mmol NaOH}$$

Now we calculate the number of millimoles of H_2SO_4 required,

$$?\ \text{mmol H}_2\text{SO}_4 = 0.440\ \text{mmol NaOH} \times \frac{1\ \text{mmol H}_2\text{SO}_4}{2\ \text{mmol NaOH}} = 0.220\ \text{mol H}_2\text{SO}_4$$

and finally the volume of H_2SO_4 solution that contains 0.220 mmol H_2SO_4.

$$?\ \text{mL} = 0.220\ \text{mol H}_2\text{SO}_4 \times \frac{1\ \text{mL}}{0.0210\ \text{mmol H}_2\text{SO}_4} = \underline{\underline{10.5\ \text{mL}}}$$

Example 11-9: What is the molarity of a solution of KOH if 35.0 mL of it neutralize 26.2 mL of 0.0100 M phosphoric acid, H_3PO_4?

3 KOH	+	H_3PO_4	$\longrightarrow$	K_3PO_4	+	3 H_2O
3 mmol		1 mmol				

The number of millimoles of H_3PO_4 neutralized is calculated first.

$$?\ \text{mmol H}_3\text{PO}_4 = 26.2\ \text{mL} \times \frac{0.0100\ \text{mmol H}_3\text{PO}_4}{\text{mL}} = 0.262\ \text{mmol H}_3\text{PO}_4$$

Now we calculate the number of millimoles of KOH that must react,

$$?\ \text{mmol KOH} = 0.262\ \text{mmol H}_3\text{PO}_4 \times \frac{3\ \text{mmol KOH}}{1\ \text{mmol H}_3\text{PO}_4} = 0.786\ \text{mmol KOH}$$

and finally the molar concentration of the KOH solution.

$$\frac{?\ \text{mmol KOH}}{\text{mL}} = \frac{0.786\ \text{mmol}}{35.0\ \text{mL}} = \underline{0.0225\ \underline{M}\ \text{KOH}}$$

2. Equivalent Weights and Normality

The concept of the equivalent weight finds application mainly in the study of acid-base reactions and oxidation-reduction (redox) reactions (Chapter 12).

One equivalent weight (eq) of an acid is the mass of the acid that produces 6.02×10^{23} hydronium ions or reacts with 6.02×10^{23} hydroxide ions. One equivalent weight of a base is the mass of the base that provides 6.02×10^{23} hydroxide ions or reacts with 6.02×10^{23} hydronium ions.

The equivalent weight of an acid or base cannot be determined just by looking at the formula of the compound (unless we assume complete neutralization). It depends upon the reaction in which the compound is used. Consider the following examples.

Example 11-10: Calculate the equivalent weights of the acids and bases in the following reactions.

(a) $HCl + H_2O \longrightarrow H_3O^+ + Cl^-$

(b) $Ba(OH)_2 \xrightarrow{H_2O} Ba^{2+} + 2\ OH^-$

(c) $H_2SO_4 + 2\ H_2O \longrightarrow 2\ H_3O^+ + SO_4^{2-}$

(d) $2\ HBr + Ca(OH)_2 \longrightarrow CaBr_2 + 2\ H_2O$

(e) $H_3PO_4 + KOH \longrightarrow KH_2PO_4 + H_2O$

	Acid		Base				
(a)	HCl	+	H_2O	$\longrightarrow$	H_3O^+	+	Cl^-
	1 mol		1 mol		1 mol		1 mol
	36.5 g		18.0 g		19.0 g		35.5 g
	6.02×10^{23}		6.02×10^{23}		6.02×10^{23}		6.02×10^{23}
	molecules		molecules		ions		ions
	1 eq		1 eq		1 eq		1 eq

Since one mole of HCl produces one mole of H_3O^+, one mol = one eq = 36.5 g HCl. Since one mole of H_2O (base) reacts with one mole of hydrogen ions, one mol = 18.0 g = 1 eq H_2O.

(b)

$Ba(OH)_2$	$\longrightarrow$	Ba^{2+}	+	$2\ OH^-$
1 mol		1 mol		2 mol
171.3 g		137.3 g		34.0 g
6.02×10^{23}		6.02×10^{23}		$2(6.02 \times 10^{23})$
formula units		ions		ions
2 eq		2 eq		2 eq

Since one mole of $Ba(OH)_2$ ionizes to produce two moles of OH^-, there are 2 eq in one mole. Or, 1 eq $Ba(OH)_2$ = 1/2 mol $Ba(OH)_2$ = 171.3 g/2 = 85.7 g.

(c)

H_2SO_4	+	$2\ H_2O$	$\longrightarrow$	$2\ H_3O^+$	+	SO_4^{2-}
1 mol		2 mol		2 mol		1 mol
98.1 g		36.0 g		38.0 g		96.1 g
6.02×10^{23}		$2(6.02 \times 10^{23})$		$2(6.02 \times 10^{23})$		6.02×10^{23}
molecules		molecules		ions		ions
2 eq		2 eq		2 eq		2 eq

One eq H_2SO_4 = 1/2 mol H_2SO_4 = 49.0 g H_2SO_4.
One eq H_2O = one mol H_2O = 18.0 g H_2O.

(d)

$2\ HBr$	+	$Ca(OH)_2$	$\longrightarrow$	$CaBr_2$	+	$2\ H_2O$
2 mol		1 mol		1 mol		2 mol
161.8 g		74.1 g		199.9 g		36.0 g
$2(6.02 \times 10^{23})$		6.02×10^{23}		6.02×10^{23}		$2(6.02 \times 10^{23})$
molecules		formula units		formula units		molecules
2 eq		2 eq		2 eq		2 eq

One mole of HBr supplies one mole of H_3O^+ and one mole of $Ca(OH)_2$ supplies two moles of OH^-. So 1 mol HBr = 1 eq HBr = 80.9 g, and 1 eq $Ca(OH)_2$ = 1/2 mol $Ca(OH)_2$ = 37.05 g.

(e)

H_3PO_4	+	KOH	$\longrightarrow$	KH_2PO_4	+	H_2O
1 mol		1 mol		1 mol		1 mol
98.0 g		56.1 g		136.1 g		18.0 g
1 eq		1 eq		1 eq		1 eq

<u>In this reaction</u>, H_3PO_4 is only one-third neutralized to form the acidic salt, KH_2PO_4.

1 eq H_3PO_4 = 1 mol H_3PO_4 = 98.0 g H_3PO_4.
1 eq KOH = 1 mol KOH = 56.1 g KOH.

Examintion of (e) shows that the equivalent weight of an acid or base depends upon the reaction in which it participates. We also note that in each case the number of eq of acid always equals the number of eq of the base with which it reacts, even though the numbers of moles may not be equal. Herein lies the usefulness of the definition of the equivalent weight.

For any acid-base reaction: <u>no. of eq of acid = no. of eq of base</u>

The normality ($\underline{N}$) of a solute is the number of eq of solute per liter of solution, or the number of milliequivalent weights (meq) of solute per milliliter of solution.

$$\underline{N} = \frac{\text{no. of gew solute}}{\text{L of sol'n}} \quad \text{or} \quad \underline{N} = \frac{\text{no. of meq solute}}{\text{mL of sol'n}}$$

Expressing concentration in terms of normality is particularly useful for acid-base (and as we shall see, redox) reactions. Since volume times concentration equals the amount of solute in a solution,

$$\text{L} \times \underline{N} = \text{no. of eq solute} \quad \text{or} \quad \text{mL} \times \underline{N} = \text{no. of meq solute}$$

For any acid-base reaction:

$$\text{no. eq acid} = \text{no. eq base}$$

so

$$L_{acid} \times \underline{N}_{acid} = L_{base} \times \underline{N}_{base}$$

or

$$mL_{acid} \times \underline{N}_{acid} = mL_{base} \times \underline{N}_{base}$$

<u>Example 11-11:</u> What are the normality and molarity of a solution of H_2SO_4 if 30.0 mL of it are required to neutralize 24.0 mL of 0.200 $\underline{M}$ NaOH solution?

$$H_2SO_4 + 2\ NaOH \longrightarrow Na_2SO_4 + 2\ H_2O$$

We know that 0.200 $\underline{M}$ NaOH = 0.200 $\underline{N}$ NaOH because 1 mol NaOH = 1 eq NaOH.

$$mL_{acid} \times \underline{N}_{acid} = mL_{base} \times \underline{N}_{base}$$

$$\underline{N}_{acid} = \frac{mL_{base} \times \underline{N}_{base}}{mL_{acid}} = \frac{24.0\text{ mL} \times 0.200\text{ N}}{30.0\text{ mL}} = \underline{\underline{0.160\ \underline{N}\ H_2SO_4}}$$

Now we calculate molarity.

$$\frac{?\ \text{mol } H_2SO_4}{L} = \frac{0.160\ \text{eq}}{L} \times \frac{1\ \text{mol}}{2\ \text{eq}} = \underline{\underline{0.080\ \underline{M}\ H_2SO_4}}$$

Note the molarity must always be equal to or less than normality since one mole of any acid or base is at least one eq. In general terms, we may write:

$$\underline{M} = \underline{N} \times \frac{1\ \text{mol}}{\text{no. of eq}} \qquad \text{or} \qquad \underline{N} = \underline{M} \times \frac{\text{no. of eq}}{1\ \text{mol}}$$

Example 11-12: How many milliliters of 0.0300 $\underline{M}$ $Ca(OH)_2$ are required to react with 25.0 mL of 0.100 $\underline{M}$ HCl?

First convert the molarity of $Ca(OH)_2$ into normality. (0.100 $\underline{M}$ HCl = 0.100 $\underline{N}$)

$$\frac{?\ \text{eq } Ca(OH)_2}{L} = \frac{0.0300\ \text{mol } Ca(OH)_2}{L} \times \frac{2\ \text{eq}}{\text{mol}} = 0.0600\ \underline{N}\ Ca(OH)_2$$

$$mL_{acid} \times \underline{N}_{acid} = mL_{base} \times \underline{N}_{base}$$

$$mL_{base} = \frac{mL_{acid} \times \underline{N}_{acid}}{\underline{N}_{base}} = \frac{25.0\ \text{mL} \times 0.100\ \underline{N}}{0.0600\ \underline{N}} = \underline{\underline{41.7\ \text{mL } Ca(OH)_2}}$$

Example 11-13: Dry sodium carbonate is often used for the standardization of solutions of acids. A volume of 25.00 mL of HNO_3 solution reacts with 21.2 mg of Na_2CO_3. What are the normality and molarity of the HNO_3 solution?

$$2\ HNO_3 + Na_2CO_3 \longrightarrow 2\ NaNO_3 + CO_2 + H_2O$$

2 mmol	1 mmol	
2 meq	2 meq	
	0.1060 g	

$$\text{no of meq } HNO_3 = mL_{HNO_3} \times \underline{N}_{HNO_3}$$

$$\text{no of meq } Na_2CO_3 = 21.2\ \text{mg } Na_2CO_3 \times \frac{2\ \text{meq } Na_2CO_3}{106\ \text{mg } Na_2CO_3}$$

Since no of meq HNO_3 = no of meq Na_2CO_3

then, $$mL_{HNO_3} \times \underline{N}_{HNO_3} = 21.2\ \text{mg } Na_2CO_3 \times \frac{2\ \text{meq } Na_2CO_3}{106\ \text{mg } NaCO_3}$$

$$= 0.400 \text{ meq } Na_2CO_3 \quad (\text{also } HNO_3)$$

$$25.00 \text{ mL} \times \underline{N}_{HNO_3} = 0.400 \text{ meq } HNO_3$$

$$\underline{N}_{HNO_3} = \frac{0.400 \text{ meq}}{25.00 \text{ mL}} = 0.0160 \ \underline{N} \ HNO_3$$

Now we calculate the molarity.

$$\frac{? \text{ mol } HNO_3}{L} = \frac{0.0160 \text{ eq } HNO_3}{L} \times \frac{1 \text{ mol } HNO_3}{1 \text{ eq } HNO_3} = 0.0160 \ \underline{M} \ HNO_3$$

EXERCISES

Draw dot formulas for reactants and products in the equations below. For species exhibiting resonance show only one dot formula. Classify the reactions as one or more of the following kinds of acid-base reactions: (1) Arrhenius, (2) Bronsted-Lowry, and (3) Lewis. Identify the acids and bases and for those that are Bronsted-Lowry acid-base reactions, indicate conjugate acid-base pairs.

1. HNO_2 (aq) + NH_3 (aq) $\longrightarrow$ NH_4NO_2 (aq)
2. $SnCl_4$ (ℓ) + 2 HCl (aq) $\longrightarrow$ H_2SnCl_6 (aq) (consists of 2 H^+ and $SnCl_6^{2-}$ ions)
3. 2 HBr (aq) + $Ca(OH)_2$ (aq) $\longrightarrow$ $CaBr_2$ (aq) + 2 H_2O (ℓ)
4. HI (aq) + H_2O (ℓ) $\longrightarrow$ H_3O^+ (aq) + I^- (aq)
5. MgO (s) + SO_2 (g) $\longrightarrow$ $MgSO_3$ (s) (not in aq. solution)

Write (a) molecular, (b) total ionic and (c) net ionic equations for the acid-base reactions (in aqueous solution) indicated below. Classify each acid and base as soluble, insoluble, strong, weak, etc. In each case, assume the product is a normal salt. (Keep the solubility rules in mind.)

6. HBr + NaOH $\longrightarrow$
7. $Mg(OH)_2$ + HNO_3 $\longrightarrow$
8. H_2SO_4 + NH_3 $\longrightarrow$
9. CH_3NH_2 + HNO_3 $\longrightarrow$
10. $HClO_4$ + $Cu(OH)_2$ $\longrightarrow$
11. $Ba(OH)_2$ + H_2SO_4 $\longrightarrow$
12. H_2CO_3 + KOH $\longrightarrow$
13. HF + NH_3 $\longrightarrow$
14. HClO + NaOH $\longrightarrow$
15. H_3PO_4 + $Ca(OH)_2$ $\longrightarrow$
16. $Sr(OH)_2$ + H_2CO_3 $\longrightarrow$

Write (a) molecular, (b) total ionic and (c) net ionic equations for acid-base reactions that produce the following salts.

17. $NaNO_3$
18. NH_4Br
19. $CuBr_2$
20. $KHCO_3$
21. NaI
22. $AlPO_4$

23. What is the molarity of the normal salt produced from the mixing of the following pairs of solutions?

    (a) 300 mL of 1.00 <u>M</u> HBr and 150 mL of 1.00 <u>M</u> $Ba(OH)_2$
    (b) 250 mL of 0.100 <u>M</u> HNO_3 and 250 mL of 0.0500 <u>M</u> $Ca(OH)_2$
    (c) 500 mL of 3.00 <u>M</u> NaOH and 100 mL of 5.00 <u>M</u> H_3PO_4

24. What are the molarities of salt and excess acid or base resulting from mixing the following solutions?

(a) 100 mL of 2.00 $\underline{M}$ NaOH and 100 mL of 1.50 $\underline{M}$ HCl
(b) 400 mL of 1.00×10^{-2} $\underline{M}$ RbOH and 250 mL of 0.100 $\underline{M}$ HNO_3
(c) 300 mL of 0.200 $\underline{M}$ KOH and 100 mL of 0.100 $\underline{M}$ H_3PO_4

25. Calculate the volume of 8.00 M KOH required to prepare 200 mL of 0.250 $\underline{M}$ KOH.

26. What volume of 0.125 M H_2SO_4 would be required to react completely with 50.0 mL of 0.100 M KOH according to the equation below?

$$2\ KOH + H_2SO_4 \longrightarrow K_2SO_4 + 2\ H_2O$$

27. A 25.0 mL sample of 0.100 M HCl solution requires 33.5 mL of $Ca(OH)_2$ solution of neutralization. What is the molarity of the $Ca(OH)_2$ solution?

28. What mass of KHP is required to just react with 30.0 mL of 0.0250 M $Ba(OH)_2$? The $Ba(OH)_2$ is to be used in the reaction below.

$$H_2SO_4 + Ba(OH)_2 \longrightarrow BaSO_4 + 2\ H_2O$$

29. A 0.5240 g sample of impure solid Na_2CO_3 reacts with 43.60 mL of 0.1077 M H_2SO_4 according to the equation below.

$$H_2SO_4 + Na_2CO_3 \longrightarrow Na_2SO_4 + CO_2 + H_2O$$

30. Calculate the equivalent weight of each acid and base in the following reactions.

(a) $2\ HI + Ba(OH)_2 \longrightarrow BaI_2 + 2\ H_2O$

(b) $H_2SO_4 + NaOH \longrightarrow NaHSO_4 + H_2O$

(c) $H_2SO_4 + 2\ NaOH \longrightarrow Na_2SO_4 + 2\ H_2O$

(d) $2\ H_3AsO_4 + 3\ Ca(OH)_2 \longrightarrow Ca_3(AsO_4)_2 + 6\ H_2O$

(e) $H_3AsO_4 + 2\ KOH \longrightarrow K_2HAsO_4 + 2\ H_2O$

(f) $H_3PO_4 + 3\ NaOH \longrightarrow Na_3PO_4 + 3\ H_2O$

(g) $H_3PO_4 + NH_3 \longrightarrow (NH_4)H_2PO_4$

31. How many equivalent weights are contained in the following?

(a) 23.0 g HI in reaction 30 (a)?
(b) 44.0 g of $Ba(OH)_2$ in reaction 30 (a)?
(c) 55.0 g of H_2SO_4 in reaction 30 (b)?
(d) 55.0 g of H_2SO_4 in reaction 30 (c)?
(e) 10.0 moles of $Ca(OH)_2$ in reaction 30 (d)?
(f) 5.0 moles of H_3AsO_4 in reaction 30 (d)?
(g) 5.0 moles of H_3AsO_4 in reaction 30 (e)?

32. Calculate the volume of 17.8 M H_2SO_4 required to prepare one liter of 0.100 N H_2SO_4 for (a) the reaction,

$$H_2SO_4 + NaOH \longrightarrow NaHSO_4 + H_2O$$

and (b) the reaction

$$H_2SO_4 + 2\,NaOH \longrightarrow Na_2SO_4 + 2\,H_2O$$

33. Calculate the volume of 6.00 M H_3PO_4 required to prepare 500 mL of 0.200 N H_3PO_4 for the reactions below:

(a) $H_3PO_4 + NaOH \longrightarrow NaH_2PO_4 + H_2O$
(b) $H_3PO_4 + 2\,NaOH \longrightarrow Na_2HPO_4 + 2\,H_2O$
(c) $H_3PO_4 + 3\,NaOH \longrightarrow Na_3PO_4 + 3\,H_2O$

34. What are the molarities of the resulting solutions of 33 (a), (b), and (c)?

35. What volume of 0.480 M H_3PO_4 would be required to react with 40.0 mL of 0.600 N NaOH according to the equation below?

$$H_3PO_4 + 3\,NaOH \longrightarrow Na_3PO_4 + 3\,H_2O$$

36. A 30.00 mL sample of an aqueous ammonia, NH_3, solution reacts with (is titrated with) 40.20 mL of 0.2104 N hydrochloric acid. What is the molarity of the aqueous ammonia solution?

37. Calculate the molarity and normality of a phosphoric acid solution if a 20.0 mL sample of it produces 0.5623 g of Ag_3PO_4. The phosphoric acid is to be used in the reaction below.

$$H_3PO_4 + 3\,NaOH \longrightarrow Na_3PO_4\,(aq) + 3\,H_2O$$

The standardization reaction is:

$$H_3PO_4 + 3\,AgNO_3 \longrightarrow Ag_3PO_4\,(s) + 3\,HNO_3$$

38. Refer to problem 37. What would be the normality if the acid were used for the reaction below?

$$H_3PO_4 + 2\,NaOH \longrightarrow Na_2HPO_4 + 2\,H_2O$$

39. Calculate the normality of an HBr solution if 31.60 mL of it reacts with exactly 0.6414 g of silver nitrate according to the equation below.

$$HBr + AgNO_3 \longrightarrow AgBr \text{ (s)} + HNO_3$$

40. Refer to the reaction of problem 39. A 27.80 mL sample of an HBr solution reacts with an excess of silver nitrate to produce 0.9762 g of insoluble silver bromide. What is the normality of the HBr solution?

ANSWERS FOR EXERCISES

1. H:Ö:N::Ö: + H:N:H $\longrightarrow$ H:N:H$^+$ + :Ö:N::Ö:$^-$

                    H   H

    acid₁   base₂   acid₂   base₁

Bronsted-Lowry and Lewis

2.   :Cl:
 :Cl:Sn:Cl: + 2 H:Cl: $\longrightarrow$ 2 H$^+$,   :Cl:
    :Cl: Cl Sn Cl

    acid   base   (Lewis only)

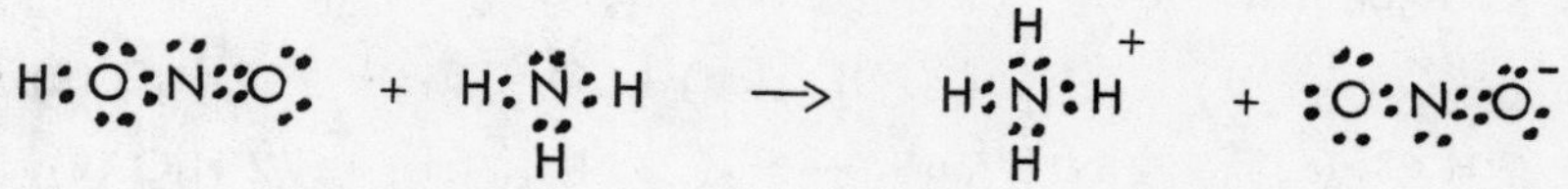

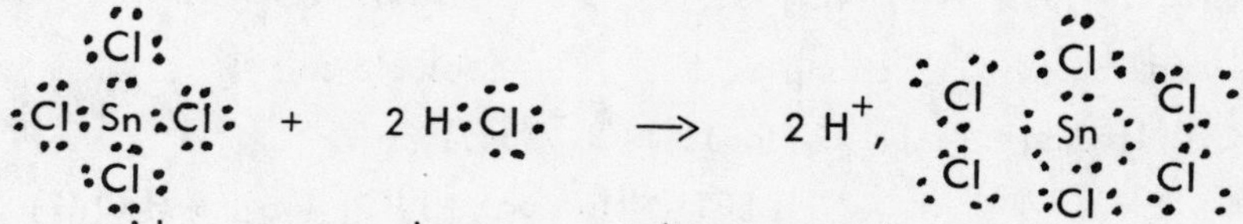

3. 2 H:Br: + Ca^{2+}, 2:Ö:H$^-$ $\longrightarrow$ Ca^{2+}, 2 :Br:$^-$ + 2 H:Ö:

                                                            H

    acid₁   base₂   base₁   acid₂

Arrhenius, Bronsted-Lowry and Lewis

4. H:I: + H:Ö: $\longrightarrow$ H:Ö:$^+$ + :I:$^-$

                    H   H

    acid₁   base₂   acid₂   base₁

Bronsted-Lowry and Lewis

5. Mg^{2+}, :Ö:$^{2-}$ + :S::Ö: $\longrightarrow$ Mg^{2+}, :Ö:S:Ö:$^{2-}$

                        :Ö: :Ö:

    base   acid   (Lewis only)

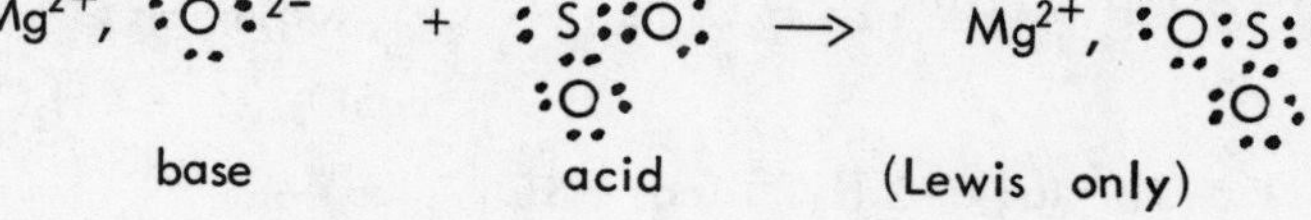

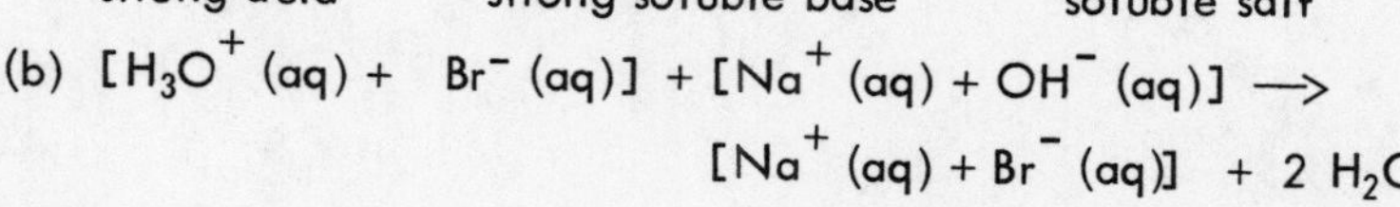

6. (a) HBr (aq) + NaOH (aq) $\longrightarrow$ NaBr (aq) + H_2O (l)

    strong acid   strong soluble base   soluble salt

    (b) [H_3O^+ (aq) + Br$^-$ (aq)] + [Na$^+$ (aq) + OH$^-$ (aq)] $\longrightarrow$

                             [Na$^+$ (aq) + Br$^-$ (aq)] + 2 H_2O (l)

6. (c) $H_3O^+ (aq) + OH^- (aq) \longrightarrow 2 H_2O (\ell)$

7. (a) $Mg(OH)_2 (s) + 2 HNO_3 (aq) \longrightarrow Mg(NO_3)_2 (aq) + 2 H_2O (\ell)$

insoluble base — strong acid — soluble salt

(b) $Mg(OH)_2 (s) + 2[H_3O^+ (aq) + NO_3^- (aq)] \longrightarrow$

$[Mg^{2+} (aq) + 2 NO_3^- (aq)] + 4 H_2O (\ell)$

(c) $Mg(OH)_2 (s) + 2 H_3O^+ (aq) \longrightarrow Mg^{2+} (aq) + 4 H_2O (\ell)$

8. (a) $H_2SO_4 (aq) + 2 NH_3 (aq) \longrightarrow (NH_4)_2SO_4 (aq)$

strong acid — weak base — soluble salt

(b) $[2 H_3O^+ (aq) + SO_4^{2-} (aq)] + 2 NH_3 (aq) \longrightarrow$

$[2 NH_4^+ (aq) + SO_4^{2-} (aq)] + 2 H_2O (\ell)$

(c) $2 H_3O^+ (aq) + 2 NH_3 (aq) \longrightarrow 2 NH_4^+ (aq) + 2 H_2O (\ell)$

or $H_3O^+ (aq) + NH_3 (aq) \longrightarrow NH_4^+ (aq) + H_2O (\ell)$

9. (a) $CH_3NH_2 (aq) + HNO_3 (aq) \longrightarrow CH_3NH_3NO_3$

weak base — strong acid — soluble salt

(b) $CH_3NH_2 (aq) + [H_3O^+ (aq) + NO_3^- (aq)] \longrightarrow$

$[CH_3NH_3^+ (aq) + NO_3^- (aq)] + H_2O (\ell)$

(c) $CH_3NH_2 (aq) + H_3O^+ (aq) \longrightarrow CH_3NH_3^+ (aq) + H_2O (\ell)$

10. (a) $2 HClO_4 (aq) + Cu(OH)_2 (s) \longrightarrow Cu(ClO_4)_2 + 2 H_2O (\ell)$

strong acid — insoluble base — soluble salt

(b) $2 [H_3O^+ (aq) + ClO_4^- (aq)] + Cu(OH)_2 (s) \longrightarrow$

$[Cu^{2+} (aq) + 2 ClO_4^- (aq)] + 4 H_2O (\ell)$

(c) $2 H_3O^+ (aq) + Cu(OH)_2 (s) \longrightarrow Cu^{2+} (aq) + 4 H_2O (\ell)$

11. (a) $Ba(OH)_2 (aq) + H_2SO_4 (aq) \longrightarrow BaSO_4 (s) + 2 H_2O (\ell)$

strong soluble base — strong acid — insoluble salt

(b) $[Ba^{2+} (aq) + 2 OH^- (aq)] + [2 H_3O^+ (aq) + SO_4^{2-} (aq)] \longrightarrow$

$BaSO_4 (s) + 4 H_2O (\ell)$

(c) $Ba^{2+} (aq) + 2 OH^- (aq) + 2 H_3O^+ (aq) + SO_4^{2-} (aq) \longrightarrow$

$BaSO_4 (s) + 4 H_2O (\ell)$

12. (a) H_2CO_3 (aq) + 2 KOH (aq) ⟶ K_2CO_3 (aq) + 2 H_2O (ℓ)

weak acid — strong soluble base — soluble salt

(b) H_2CO_3 (aq) + 2[K^+ (aq) + OH^- (aq)] ⟶

[2 K^+ (aq) + CO_3^{2-} (aq)] + 2 H_2O (ℓ)

(c) H_2CO_3 (aq) + 2 OH^- (aq) ⟶ CO_3^{2-} (aq) + 2 H_2O (ℓ)

13. (a) HF (aq) + NH_3 (aq) ⟶ NH_4F (aq)

weak acid — weak base — soluble salt

(b) HF (aq) + NH_3 (aq) ⟶ [NH_4^+ (aq) + F^- (aq)]

(c) HF (aq) + NH_3 (aq) ⟶ NH_4^+ (aq) + F^- (aq)

14. (a) HClO (aq) + NaOH (aq) ⟶ NaClO (aq) + H_2O (ℓ)

weak acid — strong soluble base — soluble salt

(b) HClO (aq) + [Na^+ (aq) + OH^- (aq)] ⟶

[Na^+ (aq) + ClO^- (aq)] + H_2O (ℓ)

(c) HClO (aq) + OH^- (aq) ⟶ ClO^- (aq) + H_2O (ℓ)

15. (a) 2 H_3PO_4 (aq) + 3 $Ca(OH)_2$ (aq) ⟶ $Ca_3(PO_4)_2$ (s) + 6 H_2O (ℓ)

weak acid — strong soluble base — insoluble salt

(b) 2 H_3PO_4 (aq) + 3 [Ca^{2+} (aq) + 2 OH^- (aq)] ⟶

$Ca_3(PO_4)_2$ (aq) + 6 H_2O (ℓ)

(c) 2 H_3PO_4 (aq) + 3 Ca^{2+} (aq) + 6 OH^- ⟶ $Ca_3(PO_4)_2$ (aq) + 6 H_2O (ℓ)

16. (a) $Sr(OH)_2$ (aq) + H_2CO_3 (aq) ⟶ $SrCO_3$ (s) + 2 H_2O (ℓ)

strong soluble base — weak acid — insoluble salt

(b) [Sr^{2+} (aq) + 2 OH^- (aq)] + H_2CO_3 (aq) ⟶ $SrCO_3$ (s) + 2 H_2O (ℓ)

(c) Sr^{2+} (aq) + 2 OH^- (aq) + H_2CO_3 (aq) ⟶ $SrCO_3$ (s) + 2 H_2O (ℓ)

17. (a) NaOH (aq) + HNO_3 (aq) ⟶ $NaNO_3$ (aq) + H_2O (ℓ)

(b) [Na^+ (aq) + OH^- (aq)] + [H_3O^+ (aq) + NO_3^- (aq)] ⟶

[Na^+ (aq) + NO_3^- (aq)] + 2 H_2O (ℓ)

(c) OH^- (aq) + H_3O^+ (aq) ⟶ 2 H_2O (ℓ)

18. (a) NH_3 (aq) + HBr (aq) $\longrightarrow$ NH_4Br (aq)

(b) NH_3 (aq) + [H_3O^+ (aq) + Br^- (aq)] $\rightarrow$ [NH_4^+ (aq) + Br^- (aq)] + H_2O (ℓ)

(c) NH_3 (aq) + H_3O^+ (aq) $\longrightarrow$ NH_4^+ (aq) + H_2O (ℓ)

19. (a) $Cu(OH)_2$ (s) + 2 HBr (aq) $\longrightarrow$ $CuBr_2$ (aq) + 2 H_2O (ℓ)

(b) $Cu(OH)_2$ (s) + 2 [H_3O^+ (aq) + Br^- (aq)] $\rightarrow$

[Cu^{2+} (aq) + 2 Br^- (aq)] + 4 H_2O (ℓ)

(c) $Cu(OH)_2$ (s) + 2 H_3O^+ (aq) $\rightarrow$ Cu^{2+} (aq) + 4 H_2O (ℓ)

20. (a) KOH (aq) + H_2CO_3 (aq) $\longrightarrow$ $KHCO_3$ (aq) + H_2O (ℓ)

(b) [K^+ (aq) + OH^- (aq)] + H_2CO_3 (aq) $\longrightarrow$ [K^+ (aq) + HCO_3^- (aq)]

+ H_2O (ℓ)

(c) OH^- (aq) + H_2CO_3 (aq) $\longrightarrow$ HCO_3^- (aq) + H_2O (ℓ)

21. (a) NaOH (aq) + HI (aq) $\rightarrow$ NaI (aq) + H_2O (ℓ)

(b) [Na^+ (aq) + OH^- (aq)] + [H_3O^+ (aq) + I^- (aq)] $\longrightarrow$

[Na^+ (aq) + I^- (aq)] + 2 H_2O (ℓ).

(c) OH^- (aq) + H_3O^+ (aq) $\longrightarrow$ 2 H_2O (ℓ)

22. (a) $Al(OH)_3$ (s) + H_3PO_4 (aq) $\longrightarrow$ $AlPO_4$ (s) + 3 H_2O (ℓ)

(b) and (c) are the same as (a) because $Al(OH)_3$ is an insoluble base, H_3PO_4 is a weak acid and $AlPO_4$ is an insoluble salt.

23. (a) 0.333 M $BaBr_2$ (b) 0.0250 M $Ca(NO_3)_2$ (c) 0.833 M Na_3PO_4

24. (a) 0.750 M NaCl, 0.250 M NaOH (b) 6.15×10^{-3} M $RbNO_3$, 0.0323 M HNO_3, (c) 0.0250 M K_3PO_4, 0.0750 M KOH 25. 6.25 mL

26. 20.0 mL 27. 0.0373 M $Ca(OH)_2$ 28. 0.306 g KHP

29. 0.4977 g Na_2CO_3, 94.89% Na_2CO_3

30. (a) 127.9 g HI/eq, 85.7 g $Ba(OH)_2$/eq

(b) 98.1 g H_2SO_4/eq, 40.0 g NaOH/eq

(c) 49.1 g H_2SO_4/eq, 40.0 g NaOH/eq

(d) 47.3 g H_3AsO_4/eq, 37.1 g $Ca(OH)_2$/eq

(e) 71.0 g H_3AsO_4/eq, 56.1 g KOH/eq

(f) 32.7 g H_3PO_4/eq, 40.0 g NaOH/eq

(g) 98.0 g H_3PO_4/eq, 17.0 g NH_3/eq

31. (a) 0.180 eq HI (b) 0.513 eq $Ba(OH)_2$ (c) 0.561 eq H_2SO_4 (d) 1.12 eq H_2SO_4 (e) 20.0 eq $Ca(OH)_2$ (f) 15.0 eq H_3AsO_4 (g) 10.0 eq H_3AsO_4

32. (a) 5.62 mL (b) 2.81 mL

33. (a) 16.7 mL (b) 8.33 mL (c) 5.56 mL

34. (a) 0.200 M H_3PO_4 (b) 0.100 M H_3PO_4 (c) 0.0667 M H_3PO_4

35. 16.7 mL

36. 0.2819 M aq NH_3

37. 0.06717 M H_3PO_4

38. 0.1343 N H_3PO_4

39. 0.1195 N HBr

40. 0.1870 N HBr

Chapter Twelve

OXIDATION-REDUCTION REACTIONS

D,G,W: Sec. 9-7.1 and Sec. 10-5

Oxidation-reduction reactions or redox reactions are those in which elements undergo changes in oxidation number. They involve, or appear to involve electron transfer. In order to deal with redox reactions it is necessary to have a firm grasp on the rules for assigning oxidation numbers (Sections 4-12 and 12-1 in the text).

12-1 Definitions

Oxidation is an algebraic increase in oxidation number. This corresponds to a loss, or apparent loss, of electrons. Reduction is an algebraic decrease in oxidation number. This corresponds to a gain, or apparent gain, of electrons. Reducing agents are the species that are oxidized, and lose, or appear to lose, electrons. Oxidizing agents are the species that are reduced, and gain, or appear to gain, electrons.

Like many other reactions, redox reactions may be represented by (1) molecular, (2) total ionic, or (3) net ionic equations. The distinctions were described in Sections 7-3, 11-6 and 12-1 in the text.

Example 12-1: Write the following molecular equations as ionic equations, if applicable. Which ones represent redox reactions? For those that do, identify the oxidizing agent and the reducing agent.

a. $2\ Na\ (s) + Cl_2\ (g) \longrightarrow 2\ NaCl\ (s)$

b. $CO_2\ (g) + H_2O\ (\ell) \longrightarrow H_2CO_3\ (aq)$

c. $2\ K\ (s) + 2\ H_2O\ (\ell) \longrightarrow 2\ KOH\ (aq) + H_2\ (g)$

d. $10\ FeSO_4\ (aq) + 8\ H_2SO_4\ (aq) + 2\ KMnO_4\ (aq) \longrightarrow$
$+ 5\ Fe_2(SO_4)_3\ (aq) + 2\ MnSO_4\ (aq) + 8\ H_2O\ (\ell) + K_2SO_4\ (aq)$

e. $BaSO_3\ (s) \xrightarrow{\Delta} BaO\ (s) + SO_2\ (g)$

f. $2\ C_2H_6\ (g) + 7\ O_2\ (g) \xrightarrow{\Delta} 4\ CO_2\ (g) + 6\ H_2O\ (g)$

The oxidation numbers are given above each element below. Net ionic equations are written where applicable.

a. $2\overset{0}{\text{Na}}\,(s) + \overset{0}{\text{Cl}_2}\,(g) \longrightarrow 2\,\overset{+1\ -1}{\text{NaCl}}\,(s)$

reducing agent — oxidizing agent

The reaction does not occur in aqueous solution, and therefore the molecular and net ionic equations are identical. This reaction is an oxidation-reduction reaction. The oxidation number of sodium increases from zero to +1; it is oxidized and is the reducing agent. The oxidation number of chlorine decreases from zero to −1; it is reduced and is the oxidizing agent.

b. $\overset{+4-2}{\text{CO}_2}\,(g) + \overset{+1-2}{\text{H}_2\text{O}}\,(\ell) \longrightarrow \overset{+1+4-2}{\text{H}_2\text{CO}_3}\,(aq)$ (not redox)

The reaction occurs in aqueous solution, but all species are predominantly molecular. (Carbonic acid is a weak acid.)

c. $2\,\overset{0}{\text{K}}\,(s) + 2\,\overset{+1-2}{\text{H}_2\text{O}}\,(\ell) \longrightarrow 2\,[\overset{+1}{\text{K}^+}\,(aq) + \overset{-2+1}{\text{OH}^-}\,(aq)] + \overset{0}{\text{H}_2}\,(g)$

reducing agent — oxidizing agent

Potassium hydroxide is a strong soluble base, and so it is written in ionized form in ionic equations. The oxidation number of potassium increases from zero to +1; it is oxidized and is the reducing agent. Water is reduced because some of its hydrogen is reduced from the +1 to the zero oxidation state (in H_2).

d. $10\,[Fe^{2+}(aq) + SO_4^{2-}(aq)] + 8\,[2\,H^+(aq) + SO_4^{2-}(aq)] +$

$2\,[K^+(aq) + MnO_4^-(aq)] \longrightarrow 5\,[2\,Fe^{3+}(aq) + 3\,SO_4^{2-}(aq)]$

$+ 2\,[Mn^{2+}(aq) + SO_4^{2-}(aq)] + 8\,H_2O\,(\ell) + [2\,K^+(aq) + SO_4^{2-}(aq)]$

(total ionic equation)

The salts above are all water-soluble and predominantly ionized in aqueous solution. Cancellation of the spectator ions (K^+ and SO_4^{2-}) from each side and division of all coefficients by 2 gives the net ionic equation.

$\overset{+7-2}{\text{MnO}_4^-} + 8\,\overset{+1}{\text{H}^+}(aq) + 5\,\overset{+2}{\text{Fe}^{2+}}(aq) \longrightarrow \overset{+2}{\text{Mn}^{2+}}(aq) + 5\,\overset{+3}{\text{Fe}^{3+}}(aq) + 4\,\overset{+1-2}{\text{H}_2\text{O}}\,(\ell)$

oxidizing agent — reducing agent

The permanganate ion, MnO_4^-, is the oxidizing agent; the oxidation number of manganese decreases from +7 to +2. Iron(II) ion, Fe^{2+}, is the reducing agent;

it is oxidized from the +2 to the +3 oxidation state.

e. $\overset{+2}{Ba}\overset{+4}{S}\overset{-2}{O_3}\,(s) \xrightarrow{\Delta} \overset{+2}{Ba}\overset{-2}{O}\,(s) + \overset{+4}{S}\overset{-2}{O_2}\,(g)$ (not redox)

f. $2\,\overset{-3}{C_2}\overset{+1}{H_6}\,(g) + 7\,\overset{0}{O_2}\,(g) \xrightarrow{\Delta} 4\,\overset{+4}{C}\overset{-2}{O_2}\,(g) + 6\,\overset{+1}{H_2}\overset{-2}{O}\,(g)$

Ethane, C_2H_6, is the reducing agent because its carbon is oxidized from the -3 to the +4 oxidation state. Oxygen is the oxidizing agent as it is reduced from the zero to the -2 oxidation state.

BALANCING OXIDATION-REDUCTION EQUATIONS

The useful methods of balancing redox equations are (1) the change-in-oxidation number method and (2) the ion-electron method. Either one (or others) can be used to balance any redox equation. Often one method is more easily applicable to a particular equation than the other. It should be kept in mind that all balanced equations have (1) atom (mass) balance and (2) charge balance.

12-2 Change-in-Oxidation Number Method

This method is based on equal total increases and decreases in oxidation numbers. The general procedure follows.

1. Write the overall unbalanced equation.
2. Assign oxidation numbers to determine which elements undergo changes in oxidation number.
3. Insert coefficients to make the total increase and decrease in oxidation numbers equal.
4. Balance the remaining atoms by inspection.

12-3 Addition of H^+, OH^-, and/or H_2O in Balancing Redox Equations

Occasionally only enough information is provided to construct an incomplete as well as unbalanced equation. In such cases atom balances may be achieved by adding, as necessary, H^+ and/or H_2O (but not OH^-) for reactions in acidic solution or OH^- and/or H_2O (but not H^+) for reactions in basic solution.

Example 12-2: Dichromate ion, $Cr_2O_7^{2-}$, oxidizes bromide ion, Br^-, to bromine, Br_2, in acidic solution and is reduced to chromium-(III) ion, Cr^{3+}. Write the balanced net ionic equation.

The steps given above are followed sequentially and labeled below.

Step 1. $Cr_2O_7^{2-} + Br^- \longrightarrow Br_2 + Cr^{3+}$

Step 2. $\overset{+6\ -2}{Cr_2O_7^{2-}}$ + $\overset{-1}{Br^-}$ $\longrightarrow$ $\overset{0}{Br_2}$ + $\overset{+3}{Cr^{3+}}$

Step 3. Chromium and bromine undergo changes in oxidation number. The oxidation number of each chromium atom increases three units and that of each bromine atom decreases one unit.*

$$Cr^{6+} \xrightarrow{3\downarrow} 2\,Cr^{3+} \qquad Cr_2 \rightarrow 2Cr$$

$$2\,Br^{-1} \xrightarrow{1\uparrow} Br_2^0 \qquad 2Br \rightarrow Br_2$$

Since chromium atoms occur in groups of two on the left side (in $Cr_2O_7^{2-}$) two Cr^{3+} ions will be produced for every $Cr_2O_7^{2-}$ ion reduced. Likewise, two Br^- ions must be oxidized for every Br_2 molecule produced. So we multiply both "helping" equations by two, which eliminates the possibility of fractional coefficients in the balanced equation.

$$2\ Cr^{+6} \xrightarrow{6\downarrow} 2\ Cr^{+3}$$

$$2\ Br^{-1} \xrightarrow{2\uparrow} Br_2^{\,0}$$

To make the total increase in oxidation numbers equal to the total decrease in oxidation numbers, the second "helping" equation is multiplied by three.

$$2\ Cr^{+6} \xrightarrow{6\downarrow} 2\ Cr^{+3} \qquad\qquad 2\ Cr^{+6} \xrightarrow{6\downarrow} 2\ Cr^{+3}$$

or

$$3(2\ Br^{-1} \xrightarrow{2\uparrow} Br_2^{\,0}) \qquad\qquad 6\ Br^{-1} \xrightarrow{6\uparrow} 3\ Br_2$$

We now insert these coefficients into the overall equation. Each $Cr_2O_7^{2-}$ ion contains 2 Cr so only one $Cr_2O_7^{2-}$ is necessary.

$$Cr_2O_7^{2-} + 6\ Br^- \longrightarrow 3\ Br_2 + 2\ Cr^{3+} \text{ (unbalanced)}$$

Step 4. To balance oxygen atoms, 7 H_2O are placed on the right, and then 14 H^+ are placed on the left to balance hydrogen atoms.

$$14\ H^+ + Cr_2O_7^{2-} + 6\ Br^- \longrightarrow 3\ Br_2 + 2\ Cr^{3+} + 7\ H_2O \quad \text{(balanced net ionic equation)}$$

→ Pay attention & reduce.

The equation is now balanced with respect to atoms (14 H, 2 Cr, 7 O, 6 Br) and charge (6+ on each side).

---

*Keep in mind that an oxidation number such as +6 for Cr in $Cr_2O_7^{2-}$ does not imply that Cr^{6+} ions are present, only that the oxidation state of chromium is +6.

Example 12-3: In basic solution metallic manganese reduces ferric ion, Fe^{3+}, to ferrous ion, Fe^{2+}, and is reduced to manganese(II) hydroxide, $Mn(OH)_2$. Write the balanced equation for this reaction.

Step 1. $Mn + Fe^{3+} \longrightarrow Fe^{2+} + Mn(OH)_2$

Step 2. $\overset{0}{Mn} + \overset{+3}{Fe^{3+}} \longrightarrow \overset{+2}{Fe^{2+}} + \overset{+2}{Mn}(\overset{-2}{O}\overset{+1}{H})_2$

Step 3. $Mn^0 \xrightarrow{2\uparrow} Mn^{2+}$

$Fe^{+3} \xrightarrow{1\downarrow} Fe^{+2}$

$1\ (Mn^0 \xrightarrow{2\uparrow} Mn^{+2})$

$2\ (Fe^{+3} \xrightarrow{1\downarrow} Fe^{+2})$

$Mn + 2\ Fe^{+3} \longrightarrow 2\ Fe^{2+} + Mn(OH)_2$ (incomplete and unbalanced)

Step 4: (basic solution)

$Mn + 2\ Fe^{3+} + 2\ OH^- \longrightarrow 2\ Fe^{2+} + Mn(OH)_2$ (balanced net ionic equation)

Atom balance: 1 Mn, 2 Fe, 2 O, 2 H. Charge balance: 4+.

Example 12-4: Chlorous acid oxidizes sulfur dioxide to sulfate ions and is reduced to chloride ions in acidic solution. Write the balanced net ionic equation for the reaction.

Step 1. $HClO_2 + SO_2 \longrightarrow Cl^- + SO_4^{2-}$

Step 2. $\overset{+1}{H}\overset{+3}{Cl}\overset{-2}{O_2} + \overset{+4}{S}\overset{-2}{O_2} \longrightarrow \overset{-1}{Cl^-} + \overset{+6}{S}\overset{-2}{O_4^{2-}}$

Step 3. $1\ (Cl^{+3} \xrightarrow{4\downarrow} Cl^{-1})$

$2\ (S^{+4} \xrightarrow{2\uparrow} S^{+6})$

$Cl^{+3} + 2\ S^{+4} \longrightarrow Cl^{-1} + 2\ S^{+6}$

$HClO_2 + 2\ SO_2 \longrightarrow Cl^- + 2\ SO_4^{2-}$ (incomplete and unbalanced)

Step 4. (acidic solution) To balance oxygen atoms (6 O left, 8 O right), we add 2 H_2O to the left side.

$$HClO_2 + 2\ SO_2 + 2\ H_2O \rightarrow Cl^- + 2\ SO_4^{2-}$$

Hydrogen atoms (5 H left, none right) are balanced by adding 5 H^+ to the right side, which gives the balanced net ionic equation.

$$HClO_2 + 2\ SO_2 + 2\ H_2O \longrightarrow Cl^- + 2\ SO_4^{2-} + 5\ H^+$$

Atom balance: 5 H, 1 Cl, 8 O, 2 S. Charge balance: zero

Example 12-5: In basic solution permanganate ion, MnO_4^-, oxidizes metallic lead to lead(II) hydroxide, $Pb(OH)_2$, and is reduced to manganese(IV) oxide. Write the balanced net ionic equation.

Step 1. $MnO_4^- + Pb \longrightarrow Pb(OH)_2 + MnO_2$

Step 2. $\overset{+7\ -2}{MnO_4^-} + \overset{0}{Pb} \longrightarrow \overset{+2\ -2\ +1}{Pb(OH)_2} + \overset{+4\ -2}{MnO_2}$

Step 3.

$$2\,(Mn^{+7} \xrightarrow{3\downarrow} Mn^{+4})$$
$$3\,(Pb^{0} \xrightarrow{2\uparrow} Pb^{+2})$$
$$2\ Mn^{+7} + 3\ Pb \longrightarrow 2\ Mn^{+4} + 3\ Pb^{+2}$$

$$2\ MnO_4^- + 3\ Pb \longrightarrow 2\ MnO_2 + 3\ Pb(OH)_2 \quad \text{(unbalanced)}$$

Step 4. (basic solution)

on left: 8 O and no H, on right: 10 O and 6 H

To balance oxygen atoms add 2 H_2O to the left side.

$$2\ MnO_4^- + 2\ H_2O + 3\ Pb \longrightarrow 2\ MnO_2 + 3\ Pb(OH)_2 \quad \text{(unbalanced)}$$

To balance hydrogen, without disturbing oxygen balance, add 2 H_2O to left side and 2 OH^- to right side.

$$2\ MnO_4^- + 4\ H_2O + 3\ Pb \longrightarrow 2\ MnO_2 + 3\ Pb(OH)_2 + 2\ OH^-$$

(balanced net ionic equation)

Atom balance: 2 Mn, 12 O, 8 H, 3 Pb. Charge balance: 2-

12-4 The Ion-Electron Method

According to this method the oxidation and reduction half-reactions are balanced separately and completely. Then the number of electrons gained and lost are made equal and the resulting half-reactions are added to give the overall balanced equation. The procedure is given below.

1. Write as much of the overall unbalanced equation as possible.

2. Construct unbalanced oxidation and reduction half-reactions (usually incomplete as well as unbalanced).
3. Balance the atoms in each half-reaction.
4. Balance the charge in each half-reaction by adding electrons.
5. Balance the electron transfer by multiplying the balanced half-reactions by appropriate integers.
6. Add the resulting half-reactions, and eliminate any common terms to obtain the balanced equation.

Example 12-6: Nitrate ions, NO_3^-, oxidize metallic copper to copper(II) ions, Cu^{2+}, and are reduced to nitric oxide, NO, in acidic solution. Write the balanced net ionic equation for this reaction.

Step 1. $NO_3^- + Cu \longrightarrow NO + Cu^{2+}$

Step 2. $NO_3^- \longrightarrow NO$ (reduction)

$Cu \longrightarrow Cu^{2+}$ (oxidation)

Step 3. $NO_3^- + 4\,H^+ \longrightarrow NO + 2\,H_2O$ (acidic solution)

$Cu \longrightarrow Cu^{2+}$

Step 4. $NO_3^- + 4\,H^+ + 3\,e^- \longrightarrow NO + 2\,H_2O$

$Cu \longrightarrow Cu^{2+} + 2\,e^-$

Step 5. $2\,(NO_3^- + 4\,H^+ + 3\,e^- \longrightarrow NO + 2\,H_2O)$

$3\,(Cu \longrightarrow Cu^{2+} + 2\,e^-)$

Step 6. $2\,NO_3^- + 8\,H^+ + 6\,e^- + 3\,Cu \longrightarrow 2\,NO + 4\,H_2O + 3\,Cu^{2+} + 6\,e^-$

$2\,NO_3^- + 8\,H^+ + 3\,Cu \longrightarrow 2\,NO + 4\,H_2O + 3\,Cu^{2+}$

(balanced net ionic equation)

Atom balance: 2 N, 6 O, 8 H, 3 Cu. Charge balance: 6+

Example 12-7: In aqueous ammonia solution the nitrate ion, NO_3^-, oxidizes metallic cadmium to cadmium ions in the tetramminecadmium ion, $[Cd(NH_3)_4]^{2+}$, and is reduced to the nitrite ion, NO_2^-. Write the balanced net ionic equation.

Step 1. $NO_3^- + Cd + NH_3 \longrightarrow [Cd(NH_3)_4]^{2+} + NO_2^-$

Step 2. $NO_3^- \longrightarrow NO_2^-$ (reduction)

$Cd + NH_3 \longrightarrow [Cd(NH_3)_4]^{2+}$ (oxidation)

Step 3. $H_2O + NO_3^- \longrightarrow NO_2^- + 2\ OH^-$ (basic solution)

$Cd + 4\ NH_3 \longrightarrow [Cd(NH_3)_4]^{2+}$

Step 4. $H_2O + NO_3^- + 2\ e^- \longrightarrow NO_2^- + 2\ OH^-$

$Cd + 4\ NH_3 \longrightarrow [Cd(NH_3)_4]^{2+} + 2\ e^-$

Step 5. Electron transfer already balanced.

Step 6. $H_2O + NO_3^- + Cd + 4\ NH_3 \longrightarrow NO_2^- + 2\ OH^- + [Cd(NH_3)_4]^{2+}$

Atom balance: 14 H, 4 O, 5 N, 1 Cd. Charge balance: 1-

Example 12-8: In basic solution ClO^- ions oxidize $Cr(OH)_4^-$ to CrO_4^{2-} ions and are reduced to Cl^- ions. Write the balanced net ionic equation for this reaction.

Step 1. $ClO^- + Cr(OH)_4^- \longrightarrow Cl^- + CrO_4^{2-}$

Step 2. $ClO^- \longrightarrow Cl^-$ (reduction)

$Cr(OH)_4^- \longrightarrow CrO_4^{2-}$ (oxidation)

Step 3. $ClO^- + H_2O \longrightarrow Cl^- + 2\ OH^-$

$Cr(OH)_4^- + 4\ OH^- \longrightarrow CrO_4^{2-} + 4\ H_2O$

Step 4. $ClO^- + H_2O + 2\ e^- \longrightarrow Cl^- + 2\ OH^-$

$Cr(OH)_4^- + 4\ OH^- \longrightarrow CrO_4^{2-} + 4\ H_2O + 3\ e^-$

Step 5. $3(ClO^- + H_2O + 2\ e^- \longrightarrow Cl^- + 2\ OH^-)$

$2\ (Cr(OH)_4^- + 4\ OH^- \longrightarrow CrO_4^{2-} + 4\ H_2O + 3\ e^-)$

Step 6. $3\ ClO^- + 3\ H_2O + 6\ e^- + 2\ Cr(OH)_4^- + 8\ OH^- \longrightarrow 3\ Cl^- + 6\ OH^- + 2\ CrO_4^{2-} + 8\ H_2O + 6\ e^-$

$3\ ClO^- + 2\ Cr(OH)_4^- + 2\ OH^- \longrightarrow 3\ Cl^- + 2\ CrO_4^{2-} + 5\ H_2O$

Atom balance: 3 Cl, 2 Cr, 13 O, 10 H. Charge balance: 7-.

Example 12-9: In acidic solution BiO_3^- oxidizes Mn^{2+} to MnO_4^- and is reduced to Bi^{3+}. Write the balanced net ionic equation for this reaction.

Step 1. $BiO_3^- + Mn^{2+} \longrightarrow Bi^{3+} + MnO_4^-$

Step 2. $BiO_3^- \longrightarrow Bi^{3+}$ (reduction)

$Mn^{2+} \longrightarrow MnO_4^-$ (oxidation)

Step 3. $BiO_3^- + 6\ H^+ \longrightarrow Bi^{3+} + 3\ H_2O$

$Mn^{2+} + 4\ H_2O \longrightarrow MnO_4^- + 8\ H^+$

Step 4. $BiO_3^- + 6\ H^+ + 2\ e^- \longrightarrow Bi^{3+} + 3\ H_2O$

$Mn^{2+} + 4\ H_2O \longrightarrow MnO_4^- + 8\ H^+ + 5\ e^-$

Step 5. $5\ (BiO_3^- + 6\ H^+ + 2\ e^- \longrightarrow Bi^{3+} + 3\ H_2O)$

$2\ (Mn^{2+} + 4\ H_2O \longrightarrow MnO_4^- + 8\ H^+ + 5\ e^-)$

Step 6. $5\ BiO_3^- + 30\ H^+ + 10\ e^- + 2\ Mn^{2+} + 8\ H_2O \longrightarrow 5\ Bi^{3+} + 15\ H_2O$

$+ 2\ MnO_4^- + 16\ H^+ + 10\ e^-$

$5\ BiO_3^- + 14\ H^+ + 2\ Mn^{2+} \longrightarrow 5\ Bi^{3+} + 7\ H_2O + 2\ MnO_4^-$

Atom balance: 5 Bi, 15 O, 14 H, 2 Mn. Charge balance: 13+.

OXIDATION-REDUCTION TITRATIONS

In a redox titration one determines the volume of a standard solution of an oxidizing agent or reducing agent required to react with a specific amount of reducing agent or oxidizing agent. Or, as in acid-base titrations, one may determine the amount of a solution required to react with an exactly known amount of a primary standard. We may express the concentrations of solutions used in redox titrations in terms of either molarity or normality. We shall illustrate both methods, beginning with the mole method and molarity.

1. The Mole Method and Molarity

Example 12-10: What volume of 0.100 M $FeSO_4$ solution is required to react with 30.0 mL of 0.0400 M $KMnO_4$ solution?

$$2\ KMnO_4 + 8\ H_2SO_4 + 10\ FeSO_4 \longrightarrow 2\ MnSO_4 + 8\ H_2O + 5\ Fe_2(SO_4)_3 + K_2SO_4$$

To solve the problem in this way we must refer to the balanced equation to determine the reaction mole ratio.

$$2\ KMnO_4 + 8\ H_2SO_4 + 10\ FeSO_4 \longrightarrow 2\ MnSO_4 + 8\ H_2O + 5\ Fe_2(SO_4)_3 + K_2SO_4$$

2 mol   10 mol

We first calculate the number of moles of $KMnO_4$ in 30.0 mL of 0.0400 $\underline{M}$ $KMnO_4$ solution.

$$\underline{?}\ \text{mol } KMnO_4 = 30.0\ \text{mL} \times \frac{0.0400\ \text{mol } KMnO_4}{1000\ \text{mL}} = 1.20 \times 10^{-3}\ \text{mol } KMnO_4$$

Then we calculate the volume of 0.100 $\underline{M}$ $FeSO_4$ solution that just reacts with this amount of $KMnO_4$.

$$\underline{?}\ \text{mL } FeSO_4\ \text{sol'n} = 1.20 \times 10^{-3}\ \text{mol } KMnO_4 \times \frac{10\ \text{mol } FeSO_4}{2\ \text{mol } KMnO_4} \times \frac{1000\ \text{mL } FeSO_4\ \text{sol'n}}{0.100\ \text{mol } FeSO_4}$$

$$= \underline{\underline{60.0\ \text{mL } FeSO_4\ \text{sol'n}}}$$

<u>Example 12-11:</u> A 0.215 gram sample of $FeSO_4$ is oxidized by 34.2 mL of a solution of $K_2Cr_2O_7$. What is the molar concentration of the $K_2Cr_2O_7$ solution?

$$K_2Cr_2O_7 + 6\ FeSO_4 + 7\ H_2SO_4 \rightarrow 3\ Fe_2(SO_4)_3 + Cr_2(SO_4)_3 + K_2SO_4 + 7\ H_2O$$

We begin by calculating how many millimoles of $FeSO_4$ have reacted.

$$?\ \text{mmol } FeSO_4 = 0.215\ \text{g } FeSO_4 \times \frac{1\ \text{mmol } FeSO_4}{0.1519\ \text{g } FeSO_4} = 1.42\ \text{mmol } FeSO_4$$

Now we can use the information given by the balanced equation to determine how many millimoles of $K_2Cr_2O_7$ have reacted.

$$?\ \text{mmol } K_2Cr_2O_7 = 1.42\ \text{mmol } FeSO_4 \times \frac{1\ \text{mmol } K_2Cr_2O_7}{6\ \text{mmol } FeSO_4} = 0.237\ \text{mmol } K_2Cr_2O_7$$

Finally, we calculate the concentration of the $K_2Cr_2O_7$ solution.

$$\frac{?\ \text{mmol } K_2Cr_2O_7}{\text{mL}} = \frac{0.237\ \text{mmol } K_2Cr_2O_7}{34.2\ \text{mL}} = \underline{\underline{0.00693\ \underline{M}\ K_2Cr_2O_7}}$$

Example 12-12: What mass of molybdenum(III) oxide, Mo_2O_3, will be completely oxidized to potassium molybdate, K_2MoO_4, by 35.0 mL of 0.200 $\underline{M}$ $KMnO_4$ solution?

$$3\ MnO_4^- + 5\ Mo^{3+} + 8\ H_2O \longrightarrow 3\ Mn^{2+} + 5\ MoO_4^{2-} + 16\ H^+$$

We first calculate the number of moles of $KMnO_4$ in 35.0 mL of 0.200 $\underline{M}$ $KMnO_4$ solution.

$$?\ \text{mol } KMnO_4 = 35.0\ \text{mL} \times \frac{0.200\ \text{mol } KMnO_4}{1000\ \text{mL}} = 7.00 \times 10^{-3}\ \text{mol } KMnO_4$$

Now we determine the weight of Mo_2O_3 that reacts with 7.00×10^{-3} mol $KMnO_4$. Again we must consult the balanced equation to determine the reaction mole ratio. It tells us that 3 moles of MnO_4^- are consumed for every 5 moles of Mo^{3+}, or 2.5 moles of Mo_2O_3.

$$?\ \text{g } Mo_2O_3 = 7.00 \times 10^{-3}\ \text{mol } KMnO_4 \times \frac{2.5\ \text{mol } Mo_2O_3}{3\ \text{mol } KMnO_4} \times \frac{239.9\ \text{g } Mo_2O_3}{1\ \text{mol } Mo_2O_3}$$

$$= \underline{\underline{1.40\ \text{g } Mo_2O_3}}$$

2. Equivalent Weights and Normality

One equivalent weight (eq) of oxidizing agent or reducing agent is the mass that gains (oxidizing agent) or loses (reducing agent) 6.02×10^{23} electrons in a particular redox reaction. Redox titrations are analogous to acid-base titrations in the sense that one equivalent of an oxidizing agent reacts with exactly one equivalent of a reducing agent. However one mole of oxidizing agent does not necessarily react with one mole of reducing agent.

$$\text{no. of eq oxidizing agent} = \text{no. of eq reducing agent}$$

Since 1000 milliequivalent weights (meq) equal one equivalent weight, we may also write

$$\text{no. of meq oxidizing agent} = \text{no. of meq reducing agent}$$

Recall that normality times volume in liters or milliliters gives the number of gew or meq of solute, respectively. Therefore, the following relationships hold for redox titrations involving two solutions where the subscripts O and R refer to oxidizing and reducing agents, respectively.

$$L_O \times \underline{N}_O = L_R \times \underline{N}_R$$

$$mL_O \times \underline{N}_O = mL_R \times \underline{N}_R$$

Example 12-13: Balance the following equation and determine the number of equivalents in one mole of oxidizing agent and one mole of reducing agent, as well as the mass of one equivalent of each. The reaction occurs in aqueous solution.

$$KClO_3 \quad + \quad NaNO_2 \longrightarrow KCl \quad + \quad NaNO_3$$

First we determine the changes in oxidation numbers.

$$\text{in } KClO_3 \longrightarrow 1\,(Cl^{+5} \xrightarrow{6\downarrow} Cl^{-1}) \longleftarrow \text{in } KCl$$

$$\text{in } NaNO_2 \longrightarrow 3\,(N^{+3} \xrightarrow{2\uparrow} N^{+5}) \longleftarrow \text{in } NaNO_3$$

Now we balance the equation.

$$KClO_3 + 3\,NaNO_2 \longrightarrow KCl \quad + \quad 3\,NaNO_3$$

Note that the oxidation number of each Cl atom decreases from +5 (in $KClO_3$) to -1 (in KCl), a change of six units.

$$Cl^{+5} \xrightarrow{6\downarrow} Cl^{-1}$$

Therefore one mole of $KClO_3$ is six equivalents, and one eq is one-sixth of a mole.

$$1 \text{ mol } KClO_3 = 6 \text{ eq } KClO_3$$

$$1 \text{ eq } KClO_3 = \frac{1 \text{ mol}}{6} = \frac{122.6 \text{ g}}{6} = 20.43 \text{ g } KClO_3$$

The oxidation number of each nitrogen atom increases from +3 (in $NaNO_2$) to +5 (in $NaNO_3$).

$$N^{+3} \xrightarrow{2\uparrow} N^{+5}$$

Therefore one mole of $NaNO_2$ is two equivalents, and one eq is one-half of a mole.

$$1 \text{ mol } NaNO_2 = 2 \text{ eq } NaNO_2$$

$$1 \text{ eq } NaNO_2 = \frac{1 \text{ mol}}{2} = \frac{69.0 \text{ g}}{2} = 34.5 \text{ g } NaNO_2$$

These calculations tell us that 20.43 g of $KClO_3$ (1 eq) reacts with 34.5 g of $NaNO_2$ (1 eq).

Example 12-14: Exactly 25.0 mL of a $KMnO_4$ solution are required for complete reaction with 40.0 mL of 0.200 $\underline{N}$ $FeSO_4$ solution according to the equation below.

(a) What are the normality and molarity of the $KMnO_4$ solution?

(b) How many grams of $KMnO_4$ are contained in 25.0 mL of the solution?

$$2\ KMnO_4 + 8\ H_2SO_4 + 10\ FeSO_4 \longrightarrow 2\ MnSO_4 + 8\ H_2O + 5\ Fe_2(SO_4)_3 + K_2SO_4$$

(a) no. of meq $KMnO_4$ = no. of meq $FeSO_4$

$$mL_{KMnO_4} \times \underline{N}_{KMnO_4} = mL_{FeSO_4} \times \underline{N}_{FeSO_4}$$

$$\underline{N}_{KMnO_4} = (mL_{FeSO_4} \times \underline{N}_{FeSO_4})/mL_{KMnO_4}$$

$$\underline{N}_{KMnO_4} = \frac{40.0\ mL \times 0.200\ meq/mL}{25.0\ mL}$$

$$\underline{N}_{KMnO_4} = 0.320\ \frac{meq}{mL} = \underline{\underline{0.320\ \underline{N}}}$$

The Mn undergoes a decrease in oxidation number of 5 units so there are 5 eq in one mole of $KMnO_4$,

$$\text{in } KMnO_4 \rightarrow Mn^{+7} \xrightarrow{5\downarrow} Mn^{+2} \leftarrow \text{in } MnSO_4$$

and so the molarity of the $KMnO_4$ solution is 1/5 its normality.

$$\frac{?\ mol}{L} = \frac{0.320\ eq}{L} \times \frac{1\ mol\ KMnO_4}{5\ eq} = \underline{\underline{0.0640\ \underline{M}}}$$

(b) One mol $KMnO_4$ (158.0 g) is 5 eq, so 1 eq = 158.0 g/5 = 31.6 in this reaction. There are 0.253 g $KMnO_4$ in 25.0 mL of the 0.320 $\underline{N}$ $KMnO_4$ solution as shown below.

$$\underline{?}\ g\ KMnO_4 = 25.0\ mL \times \frac{0.320\ eq}{1000\ mL} \times \frac{31.6\ g}{eq} = \underline{\underline{0.253\ g\ KMnO_4}}$$

Alternatively, using molarity rather than normality to express the concentration of the solution,

$$\underline{?}\ g\ KMnO_4 = 25.0\ mL \times \frac{0.0640\ mol}{1000\ mL} \times \frac{158.0\ g}{mol} = \underline{\underline{0.253\ g\ KMnO_4}}$$

Example 12-15: What volume of 0.100 M $FeSO_4$ solution is required to react exactly with 30.0 mL of 0.200 N $KMnO_4$ solution according to the reaction of Example 12-14? (This problem is the same as Example 12-10; 0.200 N, $KMnO_4$ is 0.0400 M $KMnO_4$.)

A 0.100 M $FeSO_4$ solution is also 0.100 N because each Fe^{2+} ion (in $FeSO_4$) undergoes a change in oxidation number of one in the reaction.

$$Fe^{+2} \xrightarrow{1\uparrow} Fe^{+3}$$

That is, one eq of $FeSO_4$ is also one mol of $FeSO_4$.

$$mL_{KMnO_4} \times \underline{N}_{KMnO_4} = mL_{FeSO_4} \times \underline{N}_{FeSO_4}$$

$$mL_{FeSO_4} = \frac{30.0\ mL \times 0.200\ \underline{N}}{0.100\ \underline{N}} = \underline{\underline{60.0\ mL}}$$

Example 12-16: What mass of molybdenum(III) oxide, Mo_2O_3 will be completely oxidized to potassium molybdate, K_2MoO_4, by 35.0 mL of 0.200 M $KMnO_4$ solution? Use the equivalent weight method. (This problem is the same as Example 12-11.)

$$3\ MnO_4^- + 5\ Mo^{3+} + 8\ H_2O \longrightarrow 3\ Mn^{2+} + 5\ MoO_4^{2-} + 16\ H^+$$

There are 5 meq per mmol of $KMnO_4$ because the oxidation number of each manganese atom decreases by five units. We first calculate the normality of the $KMnO_4$ solution.

$$\frac{?\ meq\ KMnO_4}{mL} = \frac{0.200\ mmol}{mL} \times \frac{5\ meq\ KMnO_4}{mmol} = \frac{1.00\ meq}{mL} = 1.00\ \underline{N}\ KMnO_4$$

The number of meq of Mo_2O_3 must be the same as the number of meq of $KMnO_4$ in 35.0 mL of 1.00 N $KMnO_4$.

$$\text{no. of meq } Mo_2O_3 = \text{no. of meq } KMnO_4$$

$$\text{no of meq } Mo_2O_3 = 35.0\ mL \times 1.00\ \underline{N}$$

$$35.0\ meq\ Mo_2O_3 = 35.0\ meq\ KMnO_4$$

The formula weight of Mo_2O_3 is 239.9 g. There are 6 meq per mmol of Mo_2O_3 because the oxidation number of each molybdenum atom increases by 3 units and there are 2 Mo per formula unit of Mo_2O_3.

in Mo_2O_3 → $2\ (Mo^{+3} \xrightarrow{3\uparrow} Mo^{+6})$ ← in MoO_4^{2-}

or $2\ Mo^{+3} \xrightarrow{6\downarrow} 2\ Mo^{+6}$

$$1 \text{ eq } Mo_2O_3 = \frac{239.9 \text{ g } Mo_2O_3}{6} = 40.0 \text{ g } Mo_2O_3$$

Now we must calculate the number of grams of Mo_2O_3 in 35.0 meq of Mo_2O_3.

$$? \text{ g } Mo_2O_3 = 35.0 \text{ meq } Mo_2O_3 \times \frac{40.0 \text{ mg } Mo_2O_3}{\text{meq } Mo_2O_3} = 1400 \text{ mg } Mo_2O_3$$

$$= \underline{\underline{1.40 \text{ g } Mo_2O_3}}$$

The sample of Mo_2O_3 required has a mass of 1.40 grams.

Example 12-17: A 0.987 gram sample of Mo_2O_3 reacts with 30.0 mL of a $KMnO_4$ solution according to the equation in Example 12-16. What is the normality of the $KMnO_4$ solution?

$$\text{no. of meq } KMnO_4 = \text{no. of meq } Mo_2O_3$$

$$mL_{KMnO_4} \times \underline{N}_{KMnO_4} = \text{no. of meq } Mo_2O_3$$

$$30.0 \text{ mL} \times \underline{N}_{KMnO_4} = 0.987 \text{ g } Mo_2O_3 \times \frac{1 \text{ meq } Mo_2O_3}{0.0400 \text{ g } Mo_2O_3}$$

$$= 24.7 \text{ meq } Mo_2O_3$$

$$= 24.7 \text{ meq } KMnO_4$$

$$\underline{N}_{KMnO_4} = \frac{24.7 \text{ meq } KMnO_4}{30.0 \text{ mL}} = \underline{\underline{0.823 \ \underline{N} \ KMnO_4}}$$

Example 12-18: A 1.500 gram sample containing some Mo_2O_3 reacts with 20.0 mL of 0.500 N $KMnO_4$ solution. How many grams of Mo_2O_3 are contained in the sample and what is the percentage of Mo_2O_3 in the sample? Assume no other components of the sample react with $KMnO_4$.

$$\text{no of meq } KMnO_4 = \text{no of meq } Mo_2O_3$$

$$20.0 \text{ mL} \times 0.500 \ \underline{N} = \text{no of meq } Mo_2O_3$$

$$10.0 \text{ meq} = \text{no of meq } Mo_2O_3$$

$$\underline{?} \text{ g } Mo_2O_3 = 10.0 \text{ meq } Mo_2O_3 \times \frac{40.0 \text{ g } Mo_2O_3}{1000 \text{ meq } Mo_2O_3}$$

$$= 0.400 \text{ g } Mo_2O_3 \text{ in the sample}$$

$$\% \ Mo_2O_3 = \frac{g\ Mo_2O_3}{g\ sample} \times 100\% = \frac{0.400\ g}{1.500\ g} \times 100\% = \underline{\underline{26.7\%\ Mo_2O_3}}$$

EXERCISES

1. Write net ionic equations for the molecular equations below, if applicable, and identify the oxidizing agent (OA) and reducing agent (RA) in each.

   (a) P_4 (s) + 6 Cl_2 (g) $\longrightarrow$ 4 PCl_3 (ℓ)

   (b) Sb_2S_3 (s) + 3 Fe (s) $\xrightarrow{\Delta}$ 2 Sb (s) + 3 FeS (s)

   (c) 2 H_2SO_4 (aq) + S (s) $\longrightarrow$ 3 SO_2 (g) + 2 H_2O (ℓ)

   (d) As_4O_6 (s) + 8 HNO_3 (aq) + 2 H_2O (ℓ) $\longrightarrow$ 4 H_3AsO_4 (aq) + 8 NO_2 (g)

   (e) $(NH_4)_2Cr_2O_7$ (s) $\xrightarrow{\Delta}$ N_2 (g) + 4 H_2O (g) + Cr_2O_3 (s)

   (f) CO (g) + H_2O (g) $\longrightarrow$ CO_2 (g) + H_2 (g)

2. Balance the following unbalanced ionic equations.

   (a) Zn (s) + H^+ (aq) $\longrightarrow$ Zn^{2+} (aq) + H_2 (g)

   (b) Fe^{3+} (aq) + Pb (s) + SO_4^{2-} (aq) $\longrightarrow$ Fe^{2+} (aq) + $PbSO_4$ (s)

   (c) $Cr(OH)_3$ (s) + OH^- (aq) + Cl_2 (g) $\longrightarrow$ CrO_4^{2-} (aq) + Cl^- (aq) + H_2O (ℓ)

   (d) H_2O_2 (aq) + MnO_4^- (aq) + H^+ (aq) $\longrightarrow$ Mn^{2+} (aq) + O_2 (g) + H_2O (ℓ)

   (e) Cu (s) + H^+ (aq) + NO_3^- (aq) $\longrightarrow$ Cu^{2+} (aq) + NO (g) + H_2O (ℓ)

3. Balance the following unbalanced molecular equations.

   (a) Mn (s) + O_2 (g) + H_2O (ℓ) $\longrightarrow$ $Mn(OH)_2$ (s)

   (b) Al (s) + HBr (aq) $\longrightarrow$ $AlBr_3$ (aq) + H_2 (g)

   (c) Hg_2Cl_2 (s) + NH_3 (aq) $\longrightarrow$ Hg (ℓ) + $HgNH_2Cl$ (s) + NH_4Cl (aq)

   (d) C (s) + HNO_3 (aq) $\longrightarrow$ NO_2 (g) + CO_2 (g) + H_2O (ℓ)

   (e) $NaClO_2$ (s) + Cl_2 (g) $\longrightarrow$ NaCl (s) + ClO_2 (g)

   (f) I_2 (s) + $HClO_3$ (aq) + H_2O (ℓ) $\longrightarrow$ HIO_3 (aq) + HCl (aq)

   (g) Pb (s) + PbO_2 (s) + H_2SO_4 (aq) $\longrightarrow$ $PbSO_4$ (s) + H_2O (ℓ)

   (h) HBr (g) + H_2SO_4 (aq) $\longrightarrow$ Br_2 (ℓ) + SO_2 (g) + H_2O (ℓ)

4. Balance the following net ionic equations. You may add H^+ (aq) and/or H_2O (ℓ) or OH^- (aq) and/or H_2O (ℓ) when appropriate.

   (a) HClO (aq) + Hg (ℓ) + Br^- (aq) $\longrightarrow$ Cl^- (aq) + $HgBr_4^{2-}$ (acidic sol'n)

   (b) ClO^- (aq) + $Ni(OH)_2$ (s) $\longrightarrow$ Cl^- (aq) + NiO_2 (s) (basic sol'n)

   (c) UF_6^- (aq) + H_2O_2 (aq) $\longrightarrow$ UO_2^{2+} (aq) + HF (aq) (acidic sol'n)

   (d) MnO_4^- (aq) + $(COOH)_2$ (aq) $\longrightarrow$ Mn^{2+} (aq) + CO_2 (g) (acidic sol'n)

4. (e) $HSnO_2^-$ (aq) + CrO_4^{2-} (aq) $\longrightarrow$ $HSnO_3^-$ (aq) + CrO_2^- (aq) (basic sol'n)
   (f) Br_2 (ℓ) $\longrightarrow$ BrO_3^- (aq) + Br^- (aq) (basic sol'n)
   (g) NO_2^- (aq) + MnO_4^- (aq) $\longrightarrow$ MnO_2 (s) + NO_3^- (aq) (basic sol'n)
   (h) $Cr_2O_7^{2-}$ + Al (s) $\longrightarrow$ Cr^{3+} + Al^{3+} (acidic sol'n)

5. Write balanced net ionic equations for the following redox reactions.
   (a) Ferrous ions reduce permanganate ions to manganese(II) ions and are oxidized to ferric ions in acidic solution.
   (b) In basic solution potassium permanganate oxidizes potassium nitrite to potassium nitrate and is reduced to manganese(IV) oxide.
   (c) Ferrous ion reduces nitrite ion to nitrogen oxide and is oxidized to ferric ion in acidic solution.
   (d) In acidic solution bromide ion reduces permanganate ion to manganese-(II) ion and is oxidized to bromine.

6. A standardized solution is 0.0404 M in sodium thiosulfate, $Na_2S_2O_3$. A 20.0 mL sample of this solution reacts with 33.4 mL of a solution of potassium triiodide, KI_3 (which was prepraed by dissolving a sample of I_2 in an excess of aqueous KI). What is the molar concentration of the KI_3 solution?

   $2\ S_2O_3^{2-} + I_3^- \longrightarrow 3\ I^- + S_4O_6^{2-}$

7. What was the mass of iodine, I_2, that dissolved in 500 mL of excess aqueous KI in problem 6? Assume that the reaction below went to completion and assume no significant change in volume due to the addition of the solid I_2.

   I_2 (s) + KI (aq) $\longrightarrow$ KI_3 (aq)

8. What volume of 0.0322 M $KMnO_4$ solution is required to oxidize 25.0 mL of a solution containing 0.816 gram of $FeSO_4$?

9. What volume of 0.0191 M potassium dichromate, $K_2Cr_2O_7$, solution is required to oxidize 30.00 mL of 0.150 M $FeSO_4$ solution?

10. What is the molar concentration of a solution of $FeSO_4$ if 25.0 mL of the solution requires 34.8 mL of 0.0180 M $K_2Cr_2O_7$ solution for complete oxidation?

11. What is the equivalent weight of the oxidizing agent and reducing agent in each of the reactions below?
   (a) $5\ (COOH)_2 + 2\ KMnO_4 + 3\ H_2SO_4 \rightarrow 2\ MnSO_4 + K_2SO_4 + 10\ CO_2 + 8\ H_2O$
   (b) $KI_3 + 2\ K_2S_2O_3 \longrightarrow 3\ KI + K_2S_4O_6$
   (c) $2\ CuCl + 2\ NaCl + Br_2 \longrightarrow 2\ CuCl_2 + 2\ NaBr$
   (d) $K_2Cr_2O_7 + 3\ H_2S + 4\ H_2SO_4 \longrightarrow Cr_2(SO_4)_3 + 3\ S + 7\ H_2O + K_2SO_4$

12. What is the equivalent weight of $KMnO_4$ in the following reactions.

(a) $5\ Sn^{2+}\ (aq) + 2\ MnO_4^-\ (aq) + 16\ H^+\ (aq) \rightarrow 5\ Sn^{4+}\ (aq) + 2\ Mn^{2+}\ (aq) + 8\ H_2O\ (\ell)$

(b) $2\ MnO_4^-\ (aq) + 3\ Mn^{2+}\ (aq) + 2\ H_2O\ (\ell) \rightarrow 5\ MnO_2\ (s) + 4\ H^+\ (aq)$

(c) $2\ MnO_4^-\ (aq) + H_2O_2\ (aq) + 2\ OH^-\ (aq) \rightarrow 2\ MnO_4^{2-}\ (aq) + O_2\ (g) + 2\ H_2O\ (\ell)$

13. What mass of As_2O_3 is required to prepare 500 mL of 0.100 N $H_2AsO_3^-$ solution to be used in the reaction below?

$H_2AsO_3^-\ (aq) + I_2\ (s) + H_2O\ (\ell) \rightarrow H_2AsO_4^{2-}\ (aq) + 2\ I^-\ (aq) + 2\ H^+\ (aq)$

14. How many g of $Na_2S_2O_3 \cdot 5\ H_2O$ must be dissolved to prepare 700 mL of 0.300 N solution? It is to be used in the reaction in which $S_2O_3^{2-}$ reduces I_2 to I^- and is oxidized to $S_4O_6^{2-}$.

15. How much of substance 2 will react with substance 1 according to the reaction listed at the right?

	Substance 1	Substance 2	Reaction
(a)	30.0 mL of 0.100 N $KMnO_4$	___mg $(COOH)_2 \cdot 2\ H_2O$	11 (a)
(b)	25.0 mL of 0.250 N $K_2S_2O_3$	___g KI_3	11 (b)
(c)	35.0 mL of 0.050 M H_2S	___mL of 0.200 N $K_2Cr_2O_7$	11 (d)
(d)	42.0 mL of 0.1042 M $K_2Cr_2O_7$	___mg H_2S	11 (d)
(e)	0.861 g of I_2	___mL of 0.200 N $H_2AsO_3^-$	13

16. What are the normality and molarity of a $K_2Cr_2O_7$ solution if 35.0 mL of it are required for complete reaction with 25.0 mL of 0.161 N $Fe(NH_4)_2(SO_4)_2$ solution?

$14\ H^+\ (aq) + Cr_2O_7^{2-}\ (aq) + 6\ Fe^{2+}\ (aq) \rightarrow 6\ Fe^{3+}\ (aq) + 2\ Cr^{3+}\ (aq) + 7\ H_2O\ (\ell)$

17. What are the normality and molarity of a $K_2Cr_2O_7$ solution if 28.0 mL of it react with 0.2046 g of $Fe(NH_4)_2(SO_4)_2 \cdot 6\ H_2O$ according to the reaction of problem 16?

18. A 1.427 g sample of impure As_2O_3 is dissolved in concentrated NaOH solution to form $NaAsO_2$ which is then reacted with HCl to produce H_3AsO_3. The resulting solution reacts with 32.6 mL of 0.749 M $Ce(SO_4)_2$ to produce $Ce_2(SO_4)_3$ and H_3AsO_4. What were the weight of As_2O_3 and percentage purity in the original sample?

$H_3AsO_3\ (aq) + 2\ Ce^{4+}\ (aq) + H_2O\ (\ell) \rightarrow H_3AsO_4\ (aq) + 2\ Ce^{3+}\ (aq) + 2\ H^+\ (aq)$

19. An impure sample of $Fe(NH_4)_2(SO_4)_2 \cdot 6\,H_2O$ weighing 13.1 g required 42.46 mL of 0.1033 M $K_2Cr_2O_7$ for complete reaction according to the equation of problem 16. Assuming no other oxidizable species were present in the sample, how many g of $Fe(NH_4)_2(SO_4)_2 \cdot 6H_2O$ were contained in the sample and what was its percentage purity?

ANSWERS FOR EXERCISES

1. (a) not applicable; OA, Cl_2, P_4; RA
   (b) not applicable; OA, Sb_2S_3; RA, Fe
   (c) $4\,H^+ (aq) + 2\,SO_4^{2-} (aq), + S (s) \longrightarrow 3\,SO_2 (g) + 2\,H_2O (\ell)$
   OA, H_2SO_4; RA, S
   (d) $As_4O_6 (s) + 8\,H^+ (aq) + 8\,NO_3^- (aq) + 2\,H_2O (\ell) \longrightarrow 4\,H_3AsO_4 (aq) + 8\,NO_2$
   OA, HNO_3; RA, As_4O_6
   (e) not applicable; $(NH_4)_2Cr_2O_7$ is OA and RA
   (f) not applicable; OA, H_2O; RA, CO

2. Designations of state and hydration have been omitted to simplify writing these equations as well as those in 3, 4, and 5.
   (a) $Zn + 2\,H^+ \longrightarrow Zn^{2+} + H_2$
   (b) $2\,Fe^{3+} + Pb + SO_4^{2-} \longrightarrow 2\,Fe^{2+} + PbSO_4 (s)$
   (c) $2\,Cr(OH)_3 + 10\,OH^- + 3\,Cl_2 \longrightarrow 2\,CrO_4^{2-} + 6\,Cl^- + 8\,H_2O$
   (d) $5\,H_2O_2 + 2\,MnO_4^- + 6\,H^+ \longrightarrow 2\,Mn^{2+} + 5\,O_2 + 8\,H_2O$
   (e) $3\,Cu + 8\,H^+ + 2\,NO_3^- \longrightarrow 3\,Cu^{2+} + 2\,NO + 4\,H_2O$

3. (a) $2\,Mn + O_2 + 2\,H_2O \longrightarrow 2\,Mn(OH)_2$
   (b) $2\,Al + 6\,HBr \longrightarrow 2\,AlBr_3 + 3\,H_2$
   (c) $Hg_2Cl_2 + 2\,NH_3 \longrightarrow Hg + HgNH_2Cl + NH_4Cl$
   (d) $C + 4\,HNO_3 \longrightarrow 4\,NO_2 + CO_2 + 2\,H_2O$
   (e) $2\,NaClO_2 + Cl_2 \longrightarrow 2\,NaCl + 2\,ClO_2$
   (f) $3\,I_2 + 5\,HClO_3 + 3\,H_2O \longrightarrow 6\,HIO_3 + 5\,HCl$
   (g) $Pb + PbO_2 + 2\,H_2SO_4 \longrightarrow 2\,PbSO_4 + 2\,H_2O$
   (h) $2\,HBr + H_2SO_4 \longrightarrow Br_2 + SO_2 + 2\,H_2O$

4. (a) $H^+ + HClO + Hg + 4\,Br^- \longrightarrow Cl^- + HgBr_4^{2-} + H_2O$
   (b) $ClO^- + Ni(OH)_2 \longrightarrow Cl^- + NiO_2 + H_2O$
   (c) $6\,H^+ + 2\,UF_6^- + H_2O_2 + 2\,H_2O \longrightarrow 2\,UO_2^{2+} + 12\,HF$
   (d) $6\,H^+ + 2\,MnO_4^- + 10\,(COOH)_2 \longrightarrow 2\,Mn^{2+} + 10\,CO_2 + 8\,H_2O$

4. (e) $3\ HSnO_2^- + 2\ CrO_4^{2-} + H_2O \longrightarrow 3\ HSnO_3^- + 2\ CrO_2^- + 2\ OH^-$

(f) $6\ OH^- + 3\ Br_2 \longrightarrow BrO_3^- + 5\ Br^- + 3\ H_2O$

(g) $3\ NO_2^- + H_2O + 2\ MnO_4^- \longrightarrow 3\ NO_3^- + 2\ MnO_2 + 2\ OH^-$

(h) $14\ H^+ + Cr_2O_7^{2-} + 2\ Al \longrightarrow 2\ Cr^{3+} + 7\ H_2O + 2\ Al^{3+}$

5. (a) $5\ Fe^{2+} + 8\ H^+ + MnO_4^- \longrightarrow 5\ Fe^{3+} + Mn^{2+} + 4\ H_2O$

(b) $2\ MnO_4^- + 3\ NO_2^- + H_2O \longrightarrow 3\ NO_3^- + 2\ MnO_2 + 2\ OH^-$

(c) $2\ H^+ + NO_2^- + Fe^{2+} \longrightarrow Fe^{3+} + NO + H_2O$

(d) $10\ Br^- + 2\ MnO_4^- + 16\ H^+ \longrightarrow 2\ Mn^{2+} + 5\ Br_2 + 8\ H_2O$

6. 0.0121 M KI_3 7. 1.54 g I_2 8. 33.4 mL

9. 39.3 mL 10. 0.150 M $FeSO_4$

11. (a) 31.6 g/eq $KMnO_4$; 45.0 g/eq $(COOH)_2$

(b) 210 g/eq KI_3; 190 g/eq $K_2S_2O_3$

(c) 79.9 g/eq Br_2; 99.0 g/eq CuCl

(d) 49.0 g/eq $K_2Cr_2O_7$; 17.0 g/eq H_2S

12. (a) 31.6 g/eq (b) 52.7 g/eq (c) 158 g/eq

13. 2.48 g As_2O_3 14. 52.1 g $Na_2S_2O_3 \cdot 5\,H_2O$ 15. (a) 189 mg $(COOH)_2 \cdot 2\,H_2O$ (b) 1.31 g KI_3 (c) 17.5 mL (d) 448 mg H_2S (e) 33.9 mL 16. 0.115 N $K_2Cr_2O_7$, 0.0192 M $K_2Cr_2O_7$, 17. 0.0186 N $K_2Cr_2O_7$, 3.10×10^{-3} M $K_2Cr_2O_7$ 18. 1.21 g As_2O_3, 84.8% As_2O_3

19. 10.3 g $Fe(NH_4)_2(SO_4)_2 \cdot 6\,H_2O$, 78.6%

Chapter Thirteen

CHEMICAL THERMODYNAMICS

D,G,W: Chap. 3; Sec. 7-11; Sec. 12- 11.3; Chap. 14

13-1 Introduction

The study of the transfers of energy accompanying physical and chemical changes is called thermodynamics. Transfers of energy between reacting systems and their surroundings are characterized by state functions, which are functions that describe the state of a system and are independent of the pathway by which a process occurs.

The internal energy, E, of the reactants of a system is the sum of the internal energies of the individual reacting species. Likewise, the internal energy of the products is the sum of the internal energies of the individual products. Internal energy is a state function and the change in internal energy, ΔE, is also a state function.

$$\Delta E = E_{products} - E_{reactants}$$

The change in internal energy is equal to the heat (q) absorbed by the system minus the work (w) done by the system during a process.

$$\Delta E = q - w$$

The conventions regarding the signs of q and w are:

q is positive if heat is absorbed by system
q is negative if heat is released by system
w is positive if work is done by system
w is negative if work is done on system

The only important type of work done during chemical and physical changes is pressure-volume work, $P\Delta V$, in which the gases that are produced by a reaction do work on the atmosphere, or the atmosphere does work on the changing system by expanding into the volume previously occupied by a (consumed) gas. For processes involving only solids and liquids* or equal numbers of moles of gases as reactants and products, $\Delta V = 0$, and therefore $P\Delta V = 0$, and no work is done. In such cases, the change in internal energy

*Solids and liquids are essentially incompressible.

is simply the heat absorbed by the system. If a bomb calorimeter is used to measure heat transfer, the volume is held constant and $\Delta V = 0$ even for changes involving the net consumption or production of gases and $\Delta E = q_V$, the heat absorbed at constant volume. However, reactions are often carried out in the laboratory at constant (atmospheric) pressure rather than at constant volume. In such cases $P\Delta V$ may be nonzero for some gaseous systems. For such cases,

$$\Delta E = q_p + P\Delta V$$

where q_p is the heat absorbed by the system at constant (atmospheric) pressure, which is also called the enthalpy change, ΔH. Thus,

$$\Delta E = \Delta H + P\Delta V$$

If $\Delta V = 0$ then $\Delta E = \Delta H$. Numerous measurements have shown that, even in cases where $\Delta V \neq 0$, $P\Delta V$ is still usually very small relative to ΔE. Thus, for practically all chemical and physical changes, ΔE is very closely approximated by ΔH, the heat absorbed by the system at constant pressure, which is also a state function.

$$\Delta E \approx \Delta H \qquad \text{(when } \Delta V \text{ is small at constant P)}$$

If heat is released by the system to the environment ΔH is negative and the reaction is exothermic. Conversely, if the system absorbs heat during a change, the sign of ΔH is positive and the reaction is endothermic.

13-2 First Law of Thermodynamics

The First Law states that the total amount of energy in the universe is constant. It can also be stated in another way: Energy is neither created nor destroyed in ordinary chemical reactions, but merely converted from one form to another. In most thermodynamically spontaneous (energetically favorable) reactions the energy associated with the chemical bonds of the products is less than that of the reactants. The energy difference is released by the reacting system to the environment (ΔH is negative). In other words exothermicity favors, but does not require, thermodynamic spontaneity.

13-3 Standard Molar Enthalpy of Formation, ΔH_f^o

Standard pressure for thermodynamic purposes is one atmosphere. A superscript of zero following a thermodynamic symbol refers to standard pressure*

*Some texts suggest that the zero superscript refers to a standard pressure (1 atm) and a standard temperature (25°C, 298 K). However, it does not refer to standard temperature and thermodynamic quantities with a superscript of zero are functions of temperature. Temperatures are indicated only by a subscript following the symbol. If no subscript appears, the temperature is assumed to be 298.15 K.

with all dissolved species at unit molarities (or strictly speaking, unit activities), all gases at unit partial pressures, and all solids or liquids in the pure state. The standard state of a substance is its stable form under standard conditions. The heat change accompanying the formation of one mole of a substance in its standard state from its constituent elements in their standard states is called the standard molar enthalpy of formation of the substance, ΔH^o_f. Several ΔH^o_f values are tabulated in Appendix K in the text. The following reaction involves the formation of one mole of solid NH_4I from its elements in their standard states.

$$\frac{1}{2} N_2 (g) \quad + \quad 2 H_2 (g) \quad + \quad \frac{1}{2} I_2 (s) \longrightarrow NH_4I (s) \quad + 201.4 \text{ kJ}$$

For this reaction $\Delta H^o_{rxn} = \Delta H^o_f$ for NH_4I (s) = -201.4 kJ standard molar enthalpies of formation of all elements in their standard states are zero.

In each of the following reactions the reactants are elements in their standard states at 25°C, so $\Delta H^o_{rxn} = \Delta H^o_f$ for the product.

$$Hg (\ell) + Cl_2 (g) \longrightarrow HgCl_2 (s) \qquad \Delta H^o = \Delta H^o_{f\, HgCl_2 (s)} = -224 \text{ kJ}$$

$$H_2 (g) + Si (s) + 3/2\, O_2 (g) \longrightarrow H_2SiO_3 (s) \qquad \Delta H^o = \Delta H^o_{f\, H_2SiO_3(s)} = -1189 \text{ kJ}$$

$$6\, C \text{ (graphite)} + 3\, H_2 (g) \longrightarrow C_6H_6 (\ell) \qquad \Delta H^o = \Delta H^o_{f\, C_6H_6 (\ell)} = 49.03 \text{ kJ}$$

13-4 Hess' Law of Heat Summation

Hess' Law of Constant Heat Summation states that the molar enthalpy change for the conversion of a given set of reactants into a given set of products is always the same whether the conversion occurs in one step or a series of steps.

Consider the reaction of tin and chlorine:

$$Sn (s) + 2\, Cl_2 (g) \longrightarrow SnCl_4 (\ell) \qquad \Delta H^o_{f\, SnCl_4 (\ell)} = -511 \text{ kJ} \quad (\Delta H^o_{rxn})$$

The reaction could also occur in two (or more) steps. The sum of the enthalpy changes for both (or more) steps is the same as that for the one-step process.

Reaction		H^o_{rxn}	
$Sn (s) + Cl_2 (g) \longrightarrow SnCl_2 (s)$	$\Delta H^o_{f\, SnCl_2 (s)}$ =	-350 kJ	(1)
$SnCl_2 (s) + Cl_2 (g) \longrightarrow SnCl_4 (\ell)$		-161 kJ	(2)
$Sn (s) + 2\, Cl_2 (g) \longrightarrow SnCl_4 (\ell)$	$\Delta H^o_{f\, SnCl_4 (\ell)}$ =	-511 kJ	(1)+(2)

In general, for an n-step process, we may write:

$$\Delta H^\circ_{rxn} = \Delta H^\circ_{(1)} + \Delta H^\circ_{(2)} + \Delta H^\circ_{(3)} + \ldots \Delta H^\circ_{(n)}$$

A corollary of Hess' Law is summarized by the equation below. For any reaction,

$$\Delta H^\circ_{rxn} = \Sigma \Delta H^\circ_{f\ (products)} - \Sigma \Delta H^\circ_{f\ (reactants)}$$

Let us use the second reaction above to illustrate this general relationship.

standard state (→ Cl_2)

	$SnCl_2\ (s)$	+	$Cl_2\ (g)$	⟶	$SnCl_4\ (\ell)$
ΔH°_f	-350 kJ		0 kJ		-511 kJ

We calculate ΔH°_{rxn} from these ΔH°_f values using the equation for Hess' Law.

$$\Delta H^\circ_{rxn} = \Delta H^\circ_{f\ SnCl_4\ (\ell)} - [\Delta H^\circ_{f\ SnCl_2\ (s)} + \Delta H^\circ_{f\ Cl_2\ (g)}]$$

$$= -511\ kJ - (-350\ kJ) - 0\ kJ$$

$$\Delta H^\circ_{rxn} = -161\ kJ \quad \text{(agrees with above)}$$

Example 13-1: Use the tabulated values for ΔH°_f to calculate ΔH°_{rxn}, in kilojoules, for the reaction below.

$$2\ SOCl_2\ (\ell) + O_2\ (g) \longrightarrow 2\ SO_2Cl_2\ (\ell) \quad \Delta H^\circ_{rxn} = ?$$

	$2\ SOCl_2\ (\ell)$	+	$O_2\ (g)$	⟶	$2\ SO_2Cl_2\ (\ell)$
ΔH°_f	-206 kJ		0 kJ		-389 kJ

$$\Delta H^\circ_{rxn} = \Sigma \Delta H^\circ_{f\ (products)} - \Sigma \Delta H^\circ_{f\ (reactants)}$$

$$= 2\Delta H^\circ_{f\ SO_2Cl_2\ (\ell)} - [2\Delta H^\circ_{f\ SOCl_2\ (\ell)} + \Delta H^\circ_{f\ O_2\ (g)}]$$

$$= 2(-389\ kJ) - 2(-206) - 0\ kJ$$

$$\underline{\underline{\Delta H^\circ_{rxn} = -366\ kJ}}$$

The reaction is exothermic and liberates 183 kJ of heat energy for each mole of SO_2Cl_2 (ℓ) formed at standard state conditions.

Example 13-2: Given the following standard molar enthalpies of formation, and ΔH^o_{rxn} for the reaction shown, calculate ΔH^o_f for H_3AsO_4 (aq). (Can you suggest a reason, based on chemical intuition, why this reaction should not be expected to occur to any significant extent at room temperature?)

$$3\ NH_4Cl\ (aq) + H_3AsO_4\ (aq) \longrightarrow (NH_4)_3AsO_4\ (aq) + 3\ HCl\ (aq)$$

$\Delta H^o_{rxn} = 29.4$ kJ

ΔH^o_f for NH_4Cl (aq) = -300.2 kJ

ΔH^o_f for $(NH_4)_3AsO_4$ (aq) = -1268 kJ

ΔH^o_f for HCl (aq) = -167.4 kJ

$$\Delta H^o_{rxn} = \Delta H^o_{f\,(NH_4)_3AsO_4\,(aq)} + 3\Delta H^o_{f\,HCl\,(aq)} - 3\Delta H^o_{f\,NH_4Cl\,(aq)} - \Delta H^o_{f\,H_3AsO_4\,(aq)}$$

$$29.4\text{ kJ} = -1268\text{ kJ} + 3(-167.4\text{ kJ}) - 3(-300.2\text{ kJ}) - H^o_{f\,H_3AsO_4\,(aq)}$$

$$\Delta H^o_{f\,H_3AsO_4\,(aq)} = -899\text{ kJ}$$

Since this equation represents the reaction of a salt of a strong acid with a weak acid, H_3AsO_4, to form a strong acid, HCl, and the salt of a weak acid, chemical intuition should tell us that the reverse reaction occurs to a much greater extent. Whether the reaction occurs to a significant extent or not in no way affects the validity of the calculation.

Example 13-3: Given the following:

	ΔH^o_{rxn}	
$H_3BO_3\ (aq) \longrightarrow HBO_2\ (aq) + H_2O\ (\ell)$	-0.02 kJ	(1)
$H_2B_4O_7\ (s) + H_2O\ (\ell) \longrightarrow 4\ HBO_2\ (aq)$	-11.3 kJ	(2)
$H_2B_4O_7\ (s) \longrightarrow 2\ B_2O_3\ (aq) + H_2O\ (\ell)$	17.5 kJ	(3)

Calculate ΔH^o_{rxn} for the following reaction.

$$2\ H_3BO_3\ (aq) \longrightarrow B_2O_3\ (s) + 3\ H_2O\ (\ell) \qquad \Delta H^o_{rxn} = ? \qquad (4)$$

If the reverse of equation (2) is added to four times equation (1) and equation (3), the result is twice equation (4).

			ΔH^o_{rxn}	
$4\ H_3BO_3\ (aq)$	$\longrightarrow$	$4\ HBO_2\ (aq)\ +\ 4\ H_2O\ (\ell)$	-0.08 kJ	4x (1)
$4\ HBO_2\ (aq)$	$\longrightarrow$	$H_2B_4O_7\ (s)\ +\ H_2O\ (\ell)$	11.3 kJ	-(2)
$H_2B_4O_7\ (s)$	$\longrightarrow$	$2\ B_2O_3\ (s)\ +\ H_2O\ (\ell)$	17.5 kJ	(3)
$4\ H_3BO_3\ (aq)$	$\longrightarrow$	$2\ B_2O_3\ (s)\ +\ 6\ H_2O\ (\ell)$	28.7 kJ	(4)

Thus for $2\ H_3BO_3\ (aq) \longrightarrow B_2O_3\ (s) + 3\ H_2O\ (\ell)$, $\Delta H^o_{rxn} = \dfrac{28.7\ kJ}{2}$

$= 14.4\ kJ$

Example 13-4: Given the following, calculate the standard molar enthalpy of formation for H_3PO_3 (aq) and ΔH^o_{rxn} for reaction (2) without consulting a table.

(1) $PCl_3\ (g) + 3\ H_2O\ (\ell) \longrightarrow H_3PO_3\ (aq) + 3\ HCl\ (aq)$ $\Delta H^o_{rxn} = -310.0\ kJ$

(2) $4\ H_3PO_3\ (aq) \longrightarrow PH_3\ (g) + 3\ H_3PO_4\ (s)$

ΔH^o_f for $H_2O\ (\ell)$ = -285.8 kJ

ΔH^o_f for PCl_3 (g) = -306.4 kJ

ΔH^o_f for HCl (aq) = -167.4 kJ

ΔH^o_f for PH_3 (g) = 5.4 kJ

ΔH^o_f for H_3PO_4 (s) = -1281 kJ

ΔH^o_f for H_3PO_3 (aq) may be evaluated from reaction (1):

$$\Delta H^o_{rxn} = \Delta H^o_{f\ H_3PO_3\ (aq)} + 3\Delta H^o_{f\ HCl\ (aq)} - \Delta H^o_{f\ PCl_3\ (g)} - 3\Delta H^o_{f\ H_2O\ (\ell)}$$

$$-310.0\ kJ = \Delta H^o_{f\ H_3PO_3\ (aq)} + 3(-167.4\ kJ) - (-306.4\ kJ) - 3(-285.8\ kJ)$$

$$\Delta H^o_{f\ H_3PO_3\ (aq)} = -971.6\ kJ$$

ΔH^o_{rxn} for reaction (2) may now be evaluated.

$$\Delta H^{o}_{rxn\ (2)} = \Delta H^{o}_{f\ PH_3\ (g)} + 3\,\Delta H^{o}_{f\ H_3PO_4\ (s)} - 4\,\Delta H^{o}_{f\ H_3PO_3\ (aq)}$$

$$= 5.4\ kJ + 3(-1283\ kJ) - 4(-971.6\ kJ)$$

$$= \underline{\underline{49\ kJ}}$$

13-5 Bond Energies

Strengths of chemical bonds are measured in terms of bond energies. Higher bond energies correspond to stronger bonds. The bond energy of a particular bond is defined as the amount of energy that must be added (endothermic reaction) to break one mole of the bonds in gaseous molecules to form (only) gaseous atoms.

$$Cl_2\ (g)\ +\ 58.2\ kcal \longrightarrow 2\ Cl\ (g)$$

or

$$Cl_2\ (g) \longrightarrow 2\ Cl\ (g) \qquad \Delta H^{o}_{rxn} = 2\,\Delta H^{o}_{f\,Cl\ (g)} = \Delta H_{Cl-Cl} = 243.4\ kJ$$

The strengths of bonds between particular elements vary slightly from compound to compound. Tables 13-2 and 13-3 in the text contain average values for bond energies.

If the bond energies of reactants and products are known, the heat of reaction for a gas phase reaction may be estimated.

Example 13-5: Use data in Tables 13-2 and 13-3 (text) to estimate the enthalpy change for the reaction below.

$$N_2\ (g) + 3\ H_2\ (g) \longrightarrow 2\ NH_3\ (g)$$

The bond energies of interest are:

Bond	Bond Energy (kJ/mol)
$N \equiv N$	946
H - H	435
N - H	389

The amount of heat released in this gas phase reaction is the amount of heat released in forming the bonds in two moles of NH_3 minus the amount of heat absorbed in breaking the bonds in one mole of N_2 and three moles of H_2.

$$\Delta H_{rxn} = \Sigma B.E._{(reactants)} - \Sigma B.E._{(products)} \quad \text{(for gas phase reactions)}$$

$$\Delta H_{rxn} = B.E._{N\equiv N} + 3\,B.E._{H-H} - 6\,B.E._{N-H}$$

$$= 946\text{ kJ} + 3(435\text{ kJ}) - 6(389\text{ kJ})$$

$$= \underline{\underline{-83\text{ kJ}}}$$

Let us compare this estimated value with the value of ΔH°_{rxn} calculated from ΔH°_f values:

$$\Delta H^\circ_{rxn} = 2\,\Delta H^\circ_{f\ NH_3\ (g)} - \Delta H^\circ_{f\ N_2\ (g)} - 3\,\Delta H^\circ_{f\ H_2\ (g)}$$

$$= 2(-46.11\text{ kJ}) - 0 - 3(0)$$

$$= \underline{\underline{-92.22\text{ kJ}}}$$

The two values agree within 10% which tells us that the use of average bond energies gives only approximate values for ΔH°_{rxn}.

Example 13-6: Oxygen and fluorine react to produce the covalent compound OF_2, oxygen difluoride. Calculate the average bond energy of the O-F bond from the following data.

$$2\,F_2\,(g) + O_2\,(g) \longrightarrow 2\,OF_2\,(g) \quad \Delta H^\circ_{rxn} = 46\text{ kJ}$$

ΔH°_f for F (g) is 78.99 kJ

ΔH°_f for O (g) is 249.2 kJ

Since this is a gas phase reaction we may write:

$$\Delta H_{rxn} = \Sigma B.E._{(reactants)} - \Sigma B.E._{(products)}$$

$$= 2\,B.E._{F-F} + B.E._{O=O} - 4\,B.E._{O-F}$$

The values of $B.E._{F-F}$ and $B.E._{O=O}$ are just equal to $2\,\Delta H^\circ_{f\ F\ (g)}$ (158.0 kJ) and $2\,\Delta H^\circ_{f\ O\ (g)}$ (498.4 kJ), respectively.

$$\Delta H_{rxn} = 2\,B.E._{F-F} + B.E._{O=O} - 4\,B.E._{O-F}$$

$$46\text{ kJ} = 2(158.0\text{ kJ}) + 498.4\text{ kJ} - 4\,B.E._{O-F}$$

$$B.E._{O-F} = \frac{-46\text{ kJ} + 316\text{ kJ} + 498.4\text{ kJ}}{4}$$

$$B.E._{O-F} = 1.9 \times 10^2\text{ kJ}$$

13-6 The Born-Haber Cycle

Enthalpy changes for some types of chemical reactions and physical changes are given special names. Most of these enthalpy changes can be measured directly in the laboratory.

Process	Reaction (Process)	ΔH_{rxn}
fusion or melting	$A\ (s) \longrightarrow A\ (\ell)$	ΔH_{fus}
solidification	$A\ (\ell) \longrightarrow A\ (s)$	$\Delta H_{sol} = -\Delta H_{fus}$
sublimation	$A\ (s) \longrightarrow A\ (g)$	ΔH_{subl}
vaporization	$A\ (\ell) \longrightarrow A\ (g)$	ΔH_{vap}
condensation	$A\ (g) \longrightarrow A\ (\ell)$	$\Delta H_{cond} = -\Delta H_{vap}$
dissociation	$A_2\ (g) \longrightarrow 2\,A\ (g)$	ΔH_{diss}
ionization (energy)	$A\ (g) \longrightarrow A^+\ (g) + e^-$	ΔH_{ie}
adding e^- (electron affinity)	$A\ (g) + e^- \longrightarrow A\ (g)^-$	ΔH_{ea}
formation of crystal lattice from gaseous ions (crystal lattice energy)	$M^+\ (g) + X^-\ (aq) \longrightarrow MX\ (s)$	ΔH_{xtal}

However, the crystal lattice energy of an ionic solid cannot be determined directly. But it may be evaluated for many ionic compounds by application of Hess' Law. This treatment is referred to as a Born-Haber cycle.

Example 13-7: Using the following experimentally obtained data (Appendix K), calculate the crystal lattice energy for $CaCl_2$ (s).

	Reaction	ΔH° (kJ)
(1)	$Ca\ (s) \longrightarrow Ca\ (g)$	$\Delta H^\circ_{subl} = 192.6$
(2)	$Ca\ (g) \longrightarrow Ca^{+}\ (g)$	$\Delta H^\circ_{ie\ (1)} = 590$
(3)	$Ca^{+}\ (g) \longrightarrow Ca^{2+}\ (g)$	$\Delta H^\circ_{ie\ (2)} = 1333$
(4)	$Cl_2\ (g) \longrightarrow 2\ Cl\ (g)$	$\Delta H^\circ_{diss} = 243.4$
(5)	$2\ Cl\ (g) \longrightarrow 2\ Cl^{-}\ (g)$	$2\,\Delta H^\circ_{ea} = -695.2$
(6)	$Ca^{2+}\ (g) + 2\ Cl^{-}\ (g) \longrightarrow CaCl_2\ (s)$	$\Delta H^\circ_{xtal} = ?$
(7)	$Ca\ (s) + Cl_2\ (g) \longrightarrow CaCl_2\ (s)$	$\Delta H^\circ_{f\,CaCl_2\ (s)} = -795$

When equations (1) through (6) are added, equation (7) results. Thus, since enthalpy changes are state functions,

$$\Delta H^\circ_{f\ CaCl_2\ (s)} = \Delta H^\circ_{subl} + \Delta H^\circ_{ie\ (1)} + \Delta H^\circ_{ie\ (2)} + \Delta H^\circ_{diss} + 2\ \Delta H^\circ_{ea} + \Delta H^\circ_{xtal}$$

$$-795 = 192.6 + 590 + 1333 + 243.2 + (-695.2) + \Delta H^\circ_{xtal}$$

$$\Delta H^\circ_{xtal} = -795 - 192.6 - 590 - 1333 - 243.2 + 695.2$$

$$\Delta H^\circ_{xtal} = -2459\ kJ$$

The more negative the value of ΔH°_{xtal} is, the more thermodynamically stable* the corresponding crystal lattice is. It is not surprising that calcium chloride is a crystalline solid with a high melting point, since the molar enthalpy of crystallization (the crystal lattice energy) is very negative. The sequence of reactions and enthalpy changes may be summarized by the following Born-Haber cycle:

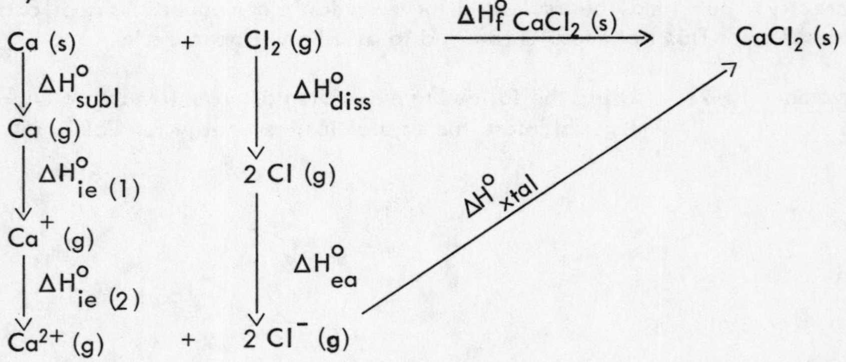

---

*A decrease in the energy of the system is favorable for the spontaneity of a process. Recall that ΔH is negative for an exothermic reaction.

13-7 Second Law of Thermodynamics

The Second Law requires that the disorder or entropy (S) of the universe must increase for a spontaneous process. If the system undergoes an increase in entropy (ΔS is positive) during a process, the spontaneity of that process is favored, but not required.

13-8 Entropy

Entropy and entropy changes are also state functions. Unfortunately, they cannot be measured directly in the laboratory. However, entropy changes (ΔS) for reactions can be evaluated by relating them to other quantities (such as enthalpy and free energy changes, electrode potentials, equilibrium constants).

13-9 Third Law of Thermodynamics

The Third Law states that the entropy of a pure perfect (infinitely ordered) crystalline substance is zero. Standard absolute entropies, S^o, for many substances are also tabulated in Appendix K. In general S^o increases in the order solids < liquids < gases, as expected, since disorder of particles increases in the same order. The Hess' Law of Constant Summation also applies to entropy changes. For a sequence of $\underline{n}$ steps in a reaction (process),

$$\Delta S_{rxn} = \Delta S_1 + \Delta S_2 + \Delta S_3 + \quad . . . \quad \Delta S_n$$

the sum of entropy changes for the individual steps equals the total entropy change for the overall reaction (process). We may also write,

$$\Delta S^o_{rxn} = \Sigma S^o_{(products)} - \Sigma S^o_{(reactants)}$$

Example 13-8: Using tabulated S^o values determine the standard entropy change for the reaction below at 25°C.

$$NH_3\,(g) + HNO_3\,(\ell) \longrightarrow N_2O\,(g) + 2\,H_2O\,(\ell)$$

$$\Delta S^o_{rxn} = S^o_{N_2O\,(g)} + 2\,S^o_{H_2O\,(\ell)} - S^o_{NH_3\,(g)} - S^o_{HNO_3\,(\ell)}$$

$$= 219.7\,\frac{J}{K} + 2(69.91\,\frac{J}{K}) - 192.3\,\frac{J}{K} = 155.6\,\frac{J}{K}$$

$$= \underline{\underline{11.6\,\frac{J}{K}}} = 0.0116\,\frac{kJ}{K}$$

Entropies and entropy changes are usually quite small in comparison to enthalpy changes. Since ΔS^o for this reaction is positive disorder is increased, a situation favorable to the spontaneity of the reaction. This, in itself, however, does not dictate that the reaction must be spontaneous under standard conditions.

13-10 Gibbs Free Energy Change

A reaction is thermodynamically spontaneous if, when products and reactants are mixed in unit molarities or partial pressures, the forward reaction proceeds to a greater extent than the reverse reaction before equilibrium is established.

The spontaneity of a process can be predicted by the Gibbs free energy change (a state function) which is related to both ΔH and ΔS for the process by the following equation, where T is the absolute temperature (K).

$$\Delta G = \Delta H - T\Delta S$$

If ΔG for a process is negative (free energy or useful energy released by the system) the process must be spontaneous under the prevailing conditions. Conversely a positive ΔG corresponds to a nonspontaneous process.* When the system is at equilibrium (neither forward nor reverse reaction spontaneous), $\Delta G = 0$. Thus, for any system at equilibrium,

$$0 = \Delta H - T\Delta S$$

and $$\Delta H = T\Delta S \qquad \text{or} \qquad T = \frac{\Delta H}{\Delta S}$$

If we know ΔH and ΔS for a given reaction we may estimate the temperature at which the system is at equilibrium using ΔH^o and ΔS^o values.** If we increase the temperature, a transition between spontaneity and non-spontaneity occurs as we pass through the equilibrium temperature.

Remember that the spontaneity of a given process is favored (but not required) if either ΔH is negative or ΔS is positive. Since ΔH and ΔS may each be either positive or negative, we can classify reactions into four categories with respect to spontaneity as shown in Table 13-6.

---

*The reverse of a spontaneous reaction, under a given set of conditions, must be nonspontaneous since $\Delta G_{forward} = -\Delta G_{reverse}$.

**This makes the assumption that ΔH^o and ΔS^o at 298 K are the same as at the temperature of interest. Strictly speaking this assumption is not absolutely true but, nevertheless, it often gives good approximations.

Table 13-1. Relationship of Thermodynamic Quantities to Spontaneity

ΔH	ΔS	ΔG	Temperature Range of Spontaneity
-	+	-	all temperatures
-	-	+ or -	only at low temperatures
+	+	+ or -	only at high temperatures
+	-	+	no temperatures

If we are interested in a process at standard conditions,

$$\Delta G^{o} = \Delta H^{o} - T\Delta S^{o}$$

The value of ΔG^{o} may be calculated by either:

(1) evaluation of ΔH^{o} and ΔS^{o} from ΔH_f^{o} and S^{o} values of the species involved, or by

(2) utilizing the relationship

$$\Delta G^{o} = \Sigma\Delta G_f^{o}\,(\text{products}) - \Sigma\Delta G_f^{o}\,(\text{reactants})$$

Standard free energy of formation values, ΔG_f^{o} are tabulated in Appendix K of the text. Elements in their standard states have ΔG_f^{o} values of zero.

Example 13-9: Use tabulated ΔG_f^{o} values to calculate the standard free energy change in kilojoules for the reaction below. Is the reaction spontaneous at 298 K?

$$2\,NH_3\,(g) + 3\,CuO\,(s) \longrightarrow 3\,H_2O\,(\ell) + N_2\,(g) + 3\,Cu\,(s)$$

$$\Delta G^{o}_{rxn} = 3\Delta G^{o}_{H_2O\,(\ell)} + \Delta G^{o}_{N_2\,(g)} + 3\Delta G^{o}_{Cu\,(s)} - 2\Delta G^{o}_{NH_3\,(g)} - 3\Delta G^{o}_{CuO\,(s)}$$

$$= 3(-237.2\text{ kJ}) + 0\text{ kJ} + 3\,(0\text{ kJ}) - 2(-16.5\text{ kJ}) - 3(-130\text{ kJ})$$

$$\Delta G^{o}_{rxn} = \underline{\underline{-289\text{ kJ}}}$$

Since ΔG^{o}_{rxn} is negative the forward reaction is spontaneous at 298 K, which tells us that the reaction can, but does not necessarily, occur at an observable rate under these conditions.

Example 13-10: Evaluate ΔG^{o}_{rxn} for the reaction in Example 13-9 using ΔH_f^{o} and S^{o} values.

	$2\ NH_3\ (g)$	+ $3\ CuO\ (s)$	$\rightarrow$ $3\ H_2O\ (\ell)$	+ $N_2\ (g)$	+ $3\ Cu\ (s)$
ΔH_f°	-46.11 kJ	-157 kJ	-285.8 kJ	0 kJ	0 kJ
S°	192.3 J/K	42.63 J/K	69.91 J/K	191.5 J/K	33.15 J/K

$$\Delta H^\circ_{rxn} = 3(-285.8\ kJ) + 0\ kJ + 3(0\ kJ) - 2(-46.11\ kJ) - 3(-157)$$

$$\Delta H^\circ_{rxn} = -294\ kJ$$

$$\Delta S^\circ_{rxn} = 3(69.91\ J/K) + 191.5\ J/K + 3(33.15\ J/K) - 2(192.3\ J/K) - 3(42.63\ J/K)$$

$$\Delta S^\circ_{rxn} = -11.81\ J/K$$

$$\Delta G^\circ_{rxn} = \Delta H^\circ_{rxn} - T\Delta S^\circ_{rxn}$$

$$= -294\ kJ - (298\ K)(-0.01181\ kJ/K)$$

$$= -294\ kJ + 3.52\ kJ$$

$$\Delta G^\circ_{rxn} = \underline{\underline{-290\ kJ}}$$ (Round-off errors account for the small difference in the results of these two calculations.)

Notice that ΔS°_{rxn} was converted to kJ to be consistent with the units of ΔH°_{rxn}.

<u>Example 13-11:</u> Estimate the temperature range over which the following reaction is spontaneous.

$$PCl_5\ (g) \longrightarrow PCl_3\ (g) + Cl_2\ (g)$$

At equilibrium $\Delta G_{rxn} = 0 = \Delta H_{rxn} - T\Delta S_{rxn}$

and,

$$\Delta H_{rxn} = T\Delta S_{rxn} \quad \text{or} \quad T = \frac{\Delta H_{rxn}}{\Delta S_{rxn}} \approx \frac{\Delta H^\circ_{rxn}}{\Delta S^\circ_{rxn}}$$

We may <u>estimate</u> the temperature at which the system is at equilibrium by evaluating ΔH°_{rxn} and ΔS°_{rxn}.

$$\Delta H^{o}_{rxn} = \Delta H^{o}_{PCl_3\,(g)} + \Delta H^{o}_{Cl_2\,(g)} - \Delta H^{o}_{PCl_5\,(g)}$$

$$= -306.4\ kJ + 0\ kJ - (-398.9\ kJ)$$

$$\Delta H^{o}_{rxn} = \underline{\underline{+92.5\ kJ}} \quad \text{(endothermic rxn)}$$

$$\Delta S^{o}_{rxn} = S^{o}_{PCl_3\,(g)} + S^{o}_{Cl_2\,(g)} - S^{o}_{PCl_5\,(g)}$$

$$= 311.7\ \frac{J}{K} + 223.0\ \frac{J}{K} - 353\ \frac{J}{K}$$

$$\Delta S^{o}_{rxn} = \underline{\underline{182\ \frac{J}{K}}}$$

$$T = \frac{\Delta H^{o}_{rxn}}{\Delta S^{o}_{rxn}} = \frac{9.25 \times 10^4\ J}{182\ J/K} = 508\ K = \underline{\underline{235^{o}C}}$$

At temperatures above 235°C the $T\Delta S^{o}$ term predominates over ΔH^{o}, ΔG^{o} is negative, and the forward reaction is spontaneous. At temperatures below 235°C, ΔH^{o} predominates over $T\Delta S^{o}$, ΔG^{o} is positive, and the forward reaction is nonspontaneous (but the reverse reaction is spontaneous).

EXERCISES

1. Calculate the change in internal energy for a system if it:
   (a) absorbs 300 joules of heat and does 200 joules of work on its surroundings.
   (b) absorbs 300 joules of heat and the surroundings do 200 joules of work on it.
   (c) releases 400 joules of heat.
   (d) does 100 joules of work on the surroundings.
   (e) does 100 joules of work on the surroundings and releases 400 joules of heat.

2. A reaction in which metallic zinc is dropped into hydrochloric acid produces hydrogen. (a) How much work is done if the dry hydrogen, collected in a 2.00 liter container, exerts a pressure of 742 torr at 30.0 °C? (b) How many moles and g of hydrogen were produced?
   1 L·atm = 101.32 J

3. Calculate the amount of heat released, in kilojoules, in the reaction of 13.4 g of magnesium with excess chlorine to produce magnesium chloride under standard state conditions.

$$Mg\ (s) \quad + \quad Cl_2\ (g) \quad \longrightarrow \quad MgCl_2\ (s)$$

4. Calculate the amount of heat released, in kilojoules, in the complete oxidation of 8.17 g of aluminum at 25°C and one atmosphere pressure to form aluminum oxide.

$$3\ O_2\ (g) \quad \longrightarrow \quad 2\ Al_2O_3\ (s)$$

5. Calculate the standard molar enthalpy change, ΔH^o, at 25°C for each of the following reactions using the data in Appendix K of the text.
   (a) $SiO_2\ (s) + Na_2CO_3\ (s) \longrightarrow Na_2SiO_3\ (s) + CO_2\ (g)$
   (b) $H_2SiF_6\ (aq) \longrightarrow 2\ HF\ (aq) + SiF_4\ (g)$
   (c) $2\ MgS\ (s) + 3\ O_2\ (g) \longrightarrow 2\ MgO\ (s) + 2\ SO_2\ (g)$
   (d) $BaSO_4\ (s) + 4\ C\ (graphite) + CaCl_2\ (s) \longrightarrow BaCl_2\ (s) + CaS\ (s) + 4\ CO\ (g)$
   (e) $2\ Pb\ (s) + 2\ H_2O\ (\ell) + O_2\ (g) \longrightarrow 2\ Pb(OH)_2\ (s)$

6. Given the following at 25°C:

   $2\ C_6H_6\ (\ell) + 15\ O_2\ (g) \longrightarrow 12\ CO_2\ (g) + 6\ H_2O\ (\ell) \qquad \Delta H^o = -6535\ kJ$

   ΔH_f^o for CO_2 (g) is −393.5 kJ/mol and for H_2O (ℓ) is −285.8 kJ/mol.

   Calculate the standard molar enthalpy of formation of benzene, C_6H_6 (ℓ).

7. Given the following at 25°C:

	ΔH°
Fe_2O_3 (s) + 3 CO (g) ⟶ 2 Fe (s) + 3 CO_2 (g)	-24.8 kJ
3 Fe_2O_3 (s) + CO (g) ⟶ 2 Fe_3O_4 (s) + CO_2 (g)	-46.4 kJ
Fe_3O_4 (s) + CO (g) ⟶ 3 FeO (s) + CO_2 (g)	19 kJ

Calculate ΔH° for the reaction:

FeO (s) + CO (g) ⟶ Fe (s) + CO_2 (g)

8. Given the following at 25°C:

P_4O_{10} (s) + 4 HNO_3 (ℓ) ⟶ 4 HPO_3 (s) + 2 N_2O_5 (g) $\Delta H^\circ = -102.3$ kJ

ΔH_f° for P_4O_{10} (s) is -2984 kJ/mol, for HNO_3 (ℓ) is -174.1 kJ/mol, for N_2O_5 (g) is 11 kJ/mol. Calculate ΔH_f° for HPO_3 (s) at 25°C.

9. Given that the H-H bond energy = 435 kJ/mol, Cl-Cl bond energy = 243 kJ/mol and ΔH_f° for HCl (g) is -92 kJ/mol, calculate the H-Cl bond energy.

10. Estimate the enthalpy change for the reaction below from the average bond energies given.

CH_4 (g) + 2 Cl_2 (g) ⟶ CH_2Cl_2 (g) + 2 HCl (g)

There are two C-Cl bonds and two C-H bonds per molecule of CH_2Cl_2.

Average Bond Energies

C-H	414 kJ/mol
Cl-Cl	243 kJ/mol
H-Cl	431 kJ/mol
C-Cl	330 kJ/mol

11. Calculate the average bond energy in kJ/mol of bonds for the C-H bond from the following data:

C (graphite) + 2 H_2 (g) ⟶ CH_4 (g) $\Delta H^\circ = -74.81$ kJ

ΔH_f° for H (g) = 218.0 kJ/mol

ΔH_f° for C (g) = 716.7 kJ/mol

12. Given the following enthalpy changes:

	ΔH°
C (graphite) + O_2 (g) ⟶ CO_2 (g)	-393.5 kJ
CO (g) + 1/2 O_2 (g) ⟶ CO_2 (g)	-283.0 kJ

Calculate ΔH_f° for CO (g).

13. Given the following information:

	ΔH°
2 C (graphite) ⟶ 2 C (g)	1433 kJ
3 H_2 (g) ⟶ 6 H (g)	1308 kJ
2 C (graphite) + 3 H_2 (g) ⟶ C_2H_6 (g)	-85 kJ

H H
C_2H_6 (g) is ethane which is H-C-C-H.
H H

The average bond energy of the C–H bond is 416 kJ/mol. Calculate the bond energy of the C–C bond in kcal/mol of bonds.

14. Given the following at 25°C.

	ΔH°
SO_3 (g) + H_2O (ℓ) ⟶ H_2SO_4 (ℓ)	-133 kJ
Pb (s) + PbO_2 (s) + 2 H_2SO_4 (ℓ) ⟶ 2 $PbSO_4$ (s) + 2 H_2O (ℓ)	-509 kJ

Calculate ΔH° for the reaction below at 25°C:

Pb (s) + PbO_2 (s) + 2 SO_3 (g) ⟶ 2 $PbSO_4$ (s)

15. Given the following information, calculate the crystal lattice energy of magnesium bromide, $MgBr_2$ (s): $\Delta H^\circ_{f\ Mg\ (g)}$ = 150 kJ/mol; first and second ionization energies for Mg (g) are 7.61 and 14.96 eV/atom, respectively; $\Delta H_{f\ Br_2\ (g)}$ = 30.91 kJ/mol; Br–Br bond energy = 192 kJ/mol; electron affinity of Br = -323 kJ/mol; ΔH°_f $MgBr_2$ (s) = -517.6 kJ/mol. One electron-volt/atom = 96.49 kJ/mol.

16. Given the following information, calculate the crystal lattice energy of aluminum fluoride, AlF_3 (s): $\Delta H^\circ_{f\ Al\ (g)}$ = 314 kJ/mol; first, second, and third ionization energies for Al (g) are 5.96 eV, 18.74 eV, and 28.31 eV/atom, respectively; F–F bond energy = 158.0 kJ/mol; electron affinity of F = -323 kJ/mol; $\Delta H_{f\ AlF_3\ (s)}$ = -1301 kJ/mol. One electron-volt/atom = 96.48 kJ/mol.

17. Calculate the standard molar entropy change, ΔS°, for each of the reactions below at 25°C using the data in Appendix K in the text.

(a) 4 NH_3 (g) + 5 O_2 (g) ⟶ 4 NO (g) + 6 H_2O (g)
(b) 3 NO_2 (g) + H_2O (ℓ) ⟶ 2 HNO_3 (aq) + NO (g)
(c) Fe_2O_3 (s) + 13 CO (g) ⟶ 2 $Fe(CO)_5$ (g) + 3 CO_2 (g)

18. Calculate ΔG° values at 25°C for each of the reactions of Problem 17 from tabulated ΔG°_f values.

19. Calculate ΔG° values at 25°C for each of the reactions of Problem 17 from tabulated ΔH°_f and S° values.

20. Estimate the normal boiling point of each of the following liquids using tabulated thermodynamic data:

(a) bromine, Br_2 (ℓ) (b) ethanol, C_2H_5OH (ℓ)

21. Calculate ΔH°, ΔS°, and ΔG° for the reaction below at 25°C.

SO_3 (g) + H_2O (ℓ) ⟶ H_2SO_4 (ℓ)

Estimate the temperature above which the reaction becomes nonspontaneous.

22. Calculate ΔS° for the reaction below from the data given.

	$PbS\ (s)$	$+\ 2\ HCl\ (g)$	$\longrightarrow$	$PbCl_2\ (s)$	$+$	$H_2S\ (g)$
ΔH_f° (kJ/mol)	-100.4	-92.31		-359.4		-20.6
ΔG_f° (kJ/mol)	-98.7	-95.30		-314.1		-33.6

23. Calculate the temperature above which the reaction of Problem 22 is nonspontaneous.

ANSWERS FOR EXERCISES

1. (a) 100 J (b) 500 J (c) -400 J (d) -100 J (e) -500 J 2. (a) 1.95 L·atm or 198 J; (b) 0.0785 mol, 0.158 g 3. 354 kJ released 4. 254 kJ 5. (a) 569 kJ (b) 57.8 kJ (c) -1103 kJ (d) 475.5 kJ (e) -460.2 kJ 6. 49 kJ/mol 7. -11 kJ 8. -951.2 kJ/mol 9. 431 kJ per mole of bonds 10. -208 kJ 11. 415.9 kJ per mol of bonds 12. -110.5 kJ 13. 330 kJ per mol of bonds (The tabulated average value for C–C bonds is 347 kJ/mol.) 14. -775 kJ 15. -2421 kJ/mol 16. -6.00×10^3 kJ/mol 17. (a) 180.8 J/K (b) -287 J/K (c) -1125 J/K 18. (a) -959.3 kJ (b) -51.1 kJ (c) -52.0 kJ 19. (a) -959.3 kJ (b) -51.3 kJ (c) -52.2 kJ 20. (a) 59°C (58.8°C is the actual b.p.) (b) 77°C (78.5°C is the actual b.p.) 21. $\Delta H^\circ = -31.65$ kcal, $\Delta S^\circ = -40.55$ cal/K, $\Delta G^\circ = -19.6$ kcal. At any temperature below 508°C the reaction is predicted to be spontaneous (since ΔH° is negative and ΔS° is negative). Therefore it is nonspontaneous above 508°C. 22. -123 J/K 23. The reaction is predicted to be nonspontaneous at temperatures above 499°C and spontaneous below this temperature.

chemical kinetics - study of rates of rxns & mechanisms by which they occur.

1) temp.
2) catalyst
3) nature of reactants
4) concen. of reactants

Chapter Fourteen

CHEMICAL KINETICS

D, G, W: Chap. 15

14-1 Introduction

The study of the rates of chemical reactions and the mechanisms, or series of steps, by which they occur is called chemical kinetics. The rate of a reaction can be defined as the amount (or concentration) of product produced or reactant consumed per unit time. Reaction rates are usually expressed in units of moles per liter per unit time. The rates of reactions depend upon four factors: (1) nature of reactants, (2) concentrations of reactants, (3) temperature, and (4) catalysts.

14-2 Nature of Reactants

The rate of reaction depends upon the substances mixed and their physical states. For reactions involving solids, rates increase as particle size decreases and surface area increases. This is because larger surface areas expose more atoms and molecules to the other reactants. Reactions between positive and negative ions in solution are also generally more rapid than those between neutral species due to coulombic attractions between oppositely charged ions and the fact that no formal chemical bonds must be broken.

14-3 Concentrations of Reactants

We can represent a hypothetical reaction as shown below where the capital letters represent formulas and the lower case letters represent coefficients in the balanced overall equation.

$$a\,A + b\,B \longrightarrow c\,C + d\,D$$

The reaction rate can be represented as the rate of decrease in concentration of a reactant or the rate of increase in concentration of a product. For example, $-\Delta[A]/\Delta t$ represents the change in concentration (moles per liter) of A per unit time interval (Δt). The negative sign signifies that the concentration of A decreases with time. The rates of decrease and increase in concentrations of reactants and products, respectively, are related by the coefficients of the balanced equation for the overall reaction.

$$\text{Rate} = -\frac{\Delta[A]}{a\Delta t} = -\frac{\Delta[B]}{b\Delta t} = \frac{\Delta[C]}{c\Delta t} = \frac{\Delta[D]}{d\Delta t}$$

The rate of a forward reaction decreases with time as reactants are consumed. The above expression describes the average rate over some arbitrary time interval (Δt). Frequently we find it useful to use another kind of expression to describe reaction rates, called the rate-law expression.

$$\text{Rate} = k[A]^x[B]^y$$

This expression describes the instantaneous rate at some arbitrary time after reactants are mixed, usually the initial rate, i.e., the rate at the instant of mixing ($t = 0$). The initial rate of a forward reaction at a given temperature is equal to a specific rate constant, k, times the concentration of the reactants each raised to a power. The powers, x and y, are experimentally determined and bear no necessary relationship to the coefficients of the balanced overall equation (a, b, c, d). As we shall see, the powers are important in deducing possible mechanisms for reactions. The value of x is said to be the order of the reaction with respect to A, and y is the order of the reaction with respect to B. The overall order of a reaction is the sum of the powers in the rate-law expression, $x + y$. The specific rate constant, k, is also experimentally determined, is different for different reactions, and, for a given reaction, increases with increasing temperature. The units of k vary with the form of the rate-law expression.

To illustrate these points consider the reaction of nitrogen oxide (nitric oxide), with hydrogen at a high temperature.

$$2\,NO\,(g) + 2\,H_2\,(g) \longrightarrow N_2\,(g) + 2\,H_2O\,(g)$$

The data in Table 14-1 show that for the reaction initiated with 0.0400 mole of NO and 0.0400 mole of H_2 in a one liter container at a certain temperature:

$$\text{Rate} = -\frac{\Delta[NO]}{2\,\Delta t} = -\frac{\Delta[H_2]}{2\,\Delta t} = \frac{\Delta[N_2]}{\Delta t} = \frac{\Delta[H_2O]}{2\,\Delta t}$$

Table 14-1

Concentrations of NO and H_2 at Various Times
(at constant temperature)

t (s)	NO (M)	H_2 (M)	N_2 (M)	H_2O (M)
0	0.0400	0.0400	0	0
5	0.0182	0.0182	0.0109	0.0218
10	0.0162	0.0162	0.0119	0.0238
15	0.0148	0.0148	0.0126	0.0252
20	0.0137	0.0137	0.0132	0.0263

Figure 14-1 shows the variations in concentrations of reactants and products versus time using the data of Table 14-1. The rate of consumption of reactants (forward reaction) decreases with time while the rate of appearance of products (reverse reaction) increases with time.

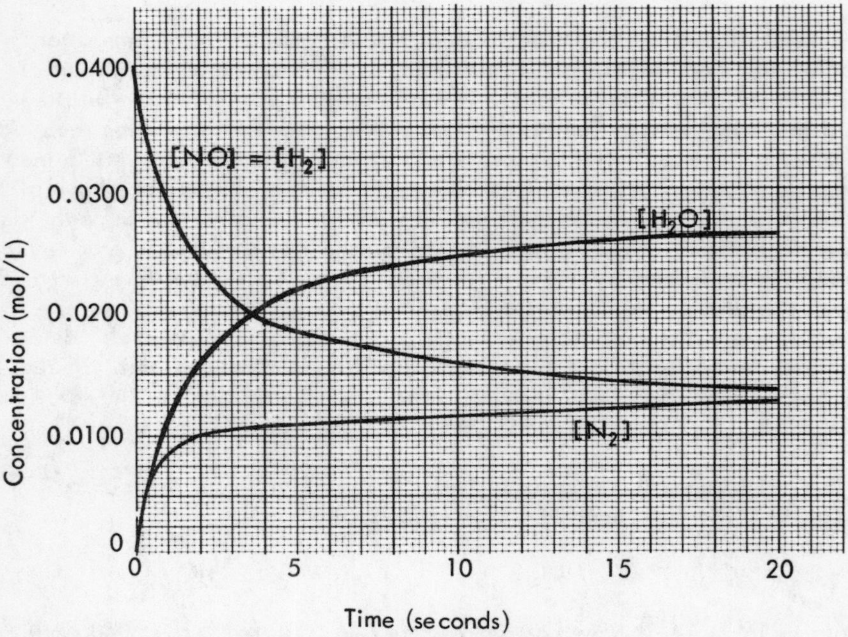

Figure 14-1. Plot of concentrations versus time for

$$2\ NO\ (g) + 2\ H_2\ (g) \longrightarrow N_2\ (g) + 2\ H_2O\ (g)$$

In general, the rate-law expression is: $\text{Rate} = k[A]^x[B]^y$

which, for this reaction, is: $\text{Rate} = k[NO]^x[H_2]^y$.

The rate-law expression for this reaction is $R = k[NO]^2[H_2]$. The powers, x and y (in this case 2 and 1, respectively), and the value of the specific rate constant, k, are determined by measuring the initial rate of reaction for different initial concentrations of reactants. Let us now develop the basic ideas.

Example 14-1: Consider the hypothetical reaction below

$$2\ A + 3\ B \longrightarrow 2\ C + D$$

at the temperature at which the initial rate data that follow were collected. Determine the rate-law expression for this reaction and evaluate k, the specific rate constant.

Experiment	Initial [A]	Initial [B]	Initial Rate of Formation of C
1	0.10 M	0.10 M	2.0×10^{-3} M·s^{-1}
2	0.10 M	0.30 M	6.0×10^{-3} M·s^{-1}
3	0.30 M	0.30 M	5.4×10^{-2} M·s^{-1}

The rate-law expression has the general form: $\text{Rate} = k[A]^x[B]^y$

Comparison of experiments 1 and 2 shows the change in rate due to changing the initial concentration of B (the concentration of A is constant). Let us divide the rate expression for experiment 2 by that for experiment 1.

$$\frac{\text{Rate}_{(2)}}{\text{Rate}_{(1)}} = \frac{k[A]^x_{(2)}[B]^y_{(2)}}{k[A]^x_{(1)}[B]^y_{(1)}}$$

The temperature does not change and so k is the same in both experiments. Also, $[A]_{(2)} = [A]_{(1)}$. So the equation simplifies to:

$$\frac{\text{Rate}_{(2)}}{\text{Rate}_{(1)}} = \frac{[B]^y_{(2)}}{[B]^y_{(1)}} \quad \text{or} \quad \frac{\text{Rate}_{(2)}}{\text{Rate}_{(1)}} = \left(\frac{[B]_{(2)}}{[B]_{(1)}}\right)^y$$

Substitution of values for [B] from Table 14-1 gives:

$$\frac{6.0 \times 10^{-3}\ \underline{M}\cdot s^{-1}}{2.0 \times 10^{-3}\ \underline{M}\cdot s^{-1}} = \left(\frac{0.30\ \underline{M}}{0.10\ \underline{M}}\right)^y \quad \text{or} \quad 3.0 = (3.0)^y, \quad \underline{y = 1}$$

This tells us that the reaction rate is directly proportional to the initial concentration of B raised to the first power. At this point we know that the rate law expression is:

$$\text{Rate} = k[A]^x[B]^1 \quad \text{or} \quad \text{Rate} = k[A]^x[B]$$

If we compare experiments 2 and 3 (initial [B] is constant), the change in initial rate must be due only to changing the initial [A]. Dividing the rate-law expression for experiment 3 by that for 2 we get:

$$\frac{\text{Rate}_{(3)}}{\text{Rate}_{(2)}} = \frac{k[A]^x_{(3)}[B]_{(3)}}{k[A]^x_{(2)}[B]_{(2)}}$$

The k and [B] terms are equal and cancel.

$$\frac{\text{Rate}_{(3)}}{\text{Rate}_{(2)}} = \left(\frac{[A]_{(3)}}{[A]_{(2)}}\right)^x$$

Substituting data from the table we are able to obtain a value for x.

$$\frac{5.4 \times 10^{-2}\ \underline{M}\cdot s^{-1}}{6.0 \times 10^{-3}\ \underline{M}\cdot s^{-1}} = \left(\frac{0.30\ \underline{M}}{0.10\ \underline{M}}\right)^x \quad \underline{\text{or}} \quad 9.0 = (3.0)^x, \quad \underline{\underline{x = 2}}$$

We have determined that the initial rate of reaction is directly proportional to the square of the initial [A] and therefore the rate-law expression is:

$$\text{Rate} = k[A]^2[B]$$

A is squared

Note that the powers to which concentrations are raised in the rate-law expression are <u>not the same as the coefficients in the balanced equation for the reaction</u>.

We may evaluate the specific rate constant, k, by using data for any one experiment. Let's use experiment 2.

$$\text{Rate}_{(2)} = k[A]^2_{(2)}\ [B]_{(2)}$$

$$6.0 \times 10^{-3}\ \underline{M}\cdot s^{-1} = k(0.10\ \underline{M})^2(0.30\ \underline{M})$$

$$k = \frac{6.0 \times 10^{-3}\ \underline{M}\cdot s^{-1}}{(0.10\ \underline{M})^2\ (0.30\ \underline{M})} = 2.0\ \underline{M}^{-2}\cdot s^{-1}$$

2

For the reaction, $2\,A + 3\,B \longrightarrow 2\,C + D$

$$\underline{\underline{\text{Rate} = 2.0\ M^{-2}\cdot s^{-1}\ [A]^2[B]}}$$

Example 14-2: What will be the rate of the reaction of Example 14-1 at the same temperature if the initial concentrations are: $[A] = 0.50\ \underline{M}$ and $[B] = 0.40\ \underline{M}$?

Since we know the rate-law expression from Example 14-1, we can determine the rate for any set of initial concentrations (at the same temperature).

$$\text{Rate} = k[A]^2[B]^1 = (2.0\ M^{-2}\cdot s^{-1})(0.50\ \underline{M})^2(0.40\ \underline{M})$$

$$\underline{\underline{\text{Rate} = 0.20\ \underline{M}\cdot s^{-1}}}$$

Example 14-3: Derive the rate-law expression for the generalized reaction

$$A + B + 3C \longrightarrow \text{Products}$$

at the temperature at which the following data were obtained.

Experiment	Initial [A]	Initial [B]	Initial [C]	Initial Rate
1	0.20 M	0.10 M	0.30 M	4.0×10^{-4} M·min^{-1}
2	0.20 M	0.20 M	0.30 M	4.0×10^{-4} M·min^{-1}
3	0.20 M	0.30 M	0.60 M	8.0×10^{-4} M·min^{-1}
4	0.60 M	0.40 M	1.20 M	1.44×10^{-2} M·min^{-1}

$$\text{Rate} = k\,[A]^x\,[B]^y\,[C]^z$$

In experiments 1 and 2, the concentrations of A and C are constant and so we use these two sets of data to determine the power to which [B] is raised in the rate-law expression.

$$\frac{\text{Rate}_{(2)}}{\text{Rate}_{(1)}} = \frac{k\,[A]^x_{(2)}\;[B]^y_{(2)}\;[C]^z_{(2)}}{k\,[A]^x_{(1)}\;[B]^y_{(1)}\;[C]^z_{(1)}}$$

Since k = k, $[A]_{(2)} = [A]_{(1)}$, and $[C]_{(2)} = [C]_{(1)}$,

$$\frac{\text{Rate}_{(2)}}{\text{Rate}_{(1)}} = \left(\frac{[B]_{(2)}}{[B]_{(1)}}\right)^y$$

$$\frac{4.0 \times 10^{-4}\ \text{M·min}^{-1}}{4.0 \times 10^{-4}\ \text{M·min}^{-1}} = \left(\frac{0.20\ \text{M}}{0.10\ \text{M}}\right)^y \qquad 1.0 = (2.0)^y,\ \ \underline{y = 0}$$

This tells us that the reaction rate is independent of the initial concentration of B. At this point we know:

$$\text{Rate} = k\,[A]^x\,[B]^0\,[C]^z \qquad [B]^0 = 1$$

$$\text{Rate} = k\,[A]^x[C]^z$$

Since the rate of the reaction is independent of [B], we may ignore the changes in [B].

The [A] is the same in experiments 2 and 3, and so we can evaluate the effect of changing [C] on the rate.

$$\frac{\text{Rate}_{(3)}}{\text{Rate}_{(2)}} = \left(\frac{[C]_{(3)}}{[C]_{(2)}}\right)^z$$

$$\frac{8.0 \times 10^{-4}\ \underline{M}\cdot\text{min}^{-1}}{4.0 \times 10^{-4}\ \underline{M}\cdot\text{min}^{-1}} = \left(\frac{0.60\ \underline{M}}{0.30\ \underline{M}}\right)^z, \quad 2.0 = (2.0)^1, \quad \underline{\underline{z = 1}}$$

$\therefore$ Rate = k $[A]^x [C]^1$ <u>or</u> Rate = k $[A]^x [C]$

Since there are no experiments in which the initial [A] changes while the initial [C] remains the same, we must evaluate x indirectly. Let us compare experiments 3 and 4. The initial [C] is doubled and this change alone should cause the rate to double because the rate is proportional to the first power of [C]. The rate actually increases by a factor of $1.44 \times 10^{-2}/8.0 \times 10^{-4} = 18$. Since a factor of 2 of the 18-fold increase is due to changing the initial [C], a factor of 18/2 = 9 must be due to changing the initial [A]. The initial [A] in experiment 4 is three times greater (it is tripled) than in experiment 3. So the [A] must be squared in the rate-law expression.

$$9.0 = (3.0)^x, \quad \text{therefore} \quad \underline{\underline{x = 2}}$$

(9.0 ↑ rate increase due to [A] alone; $(3.0)^x$ ↑ increase in initial [A])

$$\underline{\underline{\text{Rate}}} = k\,[A]^2[B]^0[C]^1 = \underline{\underline{k\,[A]^2[C]}}$$

Let's use experiment 3 to evaluate k.

$$\text{Rate}_{(3)} = k\,[A]^2[C]$$

$$8.0 \times 10^{-4}\ \underline{M}\cdot\text{min}^{-1} = k(0.20\ \underline{M})^2(0.60\ \underline{M})$$

$$k = \frac{8.0 \times 10^{-4}\ \underline{M}\cdot\text{min}^{-1}}{(0.20\ \underline{M})^2(0.60\ M)} = 3.3 \times 10^{-2}\ \underline{M}^{-2}\cdot\text{min}^{-1}$$

$$\underline{\underline{\text{Rate} = 3.3 \times 10^{-2}\ M^{-2}\cdot\text{min}^{-1}\,[A]^2[C]}}$$

14-4 Effect of Temperature

In order for a reaction to occur colliding molecules must have the proper orientations relative to each other, and they must have sufficient energy to allow their own bonds to be broken. The amount of energy necessary to initiate a reaction is called the energy of activation or more simply, activation energy, E_a. Slow reactions have high activation energies and fast reactions have low activation energies. Since molecules have higher average velocities at higher temperatures, increasing the temperature increases the rate of reaction because more molecules possess energies equal to or greater than the activation energy. A rule-of-thumb is that for every $10^{\circ}C$ rise in temperature the rate of reaction doubles. This applies strictly only when $E_a \approx 50$ kJ/mol.

Arrhenius devised a relationship between the specific rate constant, k, at a particular temperature, T, and the activation energy, E_a.

$$k = Ae^{-E_a/RT}$$

The A is the collision frequency factor which, as the name implies, is related to the number of collisions of molecules with proper relative orientations per unit time. R is the gas constant, 8.314 J/mol K.

14-5 Effect of a Catalyst

Catalysts are substances that alter the pathway of a reaction, and therefore its activation energy, but they are neither produced nor consumed in the overall process. Most catalysts are used to speed up reactions by lowering E_a, but there are inhibitory catalysts that force reactions to occur via less favorable pathways. The figure below is a plot of potential energy versus progress of a generalized reaction for a catalyzed and uncatalyzed exothermic reaction.

$$M + N \longrightarrow O + P + \text{Heat}$$

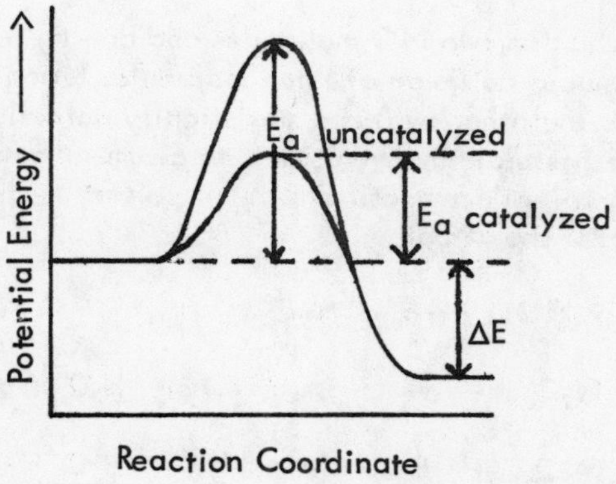

The presence of a catalyst lowers the activation energy from E_a to E_a'. The lowering of activation energy, $\Delta E_a = E_a - E_a'$, is the same for the forward and reverse (endothermic) reactions, therefore the rates of both forward and reverse reactions increase equally.

14-6 The Rate-Law Expression and Mechanisms

Recall that the mechanism of a reaction is the sequence of steps by which reactants are converted into products. The order of a reaction with respect to a particular reactant indicates the number of molecules of that substance that are reactants in the slow step (rate-determining step) or the number of molecules that are reactants in fast steps preceding the slow step. The experimental determination of the powers (orders) to which concentrations of reactants are raised in the rate law expression allows us to postulate conceivable mechanisms.

For the following reaction, the rate law expression is known to be,

$$\text{Rate} = k\,[NO]^2[H_2]$$

which tells us that the reaction is second order with respect to NO and first order with respect to H_2.

$$2\,NO\ (g) + 2\,H_2\ (g) \longrightarrow 2\,H_2O\ (g) + N_2\ (g)$$

Many mechanisms are consistent with this information including the following one.

$$\begin{array}{llr}
\text{Step 1.} & 2\,NO + H_2 \longrightarrow H_2N_2O_2 & \text{(slow)} \\
\text{Step 2.} & H_2N_2O_2 + H_2 \longrightarrow 2\,H_2O + N_2 & \text{(fast)} \\
\hline
\text{overall:} & 2\,NO + 2\,H_2 \longrightarrow 2\,H_2O + N_2 &
\end{array}$$

According to this mechanism two NO molecules and one H_2 react in the slow step. But the simultaneous collision of three molecules (each with the proper orientation relative to the other two) is a statistically unlikely event. In fact, one reasonable way to postulate mechanisms is to assume that only bimolecular collisions and unimolecular decompositions are important. The mechanism believed to be correct is shown below.

$$\begin{array}{llr}
\text{Step 1.} & 2\,NO \longrightarrow N_2O_2 & \text{(fast)} \\
\text{Step 2.} & N_2O_2 + H_2 \longrightarrow N_2O + H_2O & \text{(slow)} \\
\text{Step 3.} & N_2O + H_2 \longrightarrow N_2 + H_2O & \text{(fast)} \\
\hline
\text{overall:} & 2\,NO + 2\,H_2 \longrightarrow 2\,H_2O + N_2 &
\end{array}$$

Slow step can't occur until 1st fast one has.

Two NO molecules collide to form N_2O_2 in the first step, which is fast in comparison to step two in which the product of step one, an N_2O_2 molecule, collides with one H_2 molecule to produce one N_2O and one H_2O molecule. The N_2O molecule then rapidly combines with an H_2 molecule to yield one N_2 and another H_2O molecule. The overall rate of reaction is governed by the slow step which, of course, cannot occur until after the first (fast) step has occurred. Thus the rate is dependent upon the concentrations of reactants involved in the first two steps. Since N_2O_2 is produced in step 1 and consumed in step 2 its concentration cancels from the rate-law expression, and we find that the reaction is overall third order with 2 NO and 1 H_2 molecule being reactants in the first two steps.

Example 14-4: The reaction below is found to obey the indicated rate-law expression. Which of the mechanisms that follow are consistent with this expression?

$$2\,N_2O_5 \longrightarrow 4\,NO_2 + O_2 \qquad \text{Rate} = k[N_2O_5]$$

$K[N_2O_5]$

Possible mechanism 1:

Step 1.	$2\,N_2O_5 \longrightarrow 2\,NO_2 + 2\,NO_3$	(fast)
Step 2.	$NO_3 \longrightarrow NO + O_2$	(slow)
Step 3.	$NO + NO_3 \longrightarrow 2\,NO_2$	(fast)
overall:	$2\,N_2O_5 \longrightarrow 4\,NO_2 + O_2$	

Possible mechanism 2:

Step 1.	$N_2O_5 \longrightarrow N_2O + 2\,O_2$	(slow)
Step 2.	$N_2O_5 + N_2O \longrightarrow 2\,NO_2 + O_2 + N_2$	(fast)
Step 3.	$N_2 + 2\,O_2 \longrightarrow 2\,NO_2$	(fast)
overall:	$2\,N_2O_5 \longrightarrow 4\,NO_2 + O_2$	

Possible mechanism 3:

Step 1.	$N_2O_5 \longrightarrow NO_2 + NO_3$	(slow)
Step 2.	$N_2O_5 + NO_2 \longrightarrow N_2O_4 + NO_3$	(fast)
Step 3.	$NO_3 + N_2O_4 \longrightarrow 2\,NO_2 + NO_3$	(fast)
Step 4.	$NO_3 + NO_3 \longrightarrow 2\,NO_2 + O_2$	(fast)
overall:	$2\,N_2O_5 \longrightarrow 4\,NO_2 + O_2$	

Possible mechanism 1 is consistent with the rate law although at first it may not appear to be. The second step is the slow step for which one NO_3 molecule is the reactant ; it is not an original reactant. It is produced in step 1 from two N_2O_5 molecules. However, only one NO_2 molecule is required for the slow step. It may be "accounted for" by only one N_2O_5 molecules in the first step, although the first step actually involves a bimolecular collision, not a unimolecular decomposition. Therefore this mechanism is consistent with the fact that the reaction is first order in N_2O_5, only.

Possible mechanism 2 is unacceptable although the slow step does involve just one N_2O_5 molecule as a reactant. Step 3 involves a trimolecular collision and is unlikely to occur to a significant extent.

Possible mechanism 3 is also consistent with the rate law (as are many others) since one N_2O_5 molecule is a reactant in the first step, which is the rate-determining step. None of the steps in this mechanism involve more than two molecules as reactants.

Example 14-5: The hypothetical reaction below is found to obey the indicated rate-law expression. Which of the following mechanisms are consistent with this information?

$$2A + B + C \longrightarrow AC + AB \qquad \text{Rate} = k[A][B][C]$$

Possible mechanism 1:

Step 1.	A	+	A	$\longrightarrow$	A_2			(slow)
Step 2.	A_2	+	B	$\longrightarrow$	AB	+	A	(fast)
Step 3.	A	+	C	$\longrightarrow$	AC			(fast)
overall:	2A	+	B + C	$\longrightarrow$	AC	+	AB	

Possible mechanism 2:

Step 1.	A	+	B	$\longrightarrow$	AB			(fast)
Step 2.	AB	+	A	$\longrightarrow$	A_2B			(slow)
Step 3.	A_2B	+	C	$\longrightarrow$	AC	+	AB	(fast)
overall:	2A + B + C			$\longrightarrow$	AC	+	AB	

Possible mechanism 3:

Step 1.	A	+	C	$\rightarrow$	AC	(fast)
Step 2.	AC	+	B	$\rightarrow$	ABC	(slow)
Step 3.	ABC	+	A	$\rightarrow$	AC + AB	(fast)
overall:	2A + B + C			$\rightarrow$	AC + AB	

Mechanisms 1 and 2 are consistent with rate-law expressions of the form Rate = $k[A]^2$ and Rate = $k[A]^2[B]$, respectively. However, only mechanism 3 is consistent with: Rate = $k[A][B][C]$. Step 2, which has one B molecule and one AC molecule as reactants, is the slow step, but one AC molecule is produced from one A molecule and one C molecule in the preceding fast step.

14-7 Half-Life of a Reactant

The half-life of a reactant is the time required for half of the reactant to be consumed. If we let [A] represent the concentration of a reactant, A, after time t and $[A]_o$ represent its initial concentration, the following relationship holds for all reactions that are first order in a particular reactant, A, and first order overall.

$$\log \frac{[A]_o}{[A]} = \frac{kt}{2.303}$$

t = time elapsed since reaction was initiated

k = specific rate constant at the temperature of interest

The expression that relates the half-life, $t_{1/2}$, to the specific rate constant, k, is:

$$t_{1/2} = \frac{0.693}{k}$$

The half-life of such first order reactions is independent of the concentrations of all reactants except A, and is inversely proportional to k, the specific rate constant.

For a reaction that is second-order overall and second-order with respect to A, the half-life is

$$t_{1/2} = \frac{1}{k[A]_o} \quad \text{and} \quad \frac{1}{[A]} = \frac{1}{[A]_o} = kt$$

This relationship also holds for second-order reactions of the type

$$A + B \longrightarrow \text{Products}$$

which have rate law expressions of the form:

$$\text{Rate} = k[A][B] \quad \text{when} \quad [A]_o = [B]_o$$

Example 14-6: The compounds trans-1,2-dichloroethylene and cis-1,2-dichloroethylene are isomers of each other. The trans isomer converts to the cis isomer according to the equation below in a first order reaction with $k = 3.3 \times 10^3\ s^{-1}$ at 1000°C. What is the half-life of trans-1,2-dichloroethylene at 1000°C?

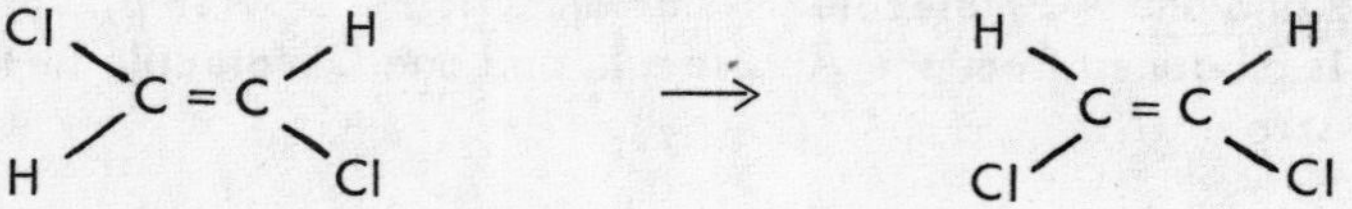

trans → cis

$$t_{1/2} = \frac{0.693}{k} = \frac{0.693}{3.3 \times 10^3\ s^{-1}} = 2.1 \times 10^{-4}\ s$$

The reaction is extremely rapid. Half of the trans isomer disappears after only 0.00021 second.

Example 14-7: Calculate the half-life for the decomposition of CO_2, which is first order with respect to CO_2 at 1000°C. $k = 1.8 \times 10^{-13} s^{-1}$

$$2\ CO_2\ (g) \longrightarrow 2\ CO\ (g) + O_2\ (g)$$

$$t_{1/2} = \frac{0.693}{k} = \frac{0.693}{1.8 \times 10^{-13}\ s^{-1}} = 3.9 \times 10^{12}\ s$$

The reaction is very slow. (3.9×10^{12} seconds is more than 120,000 years!)

Example 14-8: The rate-law expression for the following reaction is Rate = $k[NOCl]^2$. At 400 K, $k = 7.0 \times 10^{-1}\ M^{-1} \cdot s^{-1}$. What is the half-life of a reaction initiated with 0.100 mole NOCl in a one liter vessel at 400 K?

$$2\ NOCl \longrightarrow 2\ NO + Cl_2$$

$$t_{1/2} = \frac{1}{k[A]_o} = \frac{1}{(0.70\ M^{-1} \cdot s^{-1})(0.100\ \underline{M})} = 14 \text{ seconds}$$

Example 14-9: What is the half-life for the reaction of Example 14-8 for an initial NOCl concentration of 1.50 M?

$$t_{1/2} = \frac{1}{k[A]_o} = \frac{1}{(0.70\ \underline{M}^{-1}s^{-1})(1.50\ \underline{M})}$$

$$t_{1/2} = \underline{\underline{0.95 \text{ second}}}$$

Example 14-10: If the reaction of Example 14-8 is carried out at 400 K beginning with 0.100 mole of NOCl in a one liter vessel, how many moles will remain after 50.0 seconds? How many grams?

The reaction is second-order with respect to NOCl and second-order overall so the following relationship holds.

$$\frac{1}{[NOCl]} - \frac{1}{[NOCl]_o} = kt$$

$$\frac{1}{[NOCl]} - \frac{1}{0.100\ \underline{M}} = (0.70\ M^{-1}s^{-1})(50.0\ s)$$

$$\frac{0.100\ \underline{M} - [NOCl]}{(0.100\ \underline{M})[NOCl]} = 35\ \underline{M}^{-1}$$

$$0.100\ \underline{M} - [NOCl] = 3.5\ [NOCl]$$

$$0.100\ M = 4.5\ [NOCl]$$

$$[NOCl] = 0.022\ \underline{M}$$

Since the volume is one liter, 0.022 mole of NOCl remains.

$$\underline{?}\ \text{mol NOCl} = \frac{0.022\ \text{mol NOCl}}{L} \times 1.00\ L = \underline{\underline{0.022\ \text{mol NOCl}}}$$

This corresponds to 1.4 grams of NOCl.

$$\underline{?}\ \text{g NOCl} = 0.022\ \text{mol} \times \frac{65.5\ \text{g NOCl}}{\text{mol}} = \underline{\underline{1.4\ \text{g NOCl}}}$$

EXERCISES

1. The data below were obtained for the gas phase reaction shown at 25°C. Write the rate-law expression and evaluate the specific rate constant at 25°C.

$$2\,A + B \longrightarrow \text{Products}$$

Experiment	Initial [A]	Initial [B]	Initial Rate of Loss of A
1	0.20 M	0.10 M	6.0×10^{-4} M·s^{-1}
2	0.20 M	0.30 M	6.0×10^{-4} M·s^{-1}
3	0.40 M	0.30 M	1.2×10^{-3} M·s^{-1}

2. (a) What would be the rate of the reaction in Exercise 1 at 25°C if the initial concentrations of A and B were 0.60 M and 0.20 M, respectively?

   (b) What would be the rate of the same reaction at 25°C if the initial concentrations of A and B were 0.20 M and 0.60 M, respectively?

3. Consider the reaction of Exercise 2(a) in a 2.00 liter vessel. How many moles of A and B would remain after the reaction had occurred for 20.0 seconds?

4. Determine the rate-law expression for the reaction below at the temperature at which the tabulated initial rate data were obtained.

$$A + 2\,B + 3\,C \longrightarrow \text{Products}$$

Experiment	Initial [A]	Initial [B]	Initial [C]	Initial Rate of Loss of A
1	0.10 M	0.20 M	0.10 M	4.0×10^{-2} M·min^{-1}
2	0.40 M	0.20 M	0.10 M	4.0×10^{-2} M·min^{-1}
3	0.20 M	0.20 M	0.25 M	1.0×10^{-1} M·min^{-1}
4	0.20 M	0.40 M	0.10 M	1.6×10^{-1} M·min^{-1}

5. Determine the rate-law expression for the reaction below at the temperature at which the initial rate data were obtained.

$$2\,A + 3\,B \longrightarrow \text{Products}$$

Experiment	Initial [A]	Initial [B]	Initial Rate of Loss of A
1	0.25 M	0.40 M	8.0×10^{-4} M·s^{-1}
2	0.75 M	0.20 M	1.8×10^{-3} M·s^{-1}
3	0.50 M	0.40 M	3.2×10^{-3} M·s^{-1}

6. A reaction is second order in A and third order in B. If the concentration of A is reduced to one-fourth of its original value and the concentration of B is doubled, the rate of the reaction will (increase, decrease) by a factor of ________.

7. Consider the initial rate data obtained at a certain temperature for the hypothetical gas phase reaction below. Write the rate-law expression.

$$A + B + 2\,C \longrightarrow \text{Products}$$

Experiment	Initial [A]	Initial [B]	Initial [C]	Initial Rate of Loss of A
1	0.200 $\underline{M}$	0.300 $\underline{M}$	0.100 $\underline{M}$	6.40×10^{-3} $\underline{M}$•min^{-1}
2	0.300 $\underline{M}$	0.300 $\underline{M}$	0.250 $\underline{M}$	4.00×10^{-2} $\underline{M}$•min^{-1}
3	0.300 $\underline{M}$	0.600 $\underline{M}$	0.250 $\underline{M}$	4.00×10^{-2} $\underline{M}$•min^{-1}
4	0.200 $\underline{M}$	0.900 $\underline{M}$	0.400 $\underline{M}$	1.02×10^{-1} $\underline{M}$•min^{-1}

8. Consider the reaction of Exercise 7.

(a) Evaluate the specific rate constant at the temperature at which the data were obtained.

(b) What would be the initial rate of reaction in a closed 5.00 liter vessel if a reaction were initiated with 0.150 mole of A, 0.180 mole of B, and 0.260 mole of C?

(c) How many moles of A, B, and C would remain in (b) after the reaction had occurred for 3.00 minutes?

9. For each of the following pairs of reactions, which reaction is likely to occur more rapidly at room temperature? Why?

(a) $Na_2CO_3\,(s) + NiCl_2\,(s) \longrightarrow NiCO_3\,(s) + 2\,NaCl\,(s)$
$Na_2CO_3\,(aq) + NiCl_2\,(aq) \longrightarrow NiCO_3\,(s) + 2\,NaCl\,(aq)$

(b) $PCl_3\,(g) + Cl_2\,(g) \longrightarrow PCl_5\,(g)$
$AgNO_3\,(aq) + NaCl\,(aq) \longrightarrow AgCl\,(s) + NaNO_3\,(aq)$

In Exercises 10 and 11 decide which of the suggested possible mechanisms are consistent with the observed rate-law expression.

10. $2\,IBr\,(g) \longrightarrow I_2\,(g) + Br_2\,(g)$ Rate = $k\,[IBr]^2$

(a)
$IBr \longrightarrow I + Br$ slow
$I + IBr \longrightarrow I_2 + Br$ fast
$Br + Br \longrightarrow Br_2$ fast
$2\,IBr \longrightarrow I_2 + Br_2$

(b)
$IBr \longrightarrow I + Br$ fast
$I + IBr \longrightarrow I_2 + Br$ slow
$Br + Br \longrightarrow Br_2$ fast
$2\,IBr \longrightarrow I_2 + Br_2$

(c) $IBr + IBr \longrightarrow I_2Br^+ + Br^-$ slow
$I_2Br^+ \longrightarrow IBr + Br^+$ fast
$Br^- + Br^+ \longrightarrow Br_2$ fast

$2\,IBr \longrightarrow I_2 + Br_2$

(d) $IBr \longrightarrow I^+ + Br^-$ fast
$I^+ + IBr \longrightarrow IBr^+$ slow
$I_2Br^+ + Br^- \longrightarrow I_2 + Br_2$ fast

$2IBr \longrightarrow I_2 + Br_2$

(e) $IBr + IBr \longrightarrow I_2 + Br_2$ (one step)

11. $4\,PH_3\,(g) \longrightarrow P_4\,(g) + 6\,H_2\,(g)$ Rate = $k[PH_3]$ at 950°C

(a) $PH_3 + PH_3 \longrightarrow P_2H_6$ fast
$P_2H_6 + PH_3 \longrightarrow P_3 + 4H_2 + H$ slow
$PH_3 + H \longrightarrow PH_4$ fast
$PH_4 + P_3 \longrightarrow P_4 + 2\,H_2$ fast

$4\,PH_3 \longrightarrow P_4 + 6\,H_2$

(b) $PH_3 \longrightarrow PH + H_2$ slow
$PH + PH_3 \longrightarrow P_2 + 2\,H_2$ fast
$H_2 + PH_3 \longrightarrow PH_4 + H$ fast
$H + PH_3 \longrightarrow PH_4$ fast
$PH_4 + PH_4 \longrightarrow P_2 + 4\,H_2$ fast
$P_2 + P_2 \longrightarrow P_4$ fast

$4\,PH_3 \longrightarrow P_4 + 6\,H_2$

(c) same as (b) except the first step is fast and the third step is slow

(d) same as (a) except the first step is slow and the second step is fast

12. Consider the gas phase reaction $SO_2Cl_2 \longrightarrow SO_2 + Cl_2$. Propose two plausible mechanisms consistent with the rate-law expression, Rate = $k[SO_2Cl_2]$.

13. Consider the gas phase reaction $2\,N_2O \longrightarrow 2\,N_2 + O_2$. Propose two plausible mechanisms consistent with the rate-law expression, Rate = $k[N_2O]^2$.

14. The gas phase reaction below obeys the rate-law expression, Rate = $k[SO_2Cl_2]$. At 593 K the specific rate constant is $2.2 \times 10^{-5}\ s^{-1}$. A 2.0 g sample of SO_2Cl_2 is introduced into a closed 4.0 L container.

$$SO_2Cl_2 \longrightarrow SO_2 + Cl_2$$

(a) How much time must pass in order to reduce the amount of SO_2Cl_2 present to 1.8 grams?
(b) How many grams of SO_2Cl_2 remain after 3.0 hours?
(c) Refer to (b). How much Cl_2 is present after 3.0 hours?

15. The half-life for the gas phase reaction below is 9.6×10^{28} seconds at 300 K. (This is a first order reaction.)

$$C_2H_5Cl \longrightarrow C_2H_4 + HCl$$

    (a) Evaluate the specific rate constant at 300 K.
    (b) How long would it take for 1.00% of a given sample of C_2H_5Cl at 300 K to decompose into C_2H_4 and HCl?
    (c) What percentage of a 2.00 gram sample remains after one billion (10^9) years?

16. At 300 K the reaction below obeys the rate law Rate = $k[NOCl]^2$ where $k = 2.8 \times 10^{-5} M^{-1} \cdot s^{-1}$.

$$2\, NOCl \longrightarrow 2\, NO + Cl_2$$

Suppose 1.0 mole of NOCl is introduced into a 2.0 liter container at 300 K.

    (a) Evaluate the half-life of the reaction.
    (b) How much NOCl will remain after 30 minutes?
    (c) After what period of time would 0.9 mole of NOCl remain?
    (d) Refer to (c). How many moles of Cl_2 will be present?

ANSWERS TO EXERCISES

1. Rate = $k[A]$ $k = 3.0 \times 10^{-3} s^{-1}$

2. (a) Rate = $1.8 \times 10^{-3} \underline{M} \cdot s^{-1}$ (b) Rate = $6.0 \times 10^{-4} \underline{M} \cdot s^{-1}$

3. 1.13 mol A and 0.36 mol B remain   4. Rate = $k[B]^2[C]$

5. Rate = $k[A]^2[B]^2$ 6. decrease, 2   7. Rate = $k[C]^2$

8. (a) $k = 0.646 M^{-1} \cdot min^{-1}$ (b) $1.75 \times 10^{-3} \underline{M} \cdot min^{-1}$

(c) 0.124 mol A, 0.154 mol B, and 0.209 mol C remain

9. (a) Na_2CO_3 (aq) + $NiCl_2$ (aq) $\longrightarrow$ $NiCO_3$ (s) + 2 NaCl (aq)

(b) $AgNO_3$ (aq) + NaCl (aq) $\longrightarrow$ AgCl (s) + $NaNO_3$ (aq)   These reactions involve ionic compounds in aqueous solutions. Ionic or covalent bonds must be broken in the other cases.   10. (b), (c), (d), (e)   11. (b)

12. $SO_2Cl_2 \longrightarrow SO_2 + Cl_2$ (one step)

$$
\begin{array}{ll}
SO_2Cl_2 \longrightarrow SO_2Cl^+ + Cl^- & \text{slow} \\
\underline{SO_2Cl^+ + Cl^- \longrightarrow SO_2 + Cl_2} & \text{fast} \\
SO_2Cl_2 \longrightarrow SO_2 + Cl_2 &
\end{array}
$$

13. $N_2O + N_2O \longrightarrow N_2O_2 + N_2$ slow
$N_2O_2 \longrightarrow N_2 + O_2$ fast

$$2\,N_2O \longrightarrow 2\,N_2 + O_2$$

$N_2O + N_2O \longrightarrow N_3O_2 + N$ slow
$N_3O_2 + N \longrightarrow N_2 + N_2O_2$ fast
$N_2O_2 \longrightarrow N_2 + O_2$ fast

$$2\,N_2O \longrightarrow 2\,N_2 + O_2$$

14. (a) 4.8×10^3 s or 1.3 hr (b) 1.6 g (c) 0.2 g

15. (a) $7.2 \times 10^{-30}\ s^{-1}$ (b) 1.4×10^{27} s (The reaction has a very long half-life.) (c) nearly 100% 16. (a) 7.1×10^4 s (b) 0.98 mol (c) 7.9×10^3 s (d) 0.05 mol

Chapter Fifteen

CHEMICAL EQUILIBRIUM

D, G, W: Chap. 16

15-1 Introduction

Most chemical processes are reversible, i.e., occur in forward and reverse directions and do not go to completion. Double arrows indicate reversibility.

$$aA \quad + \quad bB \quad \rightleftharpoons \quad cC \quad + \quad dD$$

"Reactants" "Products"

For such processes a dynamic equilibrium is eventually established in which forward and reverse reactions occur simultaneously at the same (non-zero) rate, and some of each species is present. Unless the system is then disturbed, no further net changes in concentrations occur. The equilibrium may be approached from either direction, i.e., by starting only with "reactants", only with "products", or with a mixture of "products" and "reactants".

15-2 The Equilibrium Constant

The position of an equilibrium is related to the magnitude of the equilibrium constant (K_c). For the generalized reaction, it is defined as

$$aA \quad + \quad bB \quad \rightleftharpoons \quad cC \quad + \quad dD$$

$$K_c = \frac{[C]^c[D]^d}{[A]^a[B]^b} \begin{matrix} \leftarrow \text{"Products"} \\ \leftarrow \text{"Reactants"} \end{matrix}$$

The square brackets, [ ], refer to equilibrium concentrations. The powers to which the concentrations are raised are the coefficients in the balanced equation. By convention the species on the right are called "products" and those on the left, "reactants". "Products" always appear in the numerator of the mass action expression and "reactants" in the denominator. Values of equilibrium constants must be determined experimentally, and they change only with temperature. The larger the equilibrium constant for a given reaction, the more the forward reaction is favored.

Example 15-1: Consider the following reversible reaction. In a 3.00 liter container the following amounts are found in equilibrium at 400°C: 0.0420 mole N_2, 0.516 mole H_2, and 0.0357 mole NH_3. Evaluate K_c.

$$N_2\,(g) \quad + \quad 3\,H_2\,(g) \quad \rightleftharpoons \quad 2\,NH_3\,(g)$$

Since K_c is defined in terms of equilibrium concentrations, we must calculate these concentrations and use them in the mass action expression.

$$[N_2] = 0.0420\text{ mol}/3.00\text{ L} = 0.0140\ \underline{M}$$

$$[H_2] = 0.516\text{ mol}/3.00\text{ L} = 0.172\ \underline{M}$$

$$[NH_3] = 0.0357\text{ mol}/3.00\text{ L} = 0.0119\ \underline{M}$$

$$K_c = \frac{[NH_3]^2}{[N_2][H_2]^3} = \frac{(0.0119\ \underline{M})^2}{(0.0140\ \underline{M})(0.172\ \underline{M})^3} = \underline{1.99\ \underline{M}^{-2}}$$

Values of equilibrium constants are usually quoted without units.

15-3 Variation of K_c with the Form of the Equation

The form of the mass action expression depends upon the coefficients of the balanced equation. If we had written the equation in Example 15-1 as follows (all coefficients multiplied by 1/2),

$$1/2\ N_2\,(g) \quad + \quad 3/2\ H_2\,(g) \quad \rightleftharpoons \quad NH_3\,(g)$$

then the equilibrium constant, K_c', would be:

$$K_c' = \frac{[NH_3]}{[N_2]^{1/2}\,[H_2]^{3/2}} = \frac{(0.0119\ \underline{M})}{(0.0140\ \underline{M})^{1/2}\,(0.172\ \underline{M})^{3/2}} = 1.41 = K_c^{\ 1/2}$$

If the equation is reversed (all coefficients multiplied by -1), $K_c'' = \dfrac{1}{K_c} = K_c^{-1}$.

$$2\,NH_3\,(g) \quad \rightleftharpoons \quad N_2\,(g) \quad + \quad 3\,H_2\,(g)$$

$$K_c'' = \frac{[N_2][H_2]^3}{[NH_3]^2} = \frac{(0.0140\ \underline{M})(0.172\ \underline{M})^3}{(0.0119\ \underline{M})^2} = 0.503 = K_c^{-1}$$

In general, if the coefficients are multiplied by a number, n, the value of the original equilibrium constant is raised to the nth power.

15-4 Use of the Equilibrium Constant

Once the equilibrium constant for a given reaction is determined at a particular temperature, it may be used to determine equilibrium concentrations at that temperature for the reaction for any combination of initial concentrations.

Example 15-2: Phosgene, $COCl_2$, is a poisonous gas that decomposes into carbon monoxide and chlorine according to the following equation with $K_c = 0.083$ at 900°C. If reaction is initiated with 0.600 mole of $COCl_2$ at 900°C in 5.00 liter container, what concentrations of all species will be present after equilibrium is established? What is the percent dissociation of $COCl_2$?

$$COCl_2\ (g) \rightleftharpoons CO\ (g) + Cl_2\ (g)$$

You may find it useful to set up your calculations as follows. Indicate (1) initial concentrations, (2) changes in concentrations due to the net reaction, and (3) equilibrium concentrations, i.e., algebraic representations for any unknown concentrations. Let x = concentration of $COCl_2$ (mol/L) that decomposes.

	$COCl_2$	$\rightleftharpoons$	CO	+	Cl_2
initial:	0.12 M		0 M		0 M
change:	-x M		+x M		+x M
equil. conc'n:	(0.12 - x) M		x M		x M

The changes due to the net reaction are governed by the reaction ratio (1:1:1 in this case).

$$K_c = \frac{[CO][Cl_2]}{[COCl_2]} = 0.083$$ Substitution gives:

$$\frac{(x)(x)}{(0.12 - x)} = 0.083$$ and clearing fractions gives

$$x^2 = 0.00996 - 0.083\,x \quad \text{or} \quad x^2 + 0.083x - 0.00996 = 0$$

This is a quadratic equation, in standard form: $ax^2 + bx + c = 0$. In this case a = 1, b = 0.083, and c = -0.00996. It may be solved by the quadratic formula which invariably gives two solutions, one of which may be recognized as having no physical significance.

$$x = \frac{-b \pm \sqrt{b^2 - 4ac}}{2a} \qquad \text{or, this case,}$$

$$x = \frac{-0.083 \pm \sqrt{(0.083)^2 - 4(1)(-0.00996)}}{2(1)}$$

$$= \frac{-0.083 \pm \sqrt{0.0467}}{2} = \frac{-0.083 \pm 0.216}{2}$$

$$x = +0.067 \quad \text{and} \quad -0.15$$

Since x represents the concentration of $COCl_2$ that decomposes, its maximum value is 0.12 $\underline{M}$ (the initial concentration) and its minimum value is 0 $\underline{M}$, i.e., $0 < x < \bar{0}.12$. Thus x = -0.15 is the extraneous root, and $\underline{x = 0.067}$ is the root that has meaning. The equilibrium concentrations are:

$$[CO] = [Cl_2] = x = \underline{\underline{0.067\ \underline{M}}}$$

$$[COCl_2] = (0.12 - x) = 0.053\ \underline{M} = \underline{\underline{0.05\ \underline{M}}}$$

The percent dissociation of $COCl_2$ is given by the relationship,

$$\%\ \text{dissociation} = \frac{[COCl_2]_{\text{dissociated}}}{[COCl_2]_{\text{total}}} \times 100\%$$

Since $[COCl_2]_{\text{dissociated}} = [CO]$ (or $[Cl_2]$), we may write

$$\%\ \text{dissociation} = \frac{[CO]}{[COCl_2]_{\text{total}}} \times 100\% = \frac{0.067\ \underline{M}}{0.12\ \underline{M}} \times 100\% = \underline{\underline{56\%}}$$

Thus, 56% of the $COCl_2$ dissociates into CO and Cl_2 at 900°C.

Example 15-3: At 750°C the equilibrium constant for the following reaction is 0.771. If 0.200 mole of H_2O and 0.200 mole of CO are placed in a 1.00 liter vessel, what will the equilibrium concentrations of all species be at 750°C?

We set up the data as in Example 15-2. Let x = concentration of H_2O (or CO) that reacts:

	H_2 (g)	+	CO_2 (g)	$\rightleftharpoons$	H_2O (g)	+	CO (g)
initial:	0 $\underline{M}$		0 $\underline{M}$		0.200 $\underline{M}$	+	0.200 $\underline{M}$
change:	+x $\underline{M}$		+x $\underline{M}$		−x $\underline{M}$		−x $\underline{M}$
equil:	x $\underline{M}$		x $\underline{M}$		(0.200−x) $\underline{M}$		(0.200−x)$\underline{M}$

$$K_c = \frac{[H_2O][CO]}{[H_2][CO_2]} = 0.771$$

$$\frac{(0.200-x)(0.200-x)}{(x)(x)} = 0.771 \qquad \frac{(0.200-x)^2}{x^2} = 0.771$$

This quadratic equation is a perfect square. Taking the square root of both sides gives

$$\frac{0.200-x}{x} = 0.878 \qquad \text{or} \qquad 0.200-x = 0.878x$$

$$1.878x = 0.200 \qquad \text{so} \qquad x = \frac{0.200}{1.878} = 0.106$$

At equilibrium: $[H_2] = [CO_2] = x = \underline{\underline{0.106\ M}}$

$[H_2O] = [CO] = (0.200-x)\underline{M} = \underline{\underline{0.094\ M}}$

15-5 Reaction Quotient

The mass action expression for the reaction quotient, Q, has the same form as the equilibrium constant, except that the concentrations are not (necessarily) those at equilibrium.

$$a\,A + b\,B \rightleftharpoons c\,C + d\,D$$

$$Q = \frac{[C]^c[D]^d}{[A]^a[B]^b}$$

If	Then
Q > K	Reverse reaction favored until equilibrium is established
Q = K	Equilibrium exists
Q < K	Forward reaction favored until equilibrium is established

Evaluation of Q for a reacting system at any instant allows us to determine which concentrations must increase and which must decrease in order for equilibrium to be established.

15-6 LeChatelier's Principle

LeChatelier's Principle states that when a stress is applied to a system at equilibrium, the system responds to relieve the stress and re-establishes equilibrium. We shall consider the following types of stresses: (1) changes in concentrations, (2) changes in pressure (and volume), (3) changes in temperature, and (4) introduction of catalysts (causes no shift).

15-7 Changes in Concentration

Consider the system

$$A + B \rightleftharpoons C + D$$

at equilibrium. If the concentration of one of the species is increased (or decreased) the system will respond by temporarily favoring the reaction that consumes (or produces) that species. This illustrates LeChatelier's Principle in that the system attempts to "undo" the stress (concentration change).

Example 15-4: Consider the system below at 250°C. If 1.500 mole PCl_3 and 0.500 mole Cl_2 are placed in a 5.00 liter container, 0.390 mole PCl_5 is present after equilibrium is established. If an additional 0.100 mole of Cl_2 is then added, what will the concentrations of all species be when equilibrium is re-established? The equation is given below.

We determine the equilibrium concentrations before 0.100 mole Cl_2 (stress) is added, and the value of K_c. Since 0.0780 M PCl_5 (0.390 mol/5.00 L) is present at equilibrium, 0.0780 M of PCl_3 and Cl_2 must have been consumed.

	PCl_5 (g)	$\rightleftharpoons$	PCl_3 (g)	+	Cl_2 (g)
initial:	0 M		0.300 M		0.100 M
change:	+0.078 M		-0.078 M		0.078 M
equil:	0.078 M		0.222 M		0.022 M

$$K_c = \frac{[PCl_3][Cl_2]}{[PCl_5]} = \frac{(0.222)(0.022)}{0.078} = 0.063$$

The temperature remains constant and the value of K_c does not change. We calculate Q* at the instant Cl_2 is added, which tells us that the equilibrium shifts to the left.

	PCl_5 (g)	⇌	PCl_3 (g)	+	Cl_2 (g)	
equil:	0.078 M		0.222 M		0.022 M	
add:					+0.020 M	
new initial:	0.078 M		0.222 M		0.042 M	Q > K*
shift:	+x M		-x M		-x M	← shift
new equil:	(0.078+x)M		(0.222-x)M		(0.042-x)M	

$$0.063 = \frac{(0.222-x)(0.042-x)}{0.078+x}$$

$$0.00491 + 0.063x = 0.00932 - 0.264x + x^2$$

$$x^2 - 0.327x + 0.00441 = 0$$

Solving via the quadratic formula gives: $x = 0.313$ and $x = 0.014$. The value of x must be between 0 M and 0.042 M, thus, $x = 0.014$. The equilibrium concentrations are:

$$[PCl_5] = 0.078 + x = 0.078 + 0.014 = 0.092\ M$$

$$[PCl_3] = 0.222 - x = 0.222 - 0.014 = 0.208\ M$$

$$[Cl_2] = 0.042 - x = 0.042 - 0.014 = 0.028\ M$$

Substitution into the mass action expression gives a perfect check,

$$\frac{[PCl_3][Cl_2]}{[PCl_5]} = \frac{(0.208)(0.028)}{0.092} = 0.063 = K_c$$

because this value agrees with our original value of K_c exactly.

---

*After add'n of Cl_2 : $Q = \frac{(0.222)(0.042)}{(0.078)} = 0.12$, $Q > K$

Example 15-5: Refer to Example 15-4. If 0.100 mole of Cl_2 (0.020 M) had been removed (rather than added) what equilibrium concentrations would have resulted?

Calculation of Q* after the removal verifies our prediction from LeChatelier's Principle that the equilibrium shifts to the right.

	PCl_5 (g)	$\rightleftharpoons$ PCl_3 (g)	+ Cl_2 (g)	
orig. equil:	0.078 M	0.222 M	0.022 M	
remove:			-0.020 M	
new initial:	0.078 M	0.222 M	0.002 M	Q < K
change:	-x M	+x M	+x M	shift →
new equil:	(0.078-x)M	(0.222+x)M	(0.002+x)M	

$$0.063 = \frac{(0.222+x)(0.002+x)}{0.078-x}$$

$$0.0049 - 0.063x = 0.000444 + 0.224x + x^2$$

$$x^2 + 0.287x - 0.00446 = 0$$

Solving via the quadratic formula gives $x = -0.302$ and $+0.015$. So $x = +0.015$ M. The new equilibrium concentrations are:

$$[PCl_5] = 0.078 - x = 0.078 - 0.015 = 0.063\ M$$

$$[PCl_3] = 0.222 + x = 0.222 + 0.015 = 0.237\ M$$

$$[Cl_2] = 0.002 + x = 0.002 + 0.015 = 0.017\ M$$

Check: $$\frac{[PCl_3][Cl_2]}{[PCl_5]} = \frac{(0.237)(0.017)}{0.063} = 0.064$$

This value agrees with the previously calculated value of K_c, 0.063, within round-off error range.

*After removal of Cl_2: $Q = \frac{(0.222)(0.002)}{0.078} = 0.0057$, $Q < K$

The following tabulation summarizes Examples 15-4 and 15-5.

	Original Conc'ns	Equil. Conc'ns	K_c	Stress	New Equil. Conc'ns	Calc'd. K_c
PCl_5	0 M	0.078 M		Add	0.092 M	
PCl_3	0.300 M	0.222 M	0.063	0.020 M Cl_2	0.208 M	0.063
Cl_2	0.100 M	0.022 M		($Q>K_c$)	0.028 M	
PCl_5	0 M	0.078 M		Remove	0.063 M	
PCl_3	0.300 M	0.222 M	0.063	0.020 M Cl_2	0.237 M	0.064
Cl_2	0.100 M	0.022 M		($Q<K_c$)	0.017 M	

15-8 Changes in Volume (Pressure)

At constant temperature, a change in the volume of a gas causes a change in its pressure, and vice-versa. The pressure of a gas is directly proportional to its concentration. From the ideal gas law,

$$P = \left(\frac{n}{V}\right)RT = [\text{conc'n}]\ RT$$

As a result, pressure or volume changes can shift equilibria. For a system involving different numbers of moles of gases as products and reactants, an increase (decrease) in pressure causes a shift in the direction that produces the smaller (larger) number of moles of gases. Since pressure and volume are inversely proportional, increases in volume cause the same effect as decreases in pressure. Equilibria involving equal numbers of moles of gaseous reactants and products, and those involving only solids and/or liquids are not affected by pressure or volume changes.

Example 15-6: Refer to Example 15-4. What would the new equilibrium concentrations be if the volume of the original system (no extra Cl_2 added) were suddenly reduced to 2.00 liters with no change in temperature?

A decrease in volume (increase in pressure) shifts the equilibrium in favor of PCl_5 as our calculation of Q* at the instant of volume change shows. The new initial concentrations will be 5/2 times those given in Example 15-4 because the volume decreases from 5.00 liters to 2.00 liters.

*after volume decrease: $Q = \frac{(0.555)(0.055)}{0.195} = 0.16, \quad Q > K$

They are: $[PCl_5] = 0.078\ \underline{M} \times 5/2 = 0.195\ \underline{M}$

$[PCl_3] = 0.222\ \underline{M} \times 5/2 = 0.555\ \underline{M}$

$[Cl_2] = 0.022\ \underline{M} \times 5/2 = 0.055\ \underline{M}$

	$PCl_5\ (g)$	$\rightleftharpoons$	$PCl_3\ (g)$	+	$Cl_2\ (g)$	
new initial:	$0.195\ \underline{M}$		$0.555\ \underline{M}$	+	$0.055\ \underline{M}$	$Q > K$
shift:	$+x\ \underline{M}$		$-x\ \underline{M}$		$-x\ \underline{M}$	shift $\leftarrow$
new equil:	$(0.195+x)\underline{M}$		$(0.555-x)\underline{M}$		$(0.055-x)\underline{M}$	

$$0.063 = \frac{[PCl_3][Cl_2]}{[PCl_5]} = \frac{(0.555-x)(0.055-x)}{0.195+x}$$

$$x^2 - 0.673x + 0.0182 = 0$$

$$x = 0.65 \text{ and } 0.028$$

Since $0 < x < 0.055$, $\underline{\underline{x = 0.028\ M}}$, and when equilibrium is re-established:

$$[PCl_5] = 0.195 + x = \underline{\underline{0.223\ M}}$$
$$[PCl_3] = 0.555 - x = \underline{\underline{0.527\ M}}$$
$$[Cl_2] = 0.055 - x = \underline{\underline{0.027\ M}}$$

Check: $$\frac{[PCl_3][Cl_2]}{[PCl_5]} = \frac{(0.527)(0.027)}{0.223} = 0.064$$

If the volume had been increased (pressure decreased) the forward reaction would have been favored ($Q < K$) and the position of equilibrium would have shifted to the right.

15-9 Changes in Temperature

The magnitude of an equilibrium constant is temperature dependent. We may use LeChatelier's Principle to predict the direction of shift due to temperature changes. Consider the system,

$$A + B \rightleftharpoons C + D + \text{Heat} \qquad \text{(Exothermic Reaction)}$$

at equilibrium at a particular temperature. An increase in temperature (at constant pressure and volume) increases the amount of heat available to the system. Since the reaction to the left is endothermic, the equilibrium shifts to the left in order to consume heat. A decrease in temperature shifts this equilibrium to the right.

For the following system at equilibrium, temperature decreases shift the equilibrium to the left and temperature increases shift it to the right.

$$A + B + \text{Heat} \rightleftharpoons C + D \qquad \text{(Endothermic Reaction)}$$

15-10 Adding a Catalyst

Catalysts affect equally the rates at which both forward and reverse reactions occur and, generally, diminish the time required to establish equilibrium. A catalyst does not shift an equilibrium.

Example 15-7: Consider the following systems at equilibrium at a given temperature, and predict the effects of the following stresses on the position of equilibrium and on the relative concentration of each substance.

(1) $2\,SO_2\,(g) + O_2\,(g) \rightleftharpoons 2\,SO_3\,(g) + 197.6\text{ kJ}$

(2) $N_2\,(g) + 2\,O_2\,(g) + 66.4\text{ kJ} \rightleftharpoons 2\,NO_2\,(g)$

(3) $2\,H_2O\,(g) + 483.6\text{ kJ} \rightleftharpoons 2\,H_2\,(g) + O_2\,(g)$

(a) increasing the amount of O_2, (b) decreasing the amount of O_2, (c) raising T at constant P and V, (d) lowering T at constant P and V, (e) increasing V at constant T, (f) decreasing P at constant T, (g) decreasing V at constant T, (h) increasing P at constant T, (i) adding a catalyst.

Arrows (—> or <—) indicate shifts of the position of equilibrium to the right or left, + indicates an increase in equilibrium concentration, - a decrease, and 0 no change.

(1)

	Shift	$[SO_2]$	$[O_2]$	$[SO_3]$
(a)	—>	-	+	+
(b)	<—	+	-	-
(c)	<—	+	+	-
(d)	—>	-	-	+
(e)	<—	+	+	-
(f)	<—	+	+	-
(g)	—>	-	-	+
(h)	—>	-	-	+
(i)	0	0	0	0

(2)

Shift	$[N_2]$	$[O_2]$	$[NO_2]$
(a) —>	-	+	+
(b) <—	+	-	-
(c) —>	-	-	+
(d) <—	+	+	-
(e) <—	+	+	-
(f) <—	+	+	-
(g) —>	-	-	+
(h) —>	-	-	+
(i) 0	0	0	0

(3)

Shift	$[H_2O]$	$[H_2]$	$[O_2]$
(a) <—	+	-	+
(b) —>	-	+	-
(c) —>	-	+	+
(d) <—	+	-	-
(e) —>	-	+	+
(f) —>	-	+	+
(g) <—	+	-	-
(h) <—	+	-	-
(i) 0	0	0	0

15-11 The Equilibrium Constant in Terms of Partial Pressures

It is often convenient to measure partial pressures of gases rather than concentrations. Since the partial pressure of a gas is directly proportional to its concentration, we can also define the equilibrium constant for a reaction involving gases in terms of partial pressures. For the system,

$$a\,A\,(g) \quad + \quad b\,B\,(g) \quad \rightleftharpoons \quad c\,C\,(g) \quad + \quad d\,D\,(g)$$

$$K_p = \frac{P_C^c \times P_D^d}{P_A^a \times P_B^b} \qquad \text{and} \qquad K_c = \frac{[C]^c[D]^d}{[A]^a[B]^d}$$

For an ideal gas, $P = (\frac{n}{V})RT = [\text{conc'n}]\,RT$

At constant temperature both R and T are constants, and $P = [\text{conc'n}] \times \text{constant}$. The relationship between K_c and K_p is (where n = number of moles):

$$K_p = K_c\,(RT)^{\Delta n} \qquad\qquad \Delta n = [n_{\text{gas products}}] - [n_{\text{gas reactants}}]$$

The term Δn refers to the number of moles of gaseous products minus the number of moles of gaseous reactants. When $\Delta n = 0$, $K_p = K_c$. The magnitude of K_p is also an indicator of the position of equilibrium.

<u>Example 15-8:</u> At equilibrium at 25°C the following amounts of A, B and C were found in a 2.00 liter container: 0.200 mole of A, 0.150 mole B, and 0.0400 mole C. Calculate K_c and K_p at 25°C for the reaction, the partial pressure of each gas, and the total pressure of the system.

$$A\,(g) \quad + \quad 2\,B\,(g) \quad \rightleftharpoons \quad C\,(g)$$

Equilibrium concentrations are:

$$[A] = 0.200 \text{ mol}/2.00 \text{ L} = 0.100 \underline{M}$$

$$[B] = 0.150 \text{ mol}/2.00 \text{ L} = 0.0750 \underline{M}$$

$$[C] = 0.0400 \text{ mol}/2.00 \text{ L} = 0.0200 \underline{M}$$

$$K_c = \frac{[C]}{[A][B]^2} = \frac{0.0200 \underline{M}}{(0.100 \underline{M})(0.0750 \underline{M})^2} = \underline{\underline{35.6 \underline{M}^{-2} = K_c}}$$

$$K_p = K_c (RT)^{\Delta n} = 35.6 \underline{M}^{-2} [(0.0821 \frac{\text{L·atm}}{\text{mol·K}})(298 \text{ K})]^{-2} = \underline{\underline{0.0595 \text{ atm}^{-2}}}$$

Partial pressures can be calculated: $P_A = (\frac{n_A}{V}) RT = [A] RT$

$$P_A = [A] RT = (0.100 \frac{\text{mol}}{\text{L}})(0.0821 \frac{\text{L·atm}}{\text{mol·K}})(298 \text{ K}) = \underline{\underline{2.45 \text{ atm}}}$$

$$P_B = [B] RT = (0.0750 \frac{\text{mol}}{\text{L}})(0.0821 \frac{\text{L·atm}}{\text{mol·K}})(298 \text{ K}) = \underline{\underline{1.83 \text{ atm}}}$$

$$P_C = [C] RT = (0.0200 \frac{\text{mol}}{\text{L}})(0.0821 \frac{\text{L·atm}}{\text{mol·K}})(298 \text{ K}) = \underline{\underline{0.489 \text{ atm}}}$$

The total pressure is the sum of the partial pressures.

$$P_{Tot} = P_A + P_B + P_C = (2.45 + 1.83 + 0.49) \text{ atm} = \underline{\underline{4.77 \text{ atm}}}$$

We may check our calculations by calculating K_p directly from partial pressures.

$$K_p = \frac{P_C}{P_A \times P_B^2} = \frac{0.489 \text{ atm}}{(2.45 \text{ atm})(1.83 \text{ atm})^2} = \underline{\underline{0.0596 \text{ atm}^{-2}}}$$

This agrees with our previous calculation of K_p within round-off error.

15-12 Heterogeneous Equilibria

Heterogeneous equilibria involve species in more than one phase. Neither solids nor liquids are significantly affected by pressure changes, and the concentrations of solids and pure liquids are directly proportional to their densities, which vary only with temperature. Thus, the concentrations of solids and pure liquids are constant at a given temperature. By convention, they are not included in equilibrium constant expressions.

Consider the following equilibria and their equilibrium constant expressions.

(1) $2\ NaNO_3\ (s) \rightleftharpoons 2\ NaNO_2\ (s) + O_2\ (g)$

$$\frac{K_c'\,[NaNO_3]^2}{[NaNO_2]^2} = \text{constant} = \underline{\underline{K_c = [O_2]}} \qquad \underline{\underline{K_p = P_{O_2}}}$$

(2) $N_2H_4\ (\ell) + 2\ H_2O_2\ (\ell) \rightleftharpoons N_2\ (g) + 4\ H_2O\ (g)$

$$K_c'\,[N_2H_4][H_2O_2]^2 = \text{constant} = K_c = [N_2][H_2O]^4 \quad K_p = P_{N_2} \times P^4_{H_2O}$$

(3) $2\ NH_3\ (g) + 3\ CuO\ (s) \rightleftharpoons 3\ H_2O\ (\ell) + N_2\ (g) + 3\ Cu\ (s)$

$$\frac{K_c'\,[CuO]^3}{[H_2O]^3[Cu]^3} = \text{constant} = K_c = \frac{[N_2]}{[NH_3]^2} \qquad K_p = \frac{P_{N_2}}{P^2_{NH_3}}$$

(4) $CH_3COOH\,(aq) + H_2O\,(\ell) \rightleftharpoons H_3O^+\ (aq) + CH_3COO^-(aq)$ (dil. aq. sol'n.)

$$K_c'\,[H_2O] = \text{constant} = K_c = \frac{[H_3O^+][C_2H_3O_2^-]}{[HC_2H_3O_2]} \qquad \underline{\underline{K_p = \text{undefined}}}$$

Example 15-9: Write mass action expressions for the equilibrium constants, K_c and K_p, for the following reactions.

(1) $PCl_3 (\ell) + 3H_2O (\ell) \rightleftharpoons H_3PO_3 (s) + 3 HCl (g)$

(2)* $NH_3 (aq) + H_2O (\ell) \rightleftharpoons NH_4^+ (aq) + OH^- (aq)$

(3) $4 H_3PO_3 (\ell) \rightleftharpoons PH_3 (g) + 3 H_3PO_4 (\ell)$

(4) $2 HI (g) \rightleftharpoons H_2 (g) + I_2 (g)$

(5)* $2 H_2O_2 (aq) \rightleftharpoons 2 H_2O (\ell) + O_2 (g)$

(1) $K_c = [HCl]^3$ $\qquad K_p = P_{HCl}^{\ 3}$

(2) $K_c = \frac{[NH_4^+][OH^-]}{[NH_3]}$ $\qquad K_p =$ undefined

(3) $K_c = [PH_3]$ $\qquad K_p = P_{PH_3}$

(4) $K_c = \frac{[H_2][I_2]}{[HI]^2}$ $\qquad K_p = \frac{P_{H_2} \times P_{I_2}}{P^2_{HI}}$

(5) $K_c = \frac{[O_2]}{[H_2O_2]^2}$ $\qquad K_p = P_{O_2}$

15-13 Relationship Between ΔG^o and the Equilibrium Constant

The value of ΔG^o (or $\Delta G^o_{298\ K}$) indicates the free energy transferred in establishing equilibrium at 25°C and one atmosphere pressure when all substances, reactants and products, are mixed in unit molar quantities (or unit partial pressures, 1 atmosphere, for gases). The value of ΔG indicates the free energy transfer under any specific set of conditions except standard thermodynamic conditions. The two quantities are related to each other by the relationship

$$\Delta G = \Delta G^o + RT \ln Q$$

or, since $\ln Q = 2.303 \log Q$

$$\Delta G = \Delta G^o + 2.303\ RT \log Q$$

*Occurs in dilute aqueous solutions; $[H_2O]$ is very nearly constant (55.6 M) and is included in K_c.

where Q is the reaction quotient which has the same form as the equilibrium constant, T is the absolute temperature, and R is 8.314 J/mol K or 1.987 cal/mol K.

At equilibrium ΔG must be zero, and Q = K, so

$$\Delta G^{o} = -2.303\ RT \log K$$

where K is K_p for a gaseous system since standard states of gases are defined as unit partial pressures, and K is K_c for reactions in aqueous solution.

<u>Example 15-10:</u> Consider the reaction and evaluate K_p and K_c at 298 K from thermodynamic data in Appendix K of the text.

$$I_2\ (g) \quad + \quad Cl_2\ (g) \quad \longrightarrow \quad 2\ ICl\ (g)$$

We may calculate ΔG^{o}_{298} directly from values of ΔG^{o}_{f298}.

$$\Delta G^{o}_{298} = 2\ \Delta G^{o}_{f ICl\ (g)} - \Delta G^{o}_{f I_2\ (g)} - \Delta G^{o}_{f Cl_2\ (g)}$$

$$= 2(-5.52\ kJ) - 19.36\ kJ - 0\ kJ$$

$$\Delta G^{o}_{298} = \underline{\underline{-30.4\ kJ}} \quad \text{per mol of } I_2 \text{ (or } Cl_2\text{) consumed}$$

At equilibrium, $\Delta G^{o}_{298} = -2.303\ RT \log K_p$

$$\log K_p = \frac{\Delta G^{o}_{298}}{-2.303\ RT} = \frac{-3.04 \times 10^{4}\ J/mol}{-2.303\ (8.314\ \frac{J}{mol\ K})(298\ K)} = 5.33$$

Since $\log K_p = 5.33$, taking the antilog of both sides gives

$$K_p = 10^{+5.33} = \underline{\underline{K_p = 2.1 \times 10^{5}}} \quad \text{at 298 K}$$

Since equal numbers of moles of gases are produced and consumed ($\Delta n = 0$),

$$\underline{\underline{K_c = K_p = 2.1 \times 10^{5}}}$$

<u>Example 15-11</u> The equilibrium constant, K_c, is 4.63×10^{-3} <u>M</u> for the reaction below at 25°C and 1 atm pressure. Evaluate ΔG^{o}_{298}.

$$N_2O_4\ (g) \quad \rightleftharpoons \quad 2\ NO_2\ (g)$$

We are given K_c and we use it to calculate K_p.

$$K_p = K_c (RT)^{\Delta n} \qquad \Delta n = 1$$

$$= 4.63 \times 10^{-3} \frac{mol}{L} [(0.0821 \frac{L \cdot atm}{mol\ K})(298\ K)]^1$$

$$= 0.113 \text{ atm at } 298\ K$$

$$\Delta G^o_{298} = -2.303\ RT \log K_{p\,298}$$

$$= -2.303\ (8.314 \frac{J}{mol\ K})(298\ K) \log 0.113$$

$$= -2.303\ (8.314 \frac{J}{mol\ K})(298\ K)(-0.947)$$

$$\Delta G^o_{298} = 5403\ J/mol = 5.40\ kJ/mol \text{ of } N_2O_4 \text{ consumed}$$

15-14 Relationship between ΔH^o_{298} and Equilibrium Constants at Different Temperatures

If the assumption is made that ΔH^o is approximately temperature-independent (an inexact assumption, but a reasonable approximation), the following relationship may be derived.

$$\frac{\Delta H^o (T_2 - T_1)}{2.303\ RT_2T_1} = \log \left(\frac{K_{T_2}}{K_{T_1}}\right)$$

If ΔH^o and the equilibrium constant are known at any particular temperature, T_1, the value of the equilibrium constant at another temperature, T_2, can be estimated.

Example 15-12: For the following reaction at 25°C, $\Delta H^o_{298} = -197.6$ kJ, and $K_{p\ 298} = 7.00 \times 10^{24}\ atm^{-1}$. Estimate the equilibrium constants, K_p and K_c, for the reaction at 827°C or 1100 K.

$$2\ SO_2\ (g) \quad + \quad O_2\ (g) \quad \rightleftharpoons \quad 2\ SO_3\ (g)$$

We substitute into the relationship given above.

$$\frac{\Delta H^o_{298}\ (T_2 - T_1)}{2.303\ RT_2\ T_1} = \log \left(\frac{K_{T_2}}{K_{T_1}}\right)$$

$$\frac{-197,600(1100-298)}{2.303(8.314)(1100)(298)} = \log\left(\frac{K_{T_2}}{7.00 \times 10^{24}}\right)$$

Since $\log\left(\frac{x}{y}\right) = \log x - \log y$,

$$-25.25 = \log K_{T_2} - \log(7.00 \times 10^{24})$$

$$-25.25 = \log K_{T_2} - 24.85$$

$$-0.40 = \log K_{T_2}$$

$$K_{T_2} = 10^{-0.40} = \underline{\underline{0.40 \text{ atm}^{-1}}} \qquad (K_p \text{ at } 1100\text{ K})$$

$$K_p = K_c\,(RT)^{\Delta n} \qquad \Delta n = -1,\ T = 1100\text{ K}$$

$$K_c = \frac{K_p}{(RT)^{\Delta n}} = K_p\,(RT)^{-\Delta n}$$

$$K_c = K_p\,(RT)^1 = (0.40\text{ atm}^{-1})\left[(0.0821\ \frac{\text{L}\cdot\text{atm}}{\text{mol}\cdot\text{K}})(1100\text{ K})\right]^1$$

$$\underline{\underline{K_c = 36\ \underline{M}^{-1}}}$$

EXERCISES

1. Write mass action expressions for K_c for the following reactions.

   (a) PCl_3 (g) + Cl_2 (g) $\rightleftharpoons$ PCl_5 (g)

   (b) CO (g) + H_2O (g) $\rightleftharpoons$ CO_2 (g) + H_2 (g)

   (c) N_2O_4 (g) $\rightleftharpoons$ 2 NO_2 (g)

   (d) $MgCO_3$ (s) $\rightleftharpoons$ MgO (s) + CO_2 (g)

   (e) 2 SO_2 (g) + O_2 (g) $\rightleftharpoons$ 2 SO_3 (ℓ)

   (f) H_2 (g) + Cl_2 (g) $\rightleftharpoons$ 2 HCl (g)

   (g) 2 NO (g) + Cl_2 (g) $\rightleftharpoons$ 2 NOCl (g)

   (h) 2 $Ca_3(PO_4)_2$ (s) + 6 SiO_2 (s) + 10 C (s) $\rightleftharpoons$
   6 $CaSiO_3$ (s) + 10 CO (g) + P_4 (g)

   (i) Ca_3P_2 (s) + 6 H_2O (ℓ) $\rightleftharpoons$ 3 $Ca(OH)_2$ (s) + 2 PH_3 (g)

2. Write the mass action expression for K_p for each reaction in question 1.

3. Consider the reaction below at 350°C. The following amounts were found in a closed 5.00 liter vessel at equilibrium: 1.80 mole NO, 0.352 mole Cl_2, and 4.73 mole NOCl.

   2 NOCl (g) $\rightleftharpoons$ 2 NO (g) + Cl_2 (g)

   (a) Evaluate K_c at 350°C, (b) Evaluate K_p at 350°C

4. Evaluate K_c and K_p at 350°C for the following (see problem 3).

   (a) NOCl (g) $\rightleftharpoons$ NO (g) + 1/2 Cl_2 (g)

   (b) 2 NO (g) + Cl_2 (g) $\rightleftharpoons$ 2 NOCl (g)

   (c) 2/3 NO (g) + 1/3 Cl_2 (g) $\rightleftharpoons$ 2/3 NOCl (g)

5. The equilibrium constant, K_c, is 120 for the hypothetical reaction below at a certain temperature. What will be the equilibrium concentrations of A, B, and C in a 2.00 liter container at this temperature if the reaction is initiated with the following amounts of reactants or products?

   A (g) + B (g) $\rightleftharpoons$ 2 C (g)

   (a) 1.00×10^{-3} mole A and 1.00×10^{-3} mole B

   (b) 1.00×10^{-3} mole A and 2.00×10^{-3} mole B

   (c) 1.00×10^{-3} mole C

(d) 1.00×10^{-3} mole C and 1.00×10^{-3} mole A

6. Consider the hypothetical reaction below at a certain temperature. What will be the equilibrium concentrations of all species at this temperature in a 5.00 liter container if the reaction is initiated with the following amounts of reactants and products?

$$A\,(g) \quad + \quad B\,(g) \quad \rightleftharpoons \quad C\,(g) \quad + \quad D\,(g) \qquad K_c = 12.0$$

(a) 0.067 mole of A and 0.067 mole of B

(b) 0.040 mole of C and 0.040 mole of D

(c) 0.32 mole of A, 1.25 moles of C and 0.075 mole of D

7. Consider the hypothetical reaction below at a certain temperature. Calculate the equilibrium concentrations of all species if the reaction is initiated in a 1.00 liter container with the following amounts.

$$D\,(g) \quad + \quad E\,(g) \quad \rightleftharpoons \quad F\,(g) \quad + \quad G\,(g) \qquad K_c = 4.2 \times 10^{-2}$$

(a) 4.8×10^{-1} mole of D and 2.6×10^{-1} mole of E

(b) 2.4×10^{-2} mole of F and 5.0×10^{-2} mole of G

(c) 0.010 mole D, 0.010 mole E, and 0.010 mole G

8. $K_c = 4.62 \times 10^{-3}$ at 25°C for the decomposition of N_2O_4 to NO_2. A 1.50 gram sample of N_2O_4 is placed in a 3.00 liter container at 25°C and allowed to reach equilibrium.

$$N_2O_4\,(g) \quad \rightleftharpoons \quad 2\,NO_2\,(g)$$

(a) What masses of N_2O_4 and NO_2 are present?

(b) What is the percent dissociation of N_2O_4?

(c) What is K_p at 25°C?

(d) What are the partial pressures of N_2O_4 and NO_2?

(e) What is the total pressure?

9. At 727°C 0.500 mole of gaseous H_2S in a 500 mL container is 1.2% dissociated as indicated by the following equation.

$$2\,H_2S\,(g) \quad \rightleftharpoons \quad 2\,H_2\,(g) \quad + \quad S_2\,(g)$$

(a) Calculate K_c and K_p at this temperature.

(b) What are the concentrations of all species at equilibrium?

10. Consider the following reaction at equilibrium. In which direction will the equilibrium shift and how will the equilibrium amount of NO_2 be affected by the following changes?

$$2\,NO\,(g) \quad + \quad O_2\,(g) \quad \rightleftharpoons \quad 2\,NO_2\,(g) \quad + \quad 114.1\text{ kJ}$$

(a) lowering the temperature

(b) raising the temperature

(c) adding more O_2

(d) adding more NO

(e) removing some NO_2

(f) adding a catalyst

(g) increasing the volume of the container

(h) increasing pressure by adding an inert gas such as He

11. Consider the following reaction at equilibrium. In which direction will the equilibrium shift and how will the equilibrium amount of N_2O be affected by the following changes?

$$2\ N_2\ (g) \quad + \quad O_2\ (g) \quad + \quad 164\ kJ \quad \rightleftharpoons \quad 2\ N_2O\ (g)$$

(a) adding more N_2

(b) adding more O_2

(c) removing some N_2O

(d) adding a catalyst

(e) raising the temperature

(f) lowering the temperature

(g) decreasing the volume of the container

(h) increasing the pressure by adding neon

12. Consider the reaction below at equilibrium at 300°C. How will the equilibrium shift and how will the equilibrium amount of Cl_2 (g) be changed by the following stresses?

$$2\ Cl_2\ (g) \quad + \quad 2\ H_2O\ (g) \quad + \quad \text{Heat} \quad \rightleftharpoons \quad 4\ HCl\ (g) \quad + \quad O_2\ (g)$$

(a) removing some O_2

(b) adding some HCl

(c) adding some H_2O

(d) increasing the volume

(e) increasing the pressure by adding some He

(f) raising the temperature

(g) decreasing the temperature

13. Consider the reaction below at equilibrium at 200°C. How will the equilibrium shift and how will the equilibrium amount of Br_2 be affected by the following changes? $H_2\ (g) + Br_2\ (g) \rightleftharpoons 2\ HBr\ (g) + 109\ kJ$

(a) cooling

(b) heating

(c) removing some H_2

(d) adding some HBr

(e) adding some Br_2

(f) decreasing the volume

(g) increasing the pressure by adding argon

14. (a) Refer to Problem 8. What would be the equilibrium concentrations of N_2O_4 (g) and NO_2 (g) starting with 1.50 g of N_2O_4 in a 5.00 (rather than 3.00) liter container?

(b) Determine the percentage dissociation of 1.50 g of N_2O_4 (in 5.00 L).

(c) What masses of N_2O_4 (g) and NO_2 (g) will be present (in 5.00 L)?

(d) What will be the pressure of the system at equilibrium (in 5.00 L)?

15. For the system below, the following amounts are present in a 3.00 liter vessel at equilibrium: 0.0348 mole I_2, 0.102 mole H_2, and 0.441 mole HI. The temperature is 425°C. $2\,HI\,(g) \rightleftharpoons H_2\,(g) + I_2\,(g)$

(a) Evaluate K_c at 425°C.

(b) What would be the new equilibrium concentrations of all species if 0.0160 mole of I_2 were added to the system with no change in temperature?

(c) What would be the new equilibrium concentrations of all species if 0.0160 mole of I_2 were removed from the original system at constant temperature?

16. For the reaction below, $K_p = 6.70 \times 10^{-3}$ at 25°C.

$$COCl_2 \rightleftharpoons CO\,(g) + Cl_2\,(g)$$

(a) A sample of $COCl_2$ is placed in an enclosed 15.0 liter vessel at 25°C and it exerts a pressure of 4.65 atm before decomposition begins. What will be the partial pressures of all the species when equilibrium is established and what will the total pressure be?

(b) If the volume is increased to 30.0 liters with no change in temperature what will be the new equilibrium partial pressures and total pressure?

(c) Refer to (b). What will be the new equilibrium concentrations of the three gases?

(d) If the volume of the original system is decreased to 5.00 liters what will be the new equilibrium partial pressures of all species and the total pressure?

(e) Refer to (d). What mass (in grams) of each of the species will be present?

17. Refer to Problem 16.

(a) Calculate K_c at 25°C.

(b) Use K_c to determine the equilibrium concentrations at 25°C if the volume of the vessel is increased to 30.0 liters, i.e., solve 16(b) in terms of concentrations rather than partial pressures. Compare the results with 16(c).

(c) Convert the concentrations you just calculated in 17(b) to partial pressures. Compare the results with 16(b).

18. Using data in Appendix K, calculate ΔG° for the following reactions at 25° and then calculate K_p for each at 25°C.

(a) $H_2\,(g) + Cl_2\,(g) \rightleftharpoons 2\,HCl\,(g)$

(b) $N_2\,(g) + 1/2\,O_2\,(g) \rightleftharpoons N_2O\,(g)$

(c) $2\ NO\ (g) \rightleftharpoons O_2\ (g) + N_2\ (g)$

(d) $2\ CO\ (g) + O_2\ (g) \rightleftharpoons 2\ CO_2\ (g)$

19. Calculate K_c for each reaction of Problem 18.

20. Given the following reactions at 1000 K and their equilibrium constants, calculate ΔG^o at 1000 K for each.

(a) $2\ NH_3\ (g) \rightleftharpoons 3\ H_2\ (g) + N_2\ (g)$ $K_p = 2.9 \times 10^6$

(b) $CO_2\ (g) + CF_4\ (g) \rightleftharpoons 2\ COF_2\ (g)$ $K_p = 0.472$

21. For the reaction below, K_p is 6.25×10^3 at 1000 K. Estimate K_p at 440°C.

$$2\ NOBr\ (g) + 46.4\ kJ \rightleftharpoons 2\ NO\ (g) + Br_2\ (g)$$

ANSWERS FOR EXERCISES

1. (a) $K_c = \frac{[PCl_5]}{[PCl_3][Cl_2]}$ (b) $K_c = \frac{[CO_2][H_2]}{[CO][H_2O]}$

(c) $K_c = \frac{[NO_2]^2}{[N_2O_4]}$ (d) $K_c = [CO_2]$

(e) $K_c = \frac{1}{[SO_2]^2[O_2]}$ (f) $K_c = \frac{[HCl]^2}{[H_2][Cl_2]}$

(g) $K_c = \frac{[NOCl]^2}{[NO]^2[Cl_2]}$ (h) $K_c = [CO]^{10}[P_4]$

(i) $K_c = [PH_3]^2$

2. (a) $K_p = \frac{P_{PCl_5}}{P_{PCl_3} P_{Cl_2}}$ (b) $K_p = \frac{P_{CO_2} P_{H_2}}{P_{CO} P_{H_2O}}$

(c) $K_p = \frac{P_{NO_2}^{\ 2}}{P_{N_2O_4}}$ (d) $K_p = P_{CO_2}$

(e) $K_p = \frac{1}{P_{SO_2}^{\ 2} P_{O_2}}$ (f) $K_p = \frac{P_{HCl}^{\ 2}}{P_{H_2} P_{Cl_2}}$

(g) $K_p = \frac{P_{NOCl}^2}{P_{NO}^2 P_{Cl_2}}$ (h) $K_p = P_{CO}^{10} P_{P_4}$

(i) $K_p = P_{PH_3}^2$

3. (a) $K_c = 1.02 \times 10^{-2}$ (b) $K_p = 0.522$ 4. (a) $K_c = 0.101$, $K_p = 0.722$, (b) $K_c = 98.0$, $K_p = 1.92$ (c) $K_c = 4.61$, $K_p = 1.24$

5. (a) [A] = [B] = 7.7×10^{-5} M, [C] = 8.46×10^{-4} M
(b) [A] = 1.4×10^{-5} M, [B] = 5.14×10^{-4} M, [C] = 9.72×10^{-4} M
(c) [A] = [B] = 3.85×10^{-5} M, [C] = 4.23×10^{-4} M
(d) [A] = 5.04×10^{-4} M, [B] = 4.0×10^{-6} M, [C] = 4.92×10^{-4} M

6. (a) [A] = [B] = 3.0×10^{-3} M, [C] = [D] = 1.04×10^{-2} M
(b) [A] = [B] = 1.79×10^{-3} M, [C] = [D] = 6.21×10^{-3} M
(c) [A] = 6.8×10^{-2} M, [B] = 3.5×10^{-3} M, [C] = 0.25 M, [D] = 0.012 M

7. (a) [F] = [G] = 0.064 M, [D] = 0.42 M, [E] = 0.20 M
(b) [D] = [E] = 0.0232 M, [F] = 0.0008 M, [G] = 0.027 M
(c) [D] = [E] = 0.0095 M, [F] = 0.0005 M, [G] = 0.010 M

8. (a) 0.55 g NO_2, 0.95 g N_2O_4 (b) 36.6% (c) $K_p = 0.113$
(d) $P_{N_2O_4} = 8.42 \times 10^{-2}$ atm, $P_{NO_2} = 9.74 \times 10^{-2}$ atm (e) $P_{total} = 0.182$ atm

9. (a) $K_c = 8.9 \times 10^{-7}$, $K_p = 7.3 \times 10^{-5}$ (b) [H_2S] = 0.988 M, [H_2] = 0.012 M, [S_2] = 0.0060 M

10. (a) right, increases (b) left, decreases (c) right, increases (d) right, increases (e) right, decreases (f) none, no change (g) left, decreases (h) none, no change

11. (a) right, increases (b) right, increases (c) right, decreases (d) none, no change (e) right, increases (f) left, decreases (g) right, increases (h) none, no change

12. (a) right, decreases (b) left, increases (c) right, decreases (d) right, decreases (e) none, no change (f) right, decreases (g) left, increases

13. (a) right, decreases (b) left, increases (c) left, increases (d) left, increases (e) right, increases (f) none, no change (g) none, no change

14. (a) [NO_2] = 2.90×10^{-3} M, [N_2O_4] = 1.81×10^{-3} M (b) 44.5% (c) 0.67 g NO_2, 0.83 g N_2O_4

(d) P_{total} = 0.115 atm 15. (a) $K_c = 1.82 \times 10^{-2}$ (b) [HI] = 0.153 M, $[H_2]$ = 0.0309 M, $[I_2]$ = 0.0138 M (c) [HI] = 0.140 M, $[H_2]$ = 0.0374 M, $[I_2]$ = 0.00962 M 16. (a) $P_{CO} = P_{Cl_2}$ = 0.173 atm, P_{COCl_2} = 4.48 atm, P_{total} = 4.82 atm (b) $P_{CO} = P_{Cl_2}$ = 0.121 atm, P_{COCl_2} = 2.20 atm, P_{total} = 2.44 atm (c) [CO] = $[Cl_2]$ = 4.95×10^{-3} M, $[COCl_2]$ = 8.99×10^{-2} M (d) $P_{CO} = P_{Cl_2}$ = 0.303 atm, P_{COCl_2} = 13.7 atm, P_{total} = 14.3 atm (e) 1.74 g CO, 4.40 g Cl_2, 277 g $COCl_2$
17. (a) $K_c = 2.74 \times 10^{-4}$ (b) [CO] = $[Cl_2]$ = 4.96×10^{-3} M, $[COCl_2]$ = 8.99×10^{-2} M The concentrations are the same (within round-off accuracy) as those obtained in 16(c). (c) $P_{CO} = P_{Cl_2}$ = 0.121 atm, P_{COCl_2} = 2.20 atm. The results agree with those of 16(b). 18. (a) ΔG° = −190.6 kJ, $K_p = 2.54 \times 10^{33}$ (b) ΔG° = 104.2 kJ, $K_p = 5.5 \times 10^{-19}$ (c) ΔG° = −173.1 kJ, $K_p = 2.2 \times 10^{30}$ (d) ΔG° = −514.4 kJ, $K_p = 1.45 \times 10^{90}$
19. (a) $K_c = 2.54 \times 10^{33}$ (b) $K_c = 2.7 \times 10^{-18}$ (c) $K_c = 2.2 \times 10^{30}$ (d) $K_c = 3.55 \times 10^{91}$ 20. (a) ΔG° = −124 kJ at 1000 K (b) ΔG° = 6.24 kJ at 1000 K 21. K_p = 661 at 440°C

Chapter Sixteen

EQUILIBRIUM IN AQUEOUS SOLUTIONS-I

D, G, W: Sec. 17-1; Sec. 17-5 through 17-8; Sec. 18-1 through 18-5.1

16-1 Strong Electrolytes

Electrolytes are compounds that ionize in aqueous solution, and so their aqueous solutions conduct electric current. Nonelectrolytes do neither. Strong electrolytes are essentially completely ionized (dissociated) in dilute aqueous solutions, while weak electrolytes are only slightly ionized. The three common classes of strong electrolytes are (1) strong acids, (2) strong soluble bases, and (3) most soluble salts. Most inorganic acids are soluble in water. The names and formulas of the common strong acids are listed in the text on Table 7-1. Other common acids may be assumed to be weak. The strong soluble bases are the hydroxides of the Group IA metals and the lower members of the IIA metals (Table 7-3). Other common metal hydroxides are quite insoluble. The solubilities of salts may be predicted by application of the solubility rules (Section 7-1.5).

The concentrations of ions in solutions of strong electrolytes are simply the original concentrations of the electrolytes multiplied by the number of those ions per formula unit.

Example 16-1: What are the concentrations of $Sr(OH)_2$, Sr^{2+} and OH^- in 0.01 M $Sr(OH)_2$?

Ionization is essentially complete because $Sr(OH)_2$ is a strong soluble base.

$$Sr(OH)_2 \longrightarrow Sr^{2+} + 2\,OH^-$$

$$0.01\ \underline{M} \qquad\qquad 0.01\ \underline{M} \qquad 0.02\ \underline{M}$$

Concentrations are: $[Sr(OH)_2] = 0\ \underline{M}$, $[Sr^{2+}] = 0.01\ \underline{M}$, $[OH^-] = 0.02\ \underline{M}$

16-2 The Ionization of Water

Pure water ionizes to a slight extent according to the equation,

$$H_2O + H_2O \rightleftharpoons H_3O^+ + OH^-.$$

In pure water at 25°C, $[H_3O^+] = [OH^-] = 1.0 \times 10^{-7}$ M. In all dilute aqueous solutions the product of the concentrations of these two ions is a constant, called K_w, the ion product for water. At 25°C,

$$[H_3O^+][OH^-] = 1.0 \times 10^{-14} = K_w$$

Even though 1.0×10^{-7} M is a very low concentration, a liter of pure water contains 60.2 quadrillion H_3O^+ and 60.2 quadrillion OH^- ions.

Example 16-2: What is the $[H_3O^+]$ in a 0.40 M solution of potassium hydroxide?

$$\begin{array}{lcccc} KOH & \longrightarrow & K^+ & + & OH^- \\ 0.40\ \underline{M} & & 0.40\ \underline{M} & & 0.40\ \underline{M} \end{array}$$

$$[H_3O^+][OH^-] = 1.0 \times 10^{-14} = K_w$$

$$[H_3O^+] = \frac{1.0 \times 10^{-14}}{[OH^-]} = \frac{1.0 \times 10^{-14}}{0.40} = \underline{\underline{2.5 \times 10^{-14}\ M}}$$

16-3 The pH Scale

The pH scale provides a convenient way of expressing acidity and basicity. The "p" means "minus log of". Thus, $pH = -\log[H_3O^+]$ and $pOH = -\log[OH^-]$. If any one of the four variables, $[H_3O^+]$, $[OH^-]$, pH, or pOH is known for an aqueous solution, the other three can be calculated. The following useful relationships are valid for all aqueous solutions.

$$[H_3O^+][OH^-] = 1.00 \times 10^{-14} \quad \text{and} \quad pH + pOH = 14.000$$

It is useful to remember that as $[H_3O^+]$ increases, $[OH^-]$ decreases. As $[H_3O^+]$ increases, pH decreases, and pOH increases. The converse is also true.

Example 16-3: Calculate $[H_3O^+]$, $[OH^-]$, pH, and pOH for a 0.0500 M solution of perchloric acid, $HClO_4$.

Perchloric acid is a strong acid and ionizes completely.

$$\begin{array}{lccccccc} HClO_4 & + & H_2O & \longrightarrow & H_3O^+ & + & ClO_4^- \\ 0.0500\ \underline{M} & & & & 0.0500\ \underline{M} & & 0.0500\ \underline{M} \end{array}$$

Thus, $[H_3O^+] = \underline{\underline{0.0500\ M}}$ and we may now calculate pH.

$$pH = -\log[H_3O^+] = -\log(5.00 \times 10^{-2}) = -(0.699-2.000) = \underline{\underline{1.301}}$$

Since $[H_3O^+][OH^-] = 1.00 \times 10^{-14}$

$$[OH^-] = \frac{1.00 \times 10^{-14}}{(0.0500)} = \underline{\underline{2.00 \times 10^{-13}\ \underline{M}}}$$

$$pOH = -\log[OH^-] = -\log(2.00 \times 10^{-13}) = -(0.301 - 13^*) = \underline{\underline{12.699}}$$

pOH and $[OH^-]$ could also be calculated as follows:

$$pH + pOH = 14.000$$

$$pOH = 14.000 - pH = 14.000 - 1.301 = \underline{\underline{12.699}}$$

$$[OH^-] = 10^{-pH} = 10^{-12.699} = 10^{(0.301-13^*)}$$

$$= 10^{0.301} \times 10^{-13} = \underline{\underline{2.00 \times 10^{-13}\ \underline{M}}}$$

16-4 Ionization Constants for Weak Monoprotic Acids and Bases

The distinction between strong electrolytes and weak electrolytes is based on one criterion, extent of ionization in dilute aqueous solution. Weak acids and bases ionize only slightly. The ionization of any weak monoprotic acid, HA, can be represented as:

$$HA + H_2O \rightleftharpoons H_3O^+ + A^-$$

The ionization constant expression is: $K_a = \dfrac{[H_3O^+][A^-]}{[HA]}$

The value of K_a at a specific temperature includes the concentration of water, which for dilute solutions is essentially that for pure water, 55.6 M. The terms K_i (for $K_{ionization}$) or K_{HA} are sometimes used instead of K_a. Appendix F lists ionization constants for many common acids. The higher the value of K_a, the stronger the acid. Table 16-1 includes a few examples.

Table 16-1. Ionization Constants for Some Acids

Acid	Formula	K_a at 25°C
acetic	CH_3COOH	1.8×10^{-5}
carbonic	H_2CO_3	4.2×10^{-7} (first step)
	HCO_3^-	4.8×10^{-11} (second step)
hydrocyanic	HCN	4.0×10^{-10}
hydrofluoric	HF	7.2×10^{-4}

*Exact numbers.

If we know the value of K_a for a weak acid and the initial concentration of the acid in dilute aqueous solution, we can determine the equilibrium concentrations of all species, un-ionized HA, H_3O^+, and A^- ions, as well as percent ionization.

Example 16-4: The ionization constant for the hypothetical weak acid HA is 1.0×10^{-5}. What are the equilibrium concentrations of HA, H_3O^+, and A^- in a 0.20 M HA solution?

We write (1) the equation for the reaction, (2) the expression for the ionization constant, and then (3) set up algebraic representations of the concentrations. Let x = concentration of HA that ionizes.

	HA	+	H_2O	$\rightleftharpoons$	H_3O^+	+	A^-
initial:	0.20 M				0 M		0 M
change:	-x M				+x M		+x M
equil:	(0.20 - x) M				x M		x M

Substitution of equilibrium concentrations into the ionization constant expression gives

$$\frac{[H_3O^+][A^-]}{[HA]} = 1.0 \times 10^{-5} \qquad \frac{(x)(x)}{(0.20-x)} = 1.0 \times 10^{-5}$$

The following is a useful rule-of-thumb that may be used to avoid using the quadratic formula. Any term containing "x" added to or subtracted from a number greater than 0.05 may be simplified by deleting the "x" if the exponent of 10 is -5 or less, i.e., -5, -6, -7, -8, etc. Thus, the above expression becomes

$$\frac{x^2}{0.20} = 1.0 \times 10^{-5}, \quad x^2 = 2.0 \times 10^{-6}, \quad \underline{\underline{x = 1.4 \times 10^{-3}\ \underline{M}}}$$

$$\underline{\underline{x = [H_3O^+] = [A^-] = 1.4 \times 10^{-3}\ \underline{M}}}$$

$$[HA] = (0.20 - x) = 0.20\ \underline{M} - 0.0014\ \underline{M} \approx \underline{\underline{0.20\ \underline{M}}}$$

The species present in highest concentration is un-ionized weak acid molecules.

The basis for the assumption that $(0.20 - x) \approx 0.20$ is as follows. The "x" is defined to be the concentration of HA that ionizes. The small value of K_a, 1.0×10^{-5}, tells us that the extent of ionization is slight. It is reasonable to assume that the concentration of HA which ionizes is insignificant in comparison to the original concentration of HA (0.20 $\underline{M}$). Subtraction of "x" from 0.20 $\underline{M}$ should still give 0.20 $\underline{M}$, to two significant figures, i.e., $(0.20 - x) \approx 0.20$. The validity of this assumption should always be checked. We can do so by comparing the value of x with 0.20 $\underline{M}$.

$$(0.20 - x) = (0.20 - 0.0014)\underline{M} = 0.1986\ \underline{M} \approx 0.20\ \underline{M}$$

The assumption is valid. If such an assumption proves invalid, the "x" term which was neglected must be retained and the problem solved without simplification.

The percentage ionization is the concentration of electrolyte that ionizes divided by the concentration initially present, multiplied by 100%.

$$\%\ \text{ionization of HA} = \frac{[HA]_{ionized}}{[HA]_{total}} \times 100\%$$

In this case, $[H_3O^+] = [HA]_{ionized}$, and therefore

$$\%\ \text{ionization} = \frac{[H_3O^+]}{[HA]_{total}} \times 100\% = \frac{0.0014}{0.20} \times 100\% = \underline{\underline{0.70\%\ \text{ionized}}}$$

For a given weak acid (or base) at a given temperature, percent ionization decreases as initial concentration increases. However, concentrations of ionic species in solution increase. Table 16-2 compares the percentage ionization, $[H_3O^+]$, pH, and K_a values for 0.10 $\underline{M}$ solutions of three acids of different strengths.

Table 16-2. Comparison of Acid Strengths

Acid Solution	K_a	$[H_3O^+]$	pH	% Ionization
0.10 $\underline{M}$ HNO_3	meaningless	0.10 $\underline{M}$	1.0	100
0.10 $\underline{M}$ CH_3COOH	1.8×10^{-5}	0.0013 $\underline{M}$	2.89	1.3
0.10 $\underline{M}$ HClO	3.5×10^{-8}	0.000059 $\underline{M}$	4.23	0.059

Example 16-5: What is the percentage ionization of 0.20 $\underline{M}$ HNO_2?

The ionization reaction is:

	HNO_2	+	H_2O	$\rightleftharpoons$	H_3O^+	+	NO_2^-
start:	0.20 $\underline{M}$				0 $\underline{M}$		0 $\underline{M}$
change:	-x $\underline{M}$				+x $\underline{M}$		+x $\underline{M}$
equil:	(0.20 - x)$\underline{M}$				x $\underline{M}$		x $\underline{M}$

The ionization constant for HNO_2 is:

$$\frac{[H_3O^+][NO_2^-]}{[HNO_2]} = 4.5 \times 10^{-4}.$$ Substitution gives:

$$\frac{(x)(x)}{0.20-x} = 4.5 \times 10^{-4}$$

Since the ionization constant for HNO_2 is fairly large, we may not assume that $(0.20-x) \approx 0.20$. Therefore, we use the quadratic formula to solve the equation.

$$x^2 = 0.90 \times 10^{-4} - 4.5 \times 10^{-4}\ x$$

$$x^2 + 4.5 \times 10^{-4}\ x - 0.90 \times 10^{-4} = 0$$

Solving via the quadratic formula,

$$x = \frac{-b \pm \sqrt{b^2 - 4ac}}{2a} = \frac{-4.5 \times 10^{-4} \pm \sqrt{(4.5 \times 10^{-4})^2 - 4(1)(-0.90 \times 10^{-4})}}{2}$$

gives $\underline{\underline{x = 9.3 \times 10^{-3}\ \underline{M} = [H_3O^+] = [NO_2^-]}}$

Now we may calculate percentage ionization. Since one hydrogen (or nitrite) ion is produced by each nitrous acid molecule that ionizes, $[HNO_2]_{\text{ionized}} = [H_3O^+]$.

$$\%\ \text{ionization} = \frac{[HNO_2]_{\text{ionized}}}{[HNO_2]_{\text{total}}} \times 100\% = \frac{[H_3O^+]}{[HNO_2]_{\text{total}}} \times 100\%$$

$$= \frac{0.0093\ \underline{M}}{0.20\ \underline{M}} \times 100\% = \underline{\underline{4.6\%\ \text{ionized}}}$$

The only common weak bases are aqueous ammonia, NH_3, and its derivatives, the amines. The basicity of this class of compounds arises from the presence of the lone pair of electrons of the nitrogen atom. They make these compounds both Brønsted-Lowry bases (proton acceptors) and Lewis bases (electron pair donors). The ionization of an amine is similar to that of ammonia in water.

$$NH_3 + H_2O \rightleftharpoons NH_4^+ + OH^-$$

ammonia → ammonium ion

$$N(CH_3)_3 + H_2O \rightleftharpoons HN(CH_3)_3^+ + OH^-$$

trimethylamine → trimethylammonium ion

$$NH(C_2H_5)_2 + H_2O \rightleftharpoons NH_2(C_2H_5)_2^+ + OH^-$$

diethylamine → diethylammonium ion

Such solutions are basic since they contain an excess of OH^- ions over H_3O^+ ions. (Remember that both are present in all aqueous solutions and that a neutral solution contains equal numbers of both.) The ionization of any weak monoprotic base may be represented as

$$\text{Base} + H_2O \rightleftharpoons \text{Cation}^+ + OH^-$$

$$K_b = \frac{[\text{Cation}^+][OH^-]}{[\text{Base}]}$$

Table 16-3 lists basic ionization constants for several weak bases. Appendix G contains a longer list. These values need not be committed to memory, but their signficiance must be appreciated. Equilibrium calculations involving solutions of weak bases are entirely analogous to those of weak acids. The values of K_a for acetic acid and K_b for aqueous ammonia are coincidentally equal. This only indicates that acetic acid is as strong (or weak) an acid as ammonia is a base.

Table 16-3. Ionization Constants of Some Bases

Name	Formula	K_b
potassium hydroxide	KOH	meaningless (very large)
ammonia	NH_3	1.8×10^{-5}
methylamine	H_2NCH_3	5.0×10^{-4}
dimethylamine	$HN(CH_3)_2$	7.4×10^{-4}
trimethylamine	$N(CH_3)_3$	7.4×10^{-5}
phenylamine (aniline)	$C_6H_5NH_2$	4.2×10^{-10}

Example 16-6: Calculate the percent ionization of 0.30 M aniline solution.

The ionization reaction is:

	$C_6H_5NH_2$	+	H_2O	$\rightleftharpoons$	$C_6H_5NH_3^+$	+	OH^-
initial:	0.30 $\underline{M}$				0 $\underline{M}$		0 $\underline{M}$
change:	-x $\underline{M}$				+x $\underline{M}$		+x $\underline{M}$
equil:	(0.30-x)$\underline{M}$				x $\underline{M}$		x $\underline{M}$

The ionization constant for aniline is found in Table 16-3, and substitution gives

$$\frac{[C_6H_5NH_3^+][OH^-]}{[C_6H_5NH_2]} = 4.2 \times 10^{-10} = \frac{x^2}{0.30-x}$$

We may assume $(0.30-x) \approx 0.30$ since K_b is so small.

$$\frac{x^2}{0.30} = 4.2 \times 10^{-10} \qquad x^2 = 1.26 \times 10^{-10}$$

$$\underline{\underline{x = 1.1 \times 10^{-5}\ \underline{M} = [C_6H_5NH_3^+] = [OH^-]}}$$

Our assumption is verified since $0.30 - 1.1 \times 10^{-5} \approx 0.30$. Now we may calculate the percentage ionization.

$$\%\ \text{ionization} = \frac{[C_6H_5NH_2]_{\text{ionized}}}{[C_6H_5NH_2]_{\text{total}}} \times 100\%$$

Since $[C_6H_5NH_2]_{\text{ionized}} = [OH^-] = [C_6H_5NH_3^+]$,

$$\%\ \text{ionization} = \frac{[OH^-]}{[C_6H_5NH_2]_{\text{total}}} \times 100\% = \frac{0.000011\ \underline{M}}{0.30\ \underline{M}} \times 100\%$$

$$= \underline{\underline{0.0037\%\ \text{ionized}}} \qquad \text{(and 99.9963\% un-ionized)}$$

16-5 The Common Ion Effect and Buffer Solutions

The common ion effect is observed in solutions containing two or more compounds, both of which produce the same ion. Buffer solutions comprise a special case of the common ion effect and resist changes in pH. There are two major types of buffer solutions:

(1) A solution of a weak acid plus a soluble, ionic salt of the acid.
(2) A solution of a weak base plus a soluble, ionic salt of the base.

For such acid/salt or base/salt solutions containing reasonable concentrations of each, the presence of the relatively high concentration of anion of the acid (or cation of the base) from ionization of the salt represses ionization of the weak acid (or base). Thus, solutions of weak acids and their salts are less acidic than solutions of the same concentrations of weak acid alone. The same reasoning explains the lower basicity of solutions of weak bases and their salts as compared with solutions of identical concentrations of weak bases alone. Table 16-4 illustrates the point.

Table 16-4. Comparison of $[H_3O^+]$ and pH

Solution	$[H_3O^+]$	pH
0.10 M CH_3COOH	1.3×10^{-3} M	2.89
0.10 M CH_3COOH and 0.20 M $NaCH_3COO$	9.0×10^{-6} M	5.05

For a buffer solution which is 0.10 M in a weak acid, HA ($K_a = 1.0 \times 10^{-6}$), and 0.15 M in MA, a soluble salt, the important reactions are:

$$\underset{(0.10-x)M}{HA} + H_2O \rightleftharpoons \underset{x\,M}{H_3O^+} + \underset{x\,M}{A^-}$$

$$\underset{0.15\,M}{MA} \xrightarrow{100\%} \underset{0.15\,M}{M^+} + \underset{0.15\,M}{A^-}$$

In the equilibrium constant expression the $[A^-]$ is the sum of the concentrations of A^- from both compounds.

$$K_a = \frac{[H_3O^+][A^-]}{[HA]} = \frac{(x)(0.15+x)}{0.10-x} = 1.0 \times 10^{-6}$$

If we assume that $(0.15+x) \approx 0.15$ and $(0.10-x) \approx 0.10$ this relationship becomes:

$$\frac{(x)(0.15)}{0.10} = 1.0 \times 10^{-6} \quad \text{and} \quad x = 6.7 \times 10^{-7}\ M = [H_3O^+]$$

Note that after the simplifying assumption is made, the expression has the form

$$\frac{[H_3O^+][Salt]}{[Acid]} = K_a$$

Rearranging this expression and solving for $[H_3O^+]$ gives

$$[H_3O^+] = \frac{[Acid]}{[Salt]} \times K_a$$

Or, in a more useful form:

$$pH = pK_a + \log \frac{[Salt]}{[Acid]}$$ (Henderson-Hasselbalch equation)

In general, this expression is valid for solutions in which the original concentrations of salt and weak acid are reasonably close to each other. The [Acid] and [Salt] terms refer to original concentrations.

For buffer solutions of weak bases and their salts similar general relationships can be derived. They are:

$$[OH^-] = \frac{[Base]}{[Salt]} \times K_b$$

$$pOH = pK_b + \log \frac{[Salt]}{[Base]}$$ (Henderson-Hasselbalch equation)

The use of these two equations, where appropriate, greatly simplifies calculations of concentrations in buffer solutions. For salts containing 2 (or more) anions of the weak acid or 2 (or more) cations of the weak base per formula unit, the [Salt] must be multiplied by 2 (or more) in each of the above expressions. Examination of these equations tells us that the pH of a buffer solution can be easily adjusted simply by adjusting the ratio of salt to weak acid or salt to weak base.

Example 16-7: A buffer solution is made by mixing 250 mL of 1.00 $\underline{M}$ acetic acid with 500 mL of 0.500 $\underline{M}$ calcium acetate. Calculate the concentrations of all species in solution and the pH of the solution.

This buffer solution contains a weak acid and its salt. We first calculate the concentrations of acid and salt just after mixing the original solutions:

$$\underline{M}_{CH_3COOH} = \frac{250 \text{ mL} \times 1.00 \text{ M}}{750 \text{ mL}} = 0.333 \ \underline{M} \ CH_3COOH$$

$$\underline{M}_{Ca(CH_3COO)_2} = \frac{500 \text{ mL} \times 0.500 \text{ M}}{750 \text{ mL}} = 0.333 \ \underline{M} \ Ca(CH_3COO)_2$$

The reactions that occur and the equilibrium concentrations are:

$$Ca(CH_3COO)_2 \xrightarrow{100\%} Ca^{2+} + 2\ CH_3COO^-$$

0.333 $\underline{M}$ 0.333 $\underline{M}$ 0.666 $\underline{M}$

and, (where x = concentration of CH_3COOH that ionizes)

$$CH_3COOH \rightleftharpoons H^+ + CH_3COO^-$$
$$(0.333-x)\underline{M} \qquad x\,\underline{M} \qquad x\,\underline{M}$$

$$K_a = \frac{[H^+][CH_3COO^-]}{[CH_3COOH]} = 1.8 \times 10^{-5}$$

The $[CH_3COO^-]$ is the sum of the acetate ion concentrations produced by the calcium acetate, 0.666 M, and by acetic acid, x M.

$$\frac{(x)(x + 0.666)}{0.333-x} = 1.8 \times 10^{-5}$$

Assume $x \ll 0.333$ and obviously $x \ll 0.666$ (2 CH_3COO^- per $Ca(CH_3COO)_2$)

$$\frac{(x)(0.666)}{0.333} = 1.8 \times 10^{-5} \qquad \left(= \frac{[H^+](2)[\text{Salt}]}{[\text{Acid}]}\right)$$

$x = 9.0 \times 10^{-6}$ M $= [H^+] = [CH_3COO^-]$ from CH_3COOH ionization.

The equilibrium concentrations are:

$[H^+] = 9.0 \times 10^{-6}$ M Solution is buffered at pH = 5.05.

$$[OH^-] = \frac{K_w}{[H^+]} = \frac{1.0 \times 10^{-14}}{9.0 \times 10^{-6}} = 1.1 \times 10^{-9}\ \underline{M}$$

$[CH_3COO^-] = (0.666 + x)\,\underline{M} = (0.666 + 9.0 \times 10^{-6})\,\underline{M} = 0.666\ \underline{M}$

$[Ca^{2+}] = 0.333$ M

$[CH_3COOH] = (0.333 - x)\,\underline{M} = (0.333 - 9.0 \times 10^{-6})\,\underline{M} = 0.333\ \underline{M}$

Buffer solutions resist changes in pH upon addition of either acid or base because they are able to react with either H_3O^+ or OH^- ions. A solution of CH_3COOH and KCH_3COO is a typical weak acid/salt buffer. The CH_3COOH reacts with added base and CH_3COO^- reacts with added acid to form undissociated CH_3COOH. A solution of NH_3 and NH_4Cl is a typical weak base/salt buffer. Added H_3O^+ from acid is neutralized by the NH_3 while OH^- from added base reacts with NH_4^+ to form un-ionized NH_3 and water. The pH changes resulting from such additions are much smaller than from additions of acids or bases to unbuffered solutions.

Example 16-8: A 20.0 mL sample of 0.20 M HBr is added to 100.0 mL of a buffer solution that is 0.20 M in NH_3 and 0.20 M in NH_4Br. How much does the pH of the NH_3/NH_4Br solution change?

We must determine the pH before and after addition of the acid. Let us determine the pH of the original solution first. The appropriate equations are:

$$\begin{array}{ccccccc} NH_3 & + & H_2O & \rightleftharpoons & NH_4^+ & + & OH^- \\ (0.20-x)\,\underline{M} & & & & x\,\underline{M} & & x\,\underline{M} \end{array}$$

$$\begin{array}{ccccc} NH_4Br & \longrightarrow & NH_4^+ & + & Br^- \\ 0.20\,\underline{M} & & 0.20\,\underline{M} & & 0.20\,\underline{M} \end{array}$$

$$K_b = \frac{[NH_4^+][OH^-]}{[NH_3]} = 1.8 \times 10^{-5}$$

$$\frac{(0.20+x)(x)}{(0.20-x)} = 1.8 \times 10^{-5}$$

We may assume that $x \ll 0.20$. Thus $(0.20+x) \approx 0.20$, and $(0.20-x) \approx 0.20$. The equilibrium constant expression reduces to:

$$\frac{(0.20)(x)}{(0.20)} = 1.8 \times 10^{-5} \quad \text{or} \quad \frac{[\text{Salt}][OH^-]}{[\text{Base}]} = 1.8 \times 10^{-5}$$

$$\underline{\underline{x = 1.8 \times 10^{-5}\,\underline{M} = [OH^-]}} \quad \text{or} \quad [OH^-] = \frac{[\text{Base}]}{[\text{Salt}]}(1.8 \times 10^{-5})$$

$$[H_3O^+] = \frac{K_w}{[OH^-]} = \frac{1.0 \times 10^{-14}}{1.8 \times 10^{-5}}$$

$$[H_3O^+] = 5.6 \times 10^{-10}\,\underline{M}$$

$$pH = -\log[H_3O^+] = -\log(5.6 \times 10^{-10}) = \underline{\underline{9.25}}$$

Now let us determine pH after addition of 20.0 mL of 0.20 M HBr, a strong acid. Since the total volume of the buffer solution increases upon addition of HBr solution, we calculate the number of millimoles of each compound so that we may calculate new the molarities.

$$\underline{?}\text{ mmol HBr} = 20.0\text{ mL} \times \frac{0.20\text{ mmol HBr}}{\text{mL}} = 4.0\text{ mmol HBr}$$

$$\underline{?}\text{ mmol } NH_4Br = 100.0\text{ mL} \times \frac{0.20\text{ mmol } NH_4Br}{\text{mL}} = 20.0\text{ mmol } NH_4Br$$

$$\underline{?}\text{ mmol } NH_3 = 100.0\text{ mL} \times \frac{0.20\text{ mmol } NH_3}{\text{mL}} = 20.0\text{ mmol } NH_3$$

The added HBr reacts completely with NH_3.

	HBr	+	NH_3	$\longrightarrow$	NH_4Br
initial:	4.0 mmol		20.0 mmol		20.0 mmol
change:	-4.0 mmol		-4.0 mmol		+4.0 mmol
After rxn:	0 mmol		16.0 mmol		24.0 mmol

After reaction of 4.0 mmol of HBr with 4.0 mmol of NH_3, the concentrations are:

$$[HBr] = 0\ \underline{M}$$

$$[NH_3] = \frac{16.0\text{ mmol}}{120.0\text{ mL}} = 0.133\ \underline{M}\ NH_3$$

$$[NH_4Br] = \frac{24.0\text{ mmol}}{120.0\text{ mL}} = 0.200\ \underline{M}\ NH_4Br$$

$$K_b = \frac{[\text{Salt}][OH^-]}{[\text{Base}]} = 1.8 \times 10^{-5}$$

$$[OH^-] = \frac{(K_b)[\text{Base}]}{[\text{Salt}]} = \frac{(1.8 \times 10^{-5})(0.133)}{0.200}$$

$$[OH^-] = 1.2 \times 10^{-5}\ \underline{M} \qquad [H_3O^+][OH^-] = 1.0 \times 10^{-14}$$

$$[H_3O^+] = \frac{1.0 \times 10^{-14}}{1.2 \times 10^{-5}} = 8.3 \times 10^{-10}\ \underline{M}$$

$$pH = -\log[H_3O^+] = -\log(8.3 \times 10^{-10}) = \underline{\underline{9.08}}$$

Thus the change in pH is 9.25 - 9.08 = 0.17 pH units. If 20.0 mL of 0.20 $\underline{M}$ HBr is added to 100 mL of 0.20 $\underline{M}$ NH_3 with no NH_4Br present, the pH of the solution decreases by 1.42 pH units. You may wish to verify this by calculation.

16-6 Polyprotic Acids

Polyprotic acids are those containing two or more ionizable hydrogens which ionize in a stepwise fashion. The ionization constants are represented by K_1, K_2, K_3, ... The steps in the ionization of a triprotic acid,

H_3A, can be summarized as shown below:

Step 1: $H_3A + H_2O \rightleftharpoons H_3O^+ + H_2A^-$

$$K_1 = \frac{[H_3O^+][H_2A^-]}{[H_3A]}$$

Step 2: $H_2A^- + H_2O \rightleftharpoons H_3O^+ + HA^{2-}$

$$K_2 = \frac{[H_3O^+][HA^{2-}]}{[H_2A^-]}$$

Step 3: $HA^{2-} + H_2O \rightleftharpoons H_3O^+ + A^{3-}$

$$K_3 = \frac{[H_3O^+][A^{3-}]}{[HA^{2-}]}$$

Successive constants usually decrease by a factor of 10^4 to 10^6. The $[H_3O^+]$ produced in the second and subsequent steps is insignificant compared to that of the first step. It is important to remember that the same $[H_3O^+]$ is used in the expressions for K_1, K_2 and K_3 and that, for all practical purposes, it is equal to $[H_3O^+]$ produced in the first step alone. The anion resulting from the second step has a concentration numerically equal to K_2 as long as K_1 is at least 10^3 times greater than K_2, because $[H_3O^+] \approx [H_2A^-]$. The concentrations of all species in solution can be calculated if the initial concentration of the polyprotic acid and its ionization constants are known.

<u>Example 16-9:</u> Calculate the concentrations of all species present in 0.20 $\underline{M}$ H_3PO_4. The ionization constants for H_3PO_4 are: $K_1 = 7.5 \times 10^{-3}$, $K_2 = 6.2 \times 10^{-8}$, and $K_3 = 3.6 \times 10^{-13}$.

Each step is treated individually. Let x = concentration of H_3PO_4 that ionizes in the first step.

Step 1: $H_3PO_4 + H_2O \rightleftharpoons H_3O^+ + H_2PO_4^-$

$(0.20 - x)\,\underline{M}$ $x\,\underline{M}$ $x\,\underline{M}$

$$K_1 = \frac{[H_3O^+][H_2PO_4^-]}{[H_3PO_4]} = 7.5 \times 10^{-3} = \frac{(x)(x)}{(0.20 - x)}$$

Since K_1 is relatively large we may <u>not</u> assume that $x \ll 0.20$. Solving via the quadratic formula gives:

$x = 3.5 \times 10^{-2}\ \underline{M}\ [H_2PO_4^-] = [H_3O^+]$ from 1st step

$[H_3PO_4] = (0.20 - x)\underline{M} = 0.16\bar{5}\ \underline{M} = 0.16\ \underline{M}$

Let y = concentration of $H_2PO_4^-$ that ionizes in the second step.

Step 2: $H_2PO_4^- + H_2O \rightleftharpoons H_3O^+ + HPO_4^{2-}$

$(3.5 \times 10^{-2} - y)\underline{M}$; $y\ \underline{M}$; $y\ \underline{M}$

$$K_2 = \frac{[H_3O^+][HPO_4^{2-}]}{[H_2PO_4^-]} = 6.2 \times 10^{-8}$$

$[H^+]_{1st} + [H^+]_{2nd}$

$$\frac{(3.5 \times 10^{-2} + y)(y)}{3.5 \times 10^{-2} - y} = 6.2 \times 10^{-8}$$

Assume $y \ll 3.5 \times 10^{-2}$, $\dfrac{(3.5 \times 10^{-2})y}{3.5 \times 10^{-2}} = 6.2 \times 10^{-8}$

$y = 6.2 \times 10^{-8}\ \underline{M}\ (= K_2) = [HPO_4^{2-}] = [H_3O^+]$ from 2nd step ionization

The $[H_3O^+]$ from the second step ($6.2 \times 10^{-8}\ \underline{M}$) is much less than that from the first step ($3.5 \times 10^{-2}\ \underline{M}$). Let z = concentration of HPO_4^{2-} which ionizes in the third step.

Step 3: $HPO_4^{2-} + H_2O \rightleftharpoons H_3O^+ + PO_4^{3-}$

$(6.2 \times 10^{-8} - z)\underline{M}$; $z\ \underline{M}$; $z\ \underline{M}$

$K_3 = \dfrac{[H_3O^+][PO_4^{3-}]}{[HPO_4^{2-}]} = 3.6 \times 10^{-13}$, $[H_3O^+]_{1st} + [H_3O^+]_{2nd} + [H_3O^+]_{3rd}$

$$\frac{(3.5 \times 10^{-2} + 6.2 \times 10^{-8} + z)(z)}{6.2 \times 10^{-8} - z} = 3.6 \times 10^{-13}$$

Assume $z \ll 3.5 \times 10^{-2}$ and $z \ll 6.2 \times 10^{-8}$

$$\frac{(3.5 \times 10^{-2})(z)}{6.2 \times 10^{-8}} = 3.6 \times 10^{-13}$$

$z = 6.4 \times 10^{-19}\ \underline{M} = [PO_4^{3-}] = [H_3O^+]$ from 3rd step ionization.

The $[H_3O^+]$ from the second and third steps is so much less than from the first step that the first step ionization determines the pH of the solution.

Table 16-5 summarizes concentrations of the various species in 0.20 M H_3PO_4 and (as a matter of interest) compares them with the concentrations in 0.10 M H_3PO_4.

Table 16-5. Comparison of Concentrations in 0.10 M and 0.20 M H_3PO_4

Species	0.10 M H_3PO_4	0.20 M H_3PO_4
H_3PO_4	0.076 M	0.16 M
H_3O^+	2.4×10^{-2} M	3.5×10^{-2} M
$H_2PO_4^-$	2.4×10^{-2} M	3.5×10^{-2} M
HPO_4^{2-}	6.2×10^{-8} M	6.2×10^{-8} M
PO_4^{3-}	9.3×10^{-19} M	6.4×10^{-19} M
OH^-	4.2×10^{-13} M	2.9×10^{-13} M

16-7 Acid-Base Indicators and Titration Curves for Strong Acids and Strong Bases

Many common acid-base indicators are weak (very complex) organic acids, which we shall represent by the general formula HIn. An important characteristic of an acid-base indicator is the fact that the un-ionized acid molecules are one color, while the anion of the acid is another color.

$$HIn + H_2O \rightleftharpoons H_3O^+ + In^-$$

Color 1 (HIn) Color 2 (In^-)

$$K_{HIn} = \frac{[H_3O^+][In^-]}{[HIn]}$$

The value for the dissociation (ionization) constant for an indicator determines the pH range over which a color change occurs. (See Table 16-5 in the text.)

Let us now consider the titration of a strong acid such as hydrobromic acid, HBr, with a strong base such as potassium hydroxide, KOH.

$$HBr + KOH \longrightarrow KBr + H_2O$$

Since we are considering the reaction of a strong acid, HBr, with a strong soluble base, KOH, to form a soluble salt, KBr, there are no complications in calculating the pH as KOH solution is added to HBr solution. Consider the titration of 25.0 mL of 0.100 M HBr solution with 0.100 M KOH solution.

Before any KOH solution is added:

$$HBr + H_2O \longrightarrow H_3O^+ + Br^-$$

0.100 $\underline{M}$			0.100 $\underline{M}$	0.100 $\underline{M}$

$[H_3O^+] = 0.100\ \underline{M}$, $\underline{\underline{pH = 1.00}}$

After 10.0 mL of KOH have been added,

	HBr +	KOH →	KBr + H_2O
start:	2.50 mmol	1.00 mmol	0 mmol
change:	-1.00 mmol	-1.00 mmol	+1.00 mmol
after rxn:	1.50 mmol	0 mmol	1.00 mmol

the solution contains 1.50 mmol HBr in (25.0 + 10.0 =) 35.0 mL of solution.

$$\underline{M}_{HBr} = \frac{1.50 \text{ mmol HBr}}{35.0 \text{ mL}} = 0.0429\ \underline{M} \text{ HBr}$$

$[H_3O^+] = 0.0429\ \underline{M}$ $\underline{\underline{pH = 1.37}}$

All points up to the equivalence point on the titration curve are calculated similarly. [See the following table (16-6) for a tabulation.]

At the equivalence point, 25.0 mL of 0.100 $\underline{M}$ KOH have been added,

	HBr +	KOH →	KBr + H_2O
start:	2.50 mmol	2.50 mmol	0 mmol
change:	-2.50 mmol	-2.50 mmol	+2.50 mmol
after rxn:	0 mmol	0 mmol	2.50 mmol

and the resulting solution contains only potassium bromide, KBr, a soluble salt of a strong acid and a strong soluble base. Aqueous solutions of all such salts are neutral, i.e.,

$[H_3O^+] = [OH^-] = 1.00 \times 10^{-7}\ \underline{M}$ $\underline{\underline{pH = 7.00}}$

Beyond the equivalence point, the excess KOH determines the acidity (basicity) of the solution. After 26.0 mL of 0.100 $\underline{M}$ KOH solution have been added,

	HBr +	KOH →	KBr + H_2O
start:	2.50 mmol	2.60 mmol	0 mmol
change:	-2.50 mmol	-2.50 mmol	+2.50 mmol
after rxn:	0 mmol	0.10 mmol	2.50 mmol

$$M_{KOH} = \frac{0.10 \text{ mmol}}{51.0 \text{ mL}} = 0.0020 \underline{M} \text{ KOH}$$

Since KOH is a strong soluble base,

$$[OH^-] = 0.0020 \underline{M}, \quad pOH = 2.70, \quad \underline{\underline{pH = 11.30}}$$

All points beyond the equivalence point are calculated similarly.

The points at which we have calculated the pH of the solution are shown in Table 16-6, together with enough other points to show the shape of the titration curve definitively.

Table 16-6

Titration Data for HBr vs KOH

mL of 0.100 M KOH added	mmol KOH added	mmol excess acid or base	pH
none	0.00	2.50 H_3O^+	1.00
5.0	0.50	2.00	1.18
10.0	1.00	1.50	1.37
12.5	1.25	1.25	1.48
20.0	2.00	0.50	1.95
24.0	2.40	0.10	2.69
24.9	2.49	0.010	3.70
25.0	2.50	0.000	7.00
25.1	2.51	0.010 OH^-	10.30
26.0	2.60	0.10	11.30
27.0	2.70	0.20	11.59
30.0	3.00	0.50	11.96

Figure 16-1 shows a plot of these data.

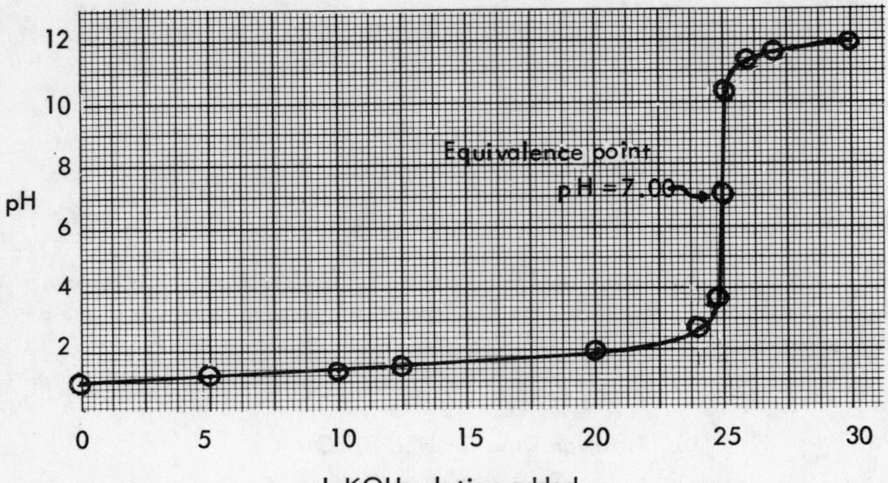

Figure 16-1. Titration curve for addition of 0.100 M KOH to 25.0 mL of 0.100 M HBr.

Note that the "vertical" portion of the curve is quite long and extends from about pH = 3.7 to pH = 10.0. Indicators that undergo changes in color within this range can be used in strong acid-strong base titrations.

Comparison of the titration curve for HBr/KOH with Figure 16-3 in the text, the titration curve for HCl/NaOH, shows that the two curves are quite similar. This is because the net reaction between any strong acid and any strong soluble base to form a soluble salt is the same.

$$H_3O^+ \quad + \quad OH^- \quad \longrightarrow \quad 2\ H_2O$$

EXERCISES

1. What are the concentrations of ions in the following solutions?

   (a) 0.25 M NaI
   (b) 0.015 M $HClO_4$
   (c) 1.40×10^{-4} M $Ba(OH)_2$
   (d) 4.5×10^{-3} M $Cr(NO_3)_3$

2. Calculate $[H_3O^+]$ and $[OH^-]$ for solutions of the following pH or pOH.

   (a) pH = 3.00
   (b) pOH = 7.00
   (c) pH = 2.15
   (d) pOH = 10.81

3. Calculate the $[H_3O^+]$, $[OH^-]$, pH and pOH of the following solutions.

   (a) 0.10 M HBr
   (b) 0.015 M LiOH
   (c) 8.1×10^{-3} M HNO_3
   (d) 3.8×10^{-5} M $Sr(OH)_2$

4. Calculate the concentrations of all species in the following aqueous solutions, as well as the pH of each.

   (a) 0.35 M HNO_2
   (b) 2.6×10^{-2} M NH_3
   (c) 4.2×10^{-3} M HClO
   (d) 0.020 M $(CH_3)_2NH$

5. Calculate the percentage ionization of each acid and base in the concentrations indicated in Problem 4.

6. How many moles of a monoprotic acid, HX, must be dissolved in water to give 2.00 liters of a solution with a pH of 5.22 if $K_a = 5.8 \times 10^{-10}$?

7. How many grams of dimethylamine must be dissolved in water to give 1.50 liters of a solution with a pH of 10.12?

8. Nitrous acid, HNO_2, is 1.6% dissociated in a particular solution. Calculate the weight of HNO_2 in 3.00 liters of this aqueous solution.

9. Calculate the pH and concentrations of all species in each of the following solutions.

   (a) 6.2×10^{-3} M acetic acid, CH_3COOH, and 4.4×10^{-3} M sodium acetate, $NaCH_3COO$.
   (b) 3.00 L of solution containing 8.1 g of hydrocyanic acid, HCN, and 19.5 g potassium cyanide, KCN.
   (c) 0.020 M methylamine, CH_3NH_2, and 0.080 M methylammonium nitrate, $CH_3NH_3NO_3$.

10. What is the pH of 500 mL of 0.020 M HNO_2 that also contains 0.851 g of KNO_2?

11. How many g of sodium hypochlorite, NaClO, must be added to 1.75 L 1.3×10^{-2} M HClO to adjust the pH to 4.95? Assume no volume change.

12. What is the pH of a solution prepared by dissolving 10.0 g of NH_4I in 100.0 mL of 0.0500 M aqueous ammonia? Assume no volume change.

13. What will be the pH of the solution resulting from the addition 50.0 mL of 0.20 M NaCN to 100.0 mL of 0.30 M HCN?

14. What is the pH of the solution resulting from the mixture of equal volumes of 7.0×10^{-2} M HClO and 3.0×10^{-2} M NaClO.

15. Calculate the pH of a solution prepared by mixing 300 mL of 0.0400 M CH_3COOH with 200 mL of 0.0300 M KOH.

16. Calculate the pH change resulting from addition of 0.0100 mole of solid KOH to one liter of each of the following. Assume no volume change due to addition of solid KOH. Compare results.

(a) pure water
(b) 0.100 M CH_3COOH solution
(c) 0.100 M CH_3COOH and 0.100 M KCH_3COO solution

17. Calculate the pH change resulting from addition of 0.0100 mole of gaseous HBr to one liter of each of the following. Assume no volume change. Compare the results with each other and with those obtained in Problem 16.

(a) pure water
(b) 0.100 M CH_3COOH solution
(c) 0.100 M CH_3COOH and 0.100 M KCH_3COO solution

18. Calculate the pH change resulting from addition of 0.0100 mole of solid KOH to one liter of each of the following solutions. Assume no volume change. Compare the results.

(a) 0.100 M aqueous NH_3
(b) 0.100 M aqueous NH_3 and 0.100 M NH_4Br

19. Calculate the pH change resulting from addition of 0.0100 mole of gaseous HBr to one liter of each of the following solutions. Assume no volume change. Compare the results.

(a) 0.100 M aqueous NH_3
(b) 0.100 M aqueous NH_3 and 0.100 M NH_4Br

20. Calculate the concentrations of all species present in 0.0100 M oxalic acid, $(COOH)_2$.

21. Calculate the concentrations of all species present in the following solutions.

(a) 0.050 M H_3PO_4
(b) 0.0050 M H_3PO_4

ANSWERS TO EXERCISES

1. (a) $[Na^+] = [I^-] = 0.25\ \underline{M}$ (b) $[H_3O^+] = [ClO_4^-] = 0.015\ \underline{M}$ (c) $[Ba^{2+}] = 1.40 \times 10^{-4}\ \underline{M}$, $[OH^-] = 2.80 \times 10^{-4}\ \underline{M}$ (d) $[Cr^{3+}] = 4.5 \times 10^{-3}\ \underline{M}$, $[NO_3^-] = 1.4 \times 10^{-2}\ \underline{M}$

2. (a) $[H_3O^+] = 1.0 \times 10^{-3}\ \underline{M}$, $[OH^-] = 1.0 \times 10^{-11}\ \underline{M}$
   (b) $[OH^-] = 1.0 \times 10^{-7}\ \underline{M}$, $[H_3O^+] = 1.0 \times 10^{-7}\ \underline{M}$
   (c) $[H_3O^+] = 7.1 \times 10^{-3}\ \underline{M}$, $[OH^-] = 1.4 \times 10^{-12}\ \underline{M}$
   (d) $[OH^-] = 1.5 \times 10^{-11}\ \underline{M}$, $[H_3O^+] = 6.7 \times 10^{-4}\ \underline{M}$

3. (a) $[H_3O^+] = 0.10\ \underline{M}$, $[OH^-] = 1.0 \times 10^{-13}$, pH = 1.00, pOH = 13.00 (b) $[OH^-] = 0.015\ \underline{M}$, $[H_3O^+] = 6.7 \times 10^{-13}$, pOH = 1.82, pH = 12.18 (c) $[H^+] = 8.1 \times 10^{-3}\ \underline{M}$, $[OH^-] = 1.2 \times 10^{-12}\ \underline{M}$, pH = 2.09, pOH = 11.91 (d) $[OH^-] = 7.6 \times 10^{-5}\ \underline{M}$, $[H^+] = 1.3 \times 10^{-10}$, pOH = 4.12, pH = 9.88

4. (a) $[H_3O^+] = [NO_2^-] = 1.2 \times 10^{-2}\ \underline{M}$, $[HNO_2] = 0.34\ \underline{M}$, $[OH^-] = 8.3 \times 10^{-13}\ \underline{M}$, pH = 1.92 (b) $[NH_4^+] = [OH^-] = 6.9 \times 10^{-4}\ \underline{M}$, $[NH_3] = 2.6 \times 10^{-2}\ \underline{M}$, $[H_3O^+] = 1.4 \times 10^{-11}\ \underline{M}$, pH = 10.85 (c) $[H_3O^+] = [ClO^-] = 1.2 \times 10^{-5}\ \underline{M}$, $[HClO] = 4.2 \times 10^{-3}\ \underline{M}$, $[OH^-] = 8.3 \times 10^{-10}\ \underline{M}$, pH = 4.92 (d) $[(CH_3)_2NH_2^+] = [OH^-] = 3.5 \times 10^{-3}\ \underline{M}$, $[(CH_3)_2NH] = 0.016\ \underline{M}$, $[H^+] = 2.9 \times 10^{-12}\ \underline{M}$, pH = 11.54 5. (a) 3.4% (b) 2.7% (c) 0.29% (d) 18% 6. 0.12 mol HX 7. 1.0×10^{-2} g $(CH_3)_2NH$ 8. 2.4×10^2 g HNO_2 9. (a) pH = 4.60, $[Na^+] = [CH_3COO^-] = 4.4 \times 10^{-3}\ \underline{M}$, $[H_3O^+] = 2.5 \times 10^{-5}\ \underline{M}$, $[CH_3COOH] = 6.2 \times 10^{-3}\ \underline{M}$, $[OH^-] = 4.0 \times 10^{-10}\ \underline{M}$, (b) pH = 9.40, $[K^+] = [CN^-] = 0.10\ \underline{M}$, $[H_3O^+] = 4.0 \times 10^{-10}\ \underline{M}$, $[HCN] = 0.10\ \underline{M}$, $[OH^-] = 2.5 \times 10^{-5}\ \underline{M}$ (c) pH = 10.11, $[CH_3NH_3^+] = [NO_3^-] = 0.080\ \underline{M}$, $[OH^-] = 1.3 \times 10^{-4}\ \underline{M}$, $[CH_3NH_2] = 0.020\ \underline{M}$, $[H_3O^+] = 7.7 \times 10^{-11}\ \underline{M}$ 10. pH = 3.35 11. 4.0×10^{-3} g NaClO (If you assumed that x is very large relative to $1.1 \times 10^{-5}\ \underline{M}$, then the result you obtained is 5.3×10^{-3} g. The correct value for x is $3.1 \times 10^{-5}\ \underline{M}$, only about 3 times $1.1 \times 10^{-5}\ \underline{M}$.) 12. pH = 8.11 13. pH = 8.92 14. pH = 7.09 15. pH = 4.74 16. (a) ΔpH = 12.00–7.00 = 5.00 unit increase (b) ΔpH = 3.80–2.89 = 0.91 unit increase

(c) $\Delta pH = 4.82-4.74 = 0.08$ unit increase The increase is much less when 0.0100 mole of KOH is added to the buffer solution (c) than the pure water or 0.100 M CH_3COOH. 17. (a) $\Delta pH = 7.00-2.00 = 5.00$ unit decrease (b) $\Delta pH = 2.89-1.99 = 0.90$ unit decrease (c) $\Delta pH = 4.74-4.66 = 0.08$ unit decrease The change in pH is much less dramatic when the HBr is added to the buffer solution (c). Note that the changes in pH in Problems 16 and 17 are the same for corresponding solutions. 18 (a) $\Delta pH = 12.00-11.11 = 0.89$ unit increase (b) $\Delta pH = 9.34-9.26 = 0.08$ unit increase The pH change is much less dramatic when the KOH is added to the buffer solution. 19. (a) $\Delta pH = 11.11-10.20 = 0.91$ unit decrease (b) $\Delta pH = 9.26-9.18 = 0.08$ unit decrease The change in pH is much less when the HBr is added to the buffer solution. 20. $[(COOH)_2] = 0.0015$ M, $[H^+] = 0.0086$ M (almost all from step 1), $[H(COO)_2^-] = 0.0084$ M, $[(COO)_2^{2-}] = 6.4 \times 10^{-5}$ M ($= K_2$), $[OH^-] = 1.2 \times 10^{-12}$ M 21. (a) $[H_3PO_4] = 0.034$ M, $[H^+] = [H_2PO_4^-] = 0.016$ M, $[HPO_4^{2-}] = 6.2 \times 10^{-8}$ M ($= K_2$), $[PO_4^{3-}] = 1.4 \times 10^{-18}$ M, $[OH^-] = 6.2 \times 10^{-13}$ M (b) $[H_3PO_4] = 1.8 \times 10^{-3}$ M, $[H^+] = [H_2PO_4^-] = 3.2 \times 10^{-3}$ M, $[HPO_4^{2-}] = 6.2 \times 10^{-8}$ M ($= K_2$), $[PO_4^{3-}] = 7.0 \times 10^{-18}$ M, $[OH^-] = 3.1 \times 10^{-12}$ M

Chapter Seventeen

EQUILIBRIA IN AQUEOUS SOLUTIONS-II

D,G,W: Sec. 17-9 and Sec. 18-5.2

17-1 Introduction

Solvolysis means reaction with the solvent. When the solvent is water the reaction is called hydrolysis. Many salts hydrolyze to produce acidic or basic solutions. Our discussion of hydrolysis will be restricted to salt solutions containing no free acid or base. Anions of weak acids and cations of weak bases hydrolyze.

It is convenient to classify normal salts into four categories on the basis of the character of their cations and anions. If you learn to recognize salts with respect to these categories, it will be much easier to decide if hydrolysis reactions occur. Each category will be illustrated with a specific example. The results can be generalized to cover all salts in the category.

17-2 Salts Derived from Strong Soluble Bases and Strong Acids

Let us consider sodium chloride, NaCl, a typical salt derived from a strong soluble base and a strong acid. Sodium chloride is soluble and ionizes completely in dilute aqueous solution. Water ionizes slightly in a reversible reaction to produce hydrogen ions and hydroxide ions. The reactions that occur when sodium chloride is dissolved in water are:

$$NaCl \xrightarrow{100\%} Na^+ + Cl^-$$

$$H_2O \rightleftharpoons OH^- + H^+$$

The Na^+ ion is derived from a strong soluble base, NaOH, and therefore it does not react with OH^- ions produced by the ionization of water. Similarly, the Cl^- ion is derived from a strong acid, HCl, and it does not react with H^+ ions from water. Since there are no possibilities for reactions between ions of the salt and the ions of water, solutions of NaCl are neutral. Similar reasoning leads to the conclusion that there is no possibility of reaction (hydrolysis) among any of the species in an aqueous solution of any salt of a strong base and strong acid. Therefore, all such solutions are neutral.

In the above example and in the examples that follow, we shall often represent H_3O^+ (aq) by H^+ to simplify writing equations.

17-3 Salts Derived from Strong Soluble Bases and Weak Acids

Consider the reactions which occur when potassium hypochlorite KClO, a typical salt of a strong soluble base and a weak acid, is dissolved in water. Potassium hypochlorite is a soluble, ionic compound.

$$KClO \xrightarrow{100\%} K^+ + ClO^-$$

$$H_2O \underset{\text{shift}}{\rightleftharpoons} \underset{\text{(excess)}}{OH^-} + H^+$$

$$ClO^- + H^+ \rightleftharpoons HClO$$

basic solution

Potassium ions and hydroxide ions do not react with each other in dilute aqueous solution because KOH is a strong base. Recall that according to Brønsted-Lowry terminology the hypochlorite ion is the conjugate base of hypochlorous acid and that the weaker the acid from which an anion is derived the stronger the anion is as a base. Thus, the hypochlorite ions react with hydrogen ions from the water to produce undissociated molecules of the weak acid, HClO. This reaction shifts the water equilibrium to the right and produces an excess of hydroxide ions and, therefore, a basic solution. The net reaction may be represented as

$$ClO^- + H_2O \rightleftharpoons HClO + OH^-$$

The equilibrium constant for the reaction is a hydrolysis constant, K_b for ClO^-.

$$K_b = \frac{[HClO][OH^-]}{[ClO^-]}$$

If we multiply both numerator and denominator of this expression by $[H^+]$, the equality will be preserved, but rearrangement allows us to relate K_b for hypochlorite ion to K_w and to K_a for HClO, the parent acid.

$$K_b = \frac{[HClO][OH^-]}{[ClO^-]} \times \frac{[H^+]}{[H^+]} = \underbrace{\frac{[HClO]}{[H^+][ClO^-]}}_{\frac{1}{K_a}} \times \underbrace{\frac{[H^+][OH^-]}{1}}_{\frac{K_w}{1}}$$

$$K_b = \frac{[HClO][OH^-]}{[ClO^-]} = \frac{K_w}{K_a} = \frac{1.0 \times 10^{-14}}{3.5 \times 10^{-8}} = 2.9 \times 10^{-7}$$

In any aqueous solution that contains a soluble salt of a strong base and weak acid, the anion of the weak acid hydrolyzes to produce a basic solution. The net reaction may be represented in general terms as:

$$\underset{\text{anion of weak acid}}{A^-} + H_2O \rightleftharpoons \underset{\text{weak acid}}{HA} + OH^-$$

The corresponding hydrolysis constant expression is:

$$K_b = \frac{[HA][OH^-]}{[A^-]} = \frac{K_w}{K_a}$$

Example 17-1: Determine the pH and concentrations of all species in 0.25 M NaCN. The ionization constant for HCN is 4.0×10^{-10}.

Sodium cyanide is a soluble ionic salt. The CN^- ions hydrolyze to form HCN and OH^-.

	CN^-	+ H_2O $\rightleftharpoons$	HCN	+	OH^-
initial:	0.25 M		0 M		0 M
change:	-x M		+x M		+x M
equil:	(0.25-x)M		x M		x M

$$K_b = \frac{[HCN][OH^-]}{[CN^-]} = \frac{K_w}{K_a} = \frac{1.0 \times 10^{-14}}{4.0 \times 10^{-10}} = 2.5 \times 10^{-5}$$

$$\frac{(x)(x)}{0.25-x} = 2.5 \times 10^{-5} \qquad \text{Assume } (0.25-x) \approx 0.25$$

$$\frac{x^2}{0.25} = 2.5 \times 10^{-5} \qquad x^2 = 6.25 \times 10^{-6}$$

$$x = 2.5 \times 10^{-3}\ M = [OH^-] = [HCN^-]$$

$$[CN^-] = (0.25-x)M = (0.25-0.0025)M \sim 0.25\ M$$

$$[H_3O^+] = \frac{1.0 \times 10^{-14}}{2.5 \times 10^{-3}} = 4.0 \times 10^{-12}\ M$$

$$pH = -\log(4.0 \times 10^{-12}) = 11.40 \qquad \text{(very basic solution)}$$

17-4 Salts Derived from Weak Bases and Strong Acids

The reactions that occur when ammonium nitrate, NH_4NO_3, a soluble ionic salt of a weak base and a strong acid, is dissolved in water are:

$$NH_4NO_3 \xrightarrow{100\%} NO_3^- + NH_4^+$$

$$H_2O \rightleftharpoons H^+ + OH^-$$

$$NH_4^+ + OH^- \rightleftharpoons NH_3 + H_2O$$

shift → excess H^+

∴ acidic solution

Since HNO_3 is a strong acid, and therefore NO_3^- is a very weak base, H^+ and NO_3^- ions do not react. But NH_4^+ ions react with OH^- ions from water to form un-ionized ammonia (a weak base) and water. This reaction occurs because NH_4^+ is the cation (conjugate acid) of the weak base, NH_3. The weaker a base is the stronger its cation is as an acid. This shifts the water equilibrium to the right and produces an acidic solution. The net reaction may be represented as:

$$NH_4^+ + H_2O \rightleftharpoons NH_3 + H_3O^+$$

$$K_a = \frac{[NH_3][H_3O^+]}{[NH_4^+]}$$

By multiplying this expression by $[OH^-]/[OH^-]$ and rearranging we can evaluate K_a in terms of K_w and K_b for NH_3.

$$K_a = \frac{[NH_3][H_3O^+]}{[NH_4^+]} \times \frac{[OH^-]}{[OH^-]} = \underbrace{\frac{[NH_3]}{[NH_4^+][OH^-]}}_{\frac{1}{K_b}} \times \underbrace{\frac{[H^+][OH^-]}{1}}_{\frac{K_w}{1}}$$

$$K_a = \frac{[NH_3][H_3O^+]}{[NH_4^+]} = \frac{K_w}{K_b} = \frac{1.0 \times 10^{-14}}{1.8 \times 10^{-5}} = 5.6 \times 10^{-10}$$

Since the cations of all weak bases hydrolyze to produce H_3O^+ ions, the net reaction is, in general terms:

$$\text{Cation}^+ + H_2O \rightleftharpoons \text{Weak Base} + H_3O^+$$

$$K_a = \frac{[\text{Weak Base}][H_3O^+]}{[\text{Cation}^+]} = \frac{K_w}{K_b}$$

Example 17-2: Determine the pH and percent hydrolysis in 0.15 M trimethylammonium nitrate, $[(CH_3)_3NH][NO_3]$. The ionization constant for trimethylamine is 7.4×10^{-5}.

Trimethylammonium nitrate is the salt of the weak base, trimethylamine, $(CH_3)_3N$, and HNO_3, a strong acid. It is a soluble, ionic salt.

$$\begin{array}{ccccccc} (CH_3)_3NH^+ & + & H_2O & \rightleftharpoons & (CH_3)N & + & H_3O^+ \\ (0.15-x)\underline{M} & & & & x\,\underline{M} & & x\,\underline{M} \end{array}$$

$$K_a = \frac{[(CH_3)_3N][H_3O^+]}{[(CH_3)_3NH^+]} = \frac{K_w}{K_b} = \frac{1.0 \times 10^{-14}}{7.4 \times 10^{-5}} = 1.4 \times 10^{-10}$$

$$\frac{x^2}{0.15-x} = 1.4 \times 10^{-10} \qquad \text{Assume } (0.15-x) \approx 0.15$$

$$\frac{x^2}{0.15} = 1.4 \times 10^{-10} \qquad x^2 = 2.1 \times 10^{-11} = 21 \times 10^{-12}$$

$$x^2 = 4.6 \times 10^{-6}\,\underline{M} = [H_3O^+] = [(CH_3)_3N]$$

$$pH = -\log(4.6 \times 10^{-6}) = \underline{5.34} \qquad \text{(acidic solution)}$$

$$\%\ \text{hydrolysis} = \frac{[(CH_3)_3NH^+]_{hydrolyzed}}{[(CH_3)_3NH^+]_{total}} \times 100\%$$

Since $[(CH_3)_3NH^+]_{hydrolyzed} = [H_3O^+]$,

$$\%\ \text{hydrolysis} = \frac{[H_3O^+]}{[(CH_3)_3NH^+]} \times 100\% = \frac{4.6 \times 10^{-6}}{0.15} \times 100\% = \underline{0.0031\%\ \text{hydrolyzed}}$$

Thus, we see that $(CH_3)_3NH^+$ ion is only very slightly hydrolyzed, yet the pH is 5.34, and the solution is distinctly acidic.

17-5 Salts Derived from Weak Bases and Weak Acids

The reactions that occur when ammonium nitrite, a soluble ionic salt, is dissolved in water are:

$$\begin{array}{lcccc} NH_4NO_2 & \xrightarrow{100\%} & NH_4^+ & + & NO_2^- \\ & & + & & + \\ H_2O & \rightleftharpoons & OH^- & + & H^+ \\ & \xrightarrow[\text{shift}]{} & \upharpoonleft\!\downharpoonright & & \upharpoonleft\!\downharpoonright \\ & & NH_3 + H_2O & & HNO_2 \end{array}$$

Ammonium nitrite is derived from aqueous ammonia, a weak base, and nitrous acid, a weak acid.

The NH_4^+ and NO_2^- ions from the NH_4NO_2 both react with the ions of water to produce un-ionized molecules of the weak base, NH_3, and the weak acid, HNO_2. The net reactions and hydrolysis constants are:

$$NH_4^+ \quad + \quad H_2O \quad \rightleftharpoons \quad NH_3 \quad + \quad H_3O^+$$

$$K_a = \frac{K_w}{K_b} = \frac{1.0 \times 10^{-14}}{1.8 \times 10^{-5}} = 5.6 \times 10^{-10} = \frac{[NH_3][H_3O^+]}{[NH_4^+]}$$

and,

$$NO_2^- \quad + \quad H_2O \quad \rightleftharpoons \quad HNO_2 \quad + \quad OH^-$$

$$K_b = \frac{K_w}{K_a} = \frac{1.0 \times 10^{-14}}{4.5 \times 10^{-4}} = 2.2 \times 10^{-11} = \frac{[HNO_2][OH^-]}{[NO_2^-]}$$

Since, in this case, NH_4^+ hydrolyzes more than NO_2^- (K_a for NH_4^+ > K_b for NO_2^-), an excess of H_3O^+ is produced and so the solution is acidic.

In general, both the cations and anions of salts of weak acids and weak bases hydrolyze. The acidity or basicity of the solution depends upon which ion hydrolyzes to the greater extent.

If	Then, solution is
$K_{a(cation)} > K_{b(anion)}$ $(K_{parent\ base} < K_{parent\ acid})$	acidic
$K_{a(cation)} = K_{b(anion)}$ $(K_{parent\ base} = K_{parent\ acid})$	neutral

If	Then, solution is
$K_{a\,(cation)} < K_{b\,(anion)}$	basic
$(K_{parent\ base} > K_{parent\ acid})$	

Since both the anions of weak acids and the cations of weak bases hydrolyze the overall reaction is obtained by adding the two hydrolysis reactions,

$$\begin{array}{lllll} \text{Cation}^+ & + & H_2O & \rightleftharpoons & \text{Weak Base} + H_3O^+ \\ \text{Anion}^- & + & H_2O & \rightleftharpoons & \text{Weak Acid} + OH^- \end{array} \quad (H_3O^+ + OH^- \rightarrow 2\,H_2O)$$

$$\text{Cation}^+ + \text{Anion}^- \rightleftharpoons \text{Weak Base} + \text{Weak Acid}^*$$

and the overall hydrolysis constant is

$$K_{overall} = \frac{[\text{Weak Acid}][\text{Weak Base}]}{[\text{Cation}^+][\text{Anion}^-]} = \frac{K_w}{K_{parent\ acid}\,K_{parent\ base}}$$

For a given initial concentration of salt, the concentrations of Weak Acid/-Anion$^-$ and Weak Base/Cation$^+$ can be calculated from hydrolysis constants. These pairs of concentrations may be used to calculate pH from the appropriate ionization constants.

We can now see that most salt solutions are not neutral. The only salts that produce neutral solutions are:

(1) salts of strong acids and strong soluble bases
(2) salts of weak acids and weak bases for which

$K_{a\,(cation)} = K_{b\,(anion)}$, a very restrictive case, in which

$K_{parent\ base} = K_{parent\ acid}$

---

*Care must be exercised to avoid the fallacious assumption that both reactions occur to the same extent. They don't except in the case in which $K_{a\,(cation)} = K_{b\,(anion)}$.

17-6 Titrations of Weak Bases with Strong Acids

In Section 16-6 in the text and Section 16-7 in this book we examined titration curves for strong acid-strong base titrations, the simplest kind of titration curves.

Let us consider the titration of 100.0 mL of 0.0100 M aqueous ammonia, a weak base ($K_b = 1.8 \times 10^{-5}$), by adding 0.0100 M HNO_3 solution. This situation is more complicated than those involving only strong acids and strong bases because after some nitric acid has been added a buffer solution consisting of NH_4NO_3 and unreacted aqueous NH_3 exists. At the equivalence point only the salt, NH_4NO_3, is present in solution, and the NH_4^+ ion hydrolyzes to produce an acidic solution. Beyond the equivalence point the excess strong acid, HNO_3, governs the pH of the solution.

Let us calculate the pH at representative points during the titration. Before any HNO_3 is added we have only 0.0100 M aqueous ammonia; its pH is determined using the mass action expression for the ionization constant of aqueous ammonia as we did in Section 16-4 in the text.

Before any HNO_3 is added:

$$\underset{(0.0100-x)\underline{M}}{NH_3} + H_2O \rightleftharpoons \underset{x\,\underline{M}}{NH_4^+} + \underset{x\,\underline{M}}{OH^-}$$

$$\frac{[NH_4^+][OH^-]}{[NH_3]} = 1.8 \times 10^{-5} = \frac{x^2}{0.0100-x} \qquad \text{assume: } x \ll 0.0100$$

$$1.8 \times 10^{-5} = \frac{x^2}{0.0100}, \qquad x^2 = 1.8 \times 10^{-7}$$

$x = 4.2 \times 10^{-4}$ M = $[OH^-]$ (assumption is valid)

pOH = 3.38, $\underline{\underline{\text{pH} = 10.62}}$

As soon as some HNO_3 is added a buffer solution (NH_4NO_3/aq NH_3) is formed. After 20.0 mL of HNO_3 are added, the pH drops to 9.86.

$$NH_3 + HNO_3 \longrightarrow NH_4NO_3$$

? mmol NH_3 = 100.0 mL × 0.0100 mmol/mL = 1.00 mmol NH_3

? mmol HNO_3 = 20.0 mL × 0.0100 mmol/mL = $\underline{0.20 \text{ mmol } HNO_3}$

After reaction: 0.80 mmol NH_3 and 0.20 mmol NH_4NO_3 in 120 mL

$$\underline{M}_{NH_3} = \frac{0.80 \text{ mmol } NH_3}{120.0 \text{ mL}} = 6.67 \times 10^{-3} \underline{M}\ NH_3$$

$$\underline{M}_{NH_4NO_3} = \frac{0.20 \text{ mmol } NH_4NO_3}{120.0 \text{ mL}} = 1.67 \times 10^{-3} \underline{M}\ NH_4NO_3$$

We calculate the pH of this solution as we did in Example 16-17 in the text. In a buffer solution the equilibrium $[NH_3] = \underline{M}_{NH_3}$ and the equilibrium $[NH_4^+] = \underline{M}_{NH_4NO_3}$.

$$\frac{[NH_4^+][OH^-]}{[NH_3]} = 1.8 \times 10^{-5} = \frac{(1.67 \times 10^{-3})[OH^-]}{6.67 \times 10^{-3}}$$

$[OH^-] = 7.2 \times 10^{-5}\ \underline{M}$, pOH = 4.14, $\underline{\underline{pH = 9.86}}$

Similar calculations at points preceding the equivalence point yield the values tabulated in Table 17-1.

At the equivalence point, hydrolysis of NH_4NO_3 produces an acidic solution, pH = 5.77).

$$NH_3 + HNO_3 \longrightarrow NH_4NO_3$$

Volume of 0.0100 $\underline{M}$ HNO_3 added = $\underline{100.0 \text{ mL}}$

$\underline{?}$ mmol NH_3 = 100.0 mL x 0.0100 mmol/mL = 1.00 mmol NH_3

$\underline{?}$ mmol HNO_3 = 100.0 mL x 0.0100 mmol/mL = 1.00 mmol HNO_3

After reaction: 1.00 mmol NH_4NO_3 in 200 mL of solution

$$\underline{M}_{NH_4NO_3} = \frac{1.00 \text{ mmol}}{200.0 \text{ mL}} = 0.0050\ \underline{M}\ NH_4NO_3$$
$$= 0.0050\ \underline{M}\ NH_4^+$$

We calculate the pH of the 0.0050 $\underline{M}$ NH_4NO_3 as we did in Example 17-2 in the text.

$$NH_4^+ + H_2O \rightleftharpoons NH_3 + H_3O^+$$
$$(0.0050-x)\underline{M} \qquad\qquad x\,\underline{M} \qquad x\,\underline{M}$$

$$K_a = \frac{[NH_3][H_3O^+]}{[NH_4^+]} = \frac{K_w}{K_b} = \frac{1.0 \times 10^{-14}}{1.8 \times 10^{-5}} = 5.6 \times 10^{-10}$$

$$5.6 \times 10^{-10} = \frac{x^2}{0.0050-x} \qquad \text{assume } x \ll 0.0050$$

$$5.6 \times 10^{-10} = \frac{x^2}{0.0050}, \qquad x^2 = 2.8 \times 10^{-12}$$

$$x = 1.7 \times 10^{-6}\ \underline{M} = [H_3O^+], \qquad \underline{pH = 5.77}$$

Beyond the equivalence point only the excess nitric acid, a strong acid, governs the pH.

The data for this titration are shown in Table 17-1 and are plotted to give the titration curve shown in Figure 17-1.

Table 17-1 Data for Aq. NH_3/HNO_3 Titration

mL 0.0100 $\underline{M}$ HNO_3 added	mmol HNO_3 added	mmol excess acid or base	pH
0	0	1.00 NH_3	10.62
20.0	0.200	0.800	9.86
30.0	0.300	0.700	9.62
50.0	0.500	0.500	9.26
60.0	0.600	0.400	9.08
70.0	0.700	0.300	8.89
80.0	0.800	0.200	8.65
90.0	0.900	0.100	8.30
99.0	0.990	0.010	7.26
99.5	0.995	0.005	6.96
100.0	1.00	0.00	5.77
100.5	1.005	0.005 H_3O^+	4.60
110.0	1.10	0.10	3.32
130.0	1.30	0.30	2.88
150.0	1.50	0.50	2.70

Note that the change in pH as a small volume of titrant is added is less dramatic than in the titration of a strong base with a strong acid. The best indicator for this titration tabulated in Table 16-5 in the text, would be methyl red (at equivalence point, pH = 5.77).

17-7 Titrations of Weak Acids with Strong Bases

A titration curve for the titration of a weak acid by adding a strong base is similar to the inverted image of the curve in Figure 17-1, beginning at a low pH, having an equivalence point at pH > 7.00 due to hydrolysis of the anion of the salt produced, and ending at a high pH. See Figure 17-1 in the text. It shows the titration curve for $CH_3COOH/NaOH$.

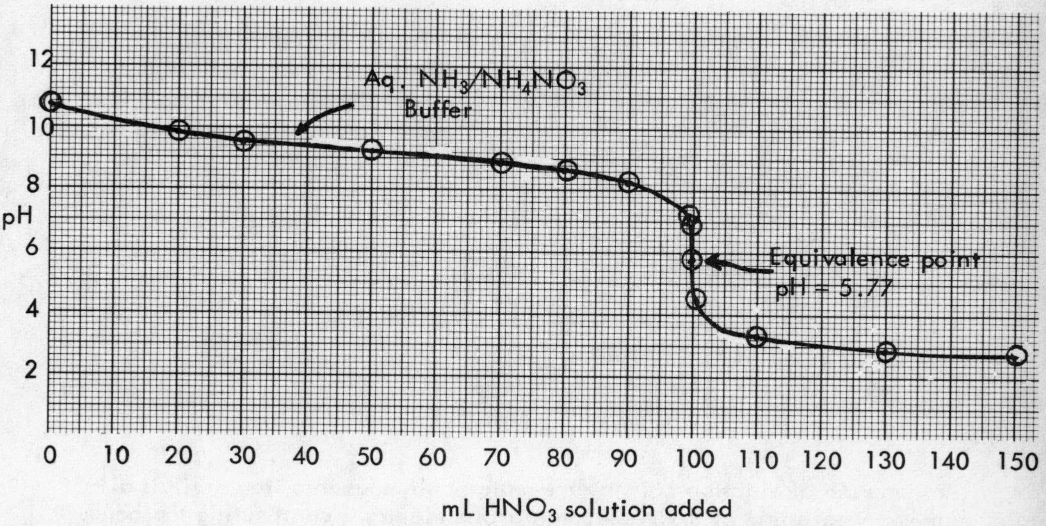

Figure 17-1. Titration curve for addition of 0.0100 M HNO_3 to 100.0 mL of 0.0100 M aqueous NH_3.

17-8 Titrations of Weak Acids and Weak Bases

Such titrations are impossible to carry out using visual indicators because the color change is not distinct enough. This is because the solution is buffered both before and after the equivalence point. At the equivalence point the cation and anion of the salt both hydrolyze and, as a result, the change in pH is gradual throughout the entire titration. See Figure 17-3 in the text for the titration curve for aqueous NH_3/CH_3COOH.

17-9 Hydrolysis of Small Highly-Charged Cations

Salts containing small, highly-charged cations also hydrolyze to produce acidic solutions. Such cations are almost exclusively the metal ions that form insoluble hydroxides. Although not often represented as such, metal ions dissolved in aqueous solution are actually hydrated, usually with four or six water molecules bonded to the metal ion through coordinate covalent bonds.

The ion that we commonly designate as Fe^{3+} (aq) is actually $[Fe(OH_2)_6]^{3+}$, the complex ion depicted below.

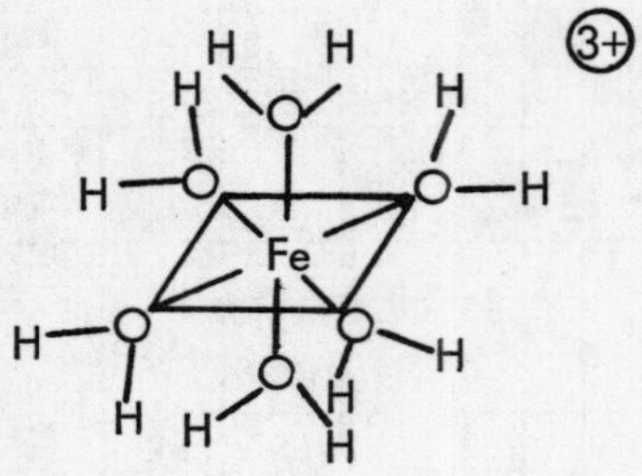

Fe(III) is sp^3d^2 hybridized and each of six octahedrally-coordinated water molecules donate two electrons into vacant Fe(III) orbitals.

The donation of electron pairs by the water molecules into iron orbitals also causes a weakening of the H-O bonds as the electrons constituting the bonds are shifted toward the oxygen atoms. As a consequence, it is easier for coordinated water molecules to act as acids and to produce H_3O^+ ions than for uncoordinated water molecules to ionize. Such a hydrolysis reaction can be represented as:

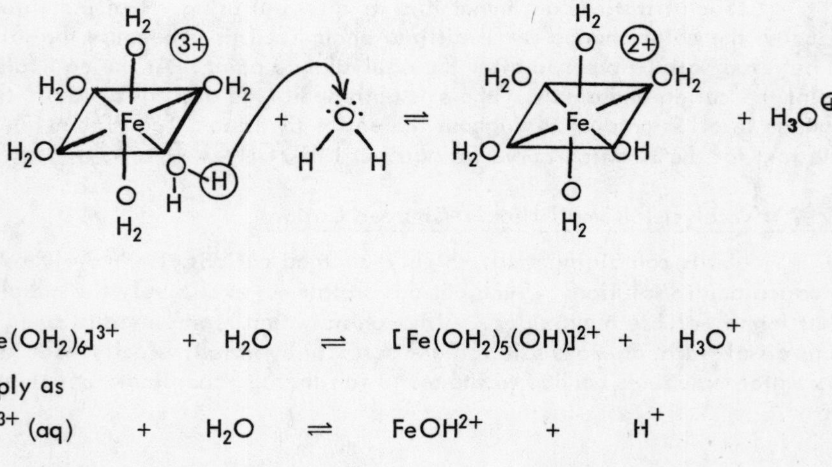

or

$$[Fe(OH_2)_6]^{3+} + H_2O \rightleftharpoons [Fe(OH_2)_5(OH)]^{2+} + H_3O^+$$

or, simply as

$$Fe^{3+}(aq) + H_2O \rightleftharpoons FeOH^{2+} + H^+$$

The hydrolysis constant for the reaction has the usual form,

$$K_a = \frac{[[Fe(OH_2)_5(OH)]^{2+}][H_3O^+]}{[Fe(OH_2)_6{}^{3+}]} = 4.0 \times 10^{-3}$$

and is often written in simplied form as:

$$K_a = \frac{[Fe(OH)^{2+}][H^+]}{[Fe^{3+}]} = 4.0 \times 10^{-3}$$

The strength of metal-oxygen bonds in hydrated ions, and therefore the degree of hydrolysis, generally increases as the positive charge on the cation increases and as its size decreases. This is shown in Table 17-3 in the text. Hydrolysis reactions of small highly-charged metal ions, M^{n+}, may be generalized as:

$$M^{n+}(aq) + H_2O \rightleftharpoons M(OH)^{(n-1)+}(aq) + H_3O^+$$

$$K_a = \frac{[M(OH)^{(n-1)+}][H_3O^+]}{[M^{n+}]}$$

If K_a is known the pH of a given solution can be calculated, or if the pH of a given solution is known, then K_a can be calculated.

Example 17-3: Calculate the pH and concentrations of all species in 0.15 M $CoBr_2$. K_a for $[Co(OH_2)_6]^{2+} = 5.0 \times 10^{-10}$

The Co^{2+} (aq) ions hydrolyze as shown.

$$\underset{(0.15-x)\,M}{[Co(OH_2)_6]^{2+}} + H_2O \rightleftharpoons \underset{x\,M}{[Co(OH_2)_5(OH)]^+} + \underset{x\,M}{H_3O^+}$$

$$K_a = \frac{[[Co(OH_2)_5(OH)]^+][H_3O^+]}{[[Co(OH_2)_6]^{2+}]} = 5.0 \times 10^{-10}$$

$$\frac{(x)(x)}{0.15-x} = 5.0 \times 10^{-10} \qquad \text{Assume } (0.15-x) \sim 0.15$$

$$\frac{x^2}{0.15} = 5.0 \times 10^{-10} \qquad x^2 = 7.5 \times 10^{-11}$$

$$x = 8.7 \times 10^{-6}\ M = [H_3O^+] = [[Co(OH_2)_5(OH)]^+], \ pH = 5.06.$$

The solution is acidic because the concentration of H_3O^+ is 87 times greater than in pure water.

$$[[Co(OH_2)_6]^{2+}] = (0.15-x)\,M \sim 0.15\ M$$

$$[Br^-] = 0.30\ M$$

EXERCISES

You may consult Appendices F and G and Table 17-3 in the text to obtain ionization and hydrolysis constants, as necessary.

1. Indicate whether each salt below hydrolyzes, and for those that do, whether they give acidic, basic, or neutral solutions.

   (a) KBr (b) NH_4I (c) NaCN (d) $NaNO_2$ (e) $(CH_3)_3NHCl$
   (f) RbCl (g) NH_4CH_3COO (h) $LiCH_3COO$ (i) NH_4NO_3

2. Write the equation for the hydrolysis (if any) of each salt of Problem 1. Also write the mass action expression and evaluate the hydrolysis constant from tabulated acid and base ionization constants.

3. Arrange the following salts in order of increasing basicity.

   (a) KNO_2 (b) KCN (c) K_2S (d) KCl (e) KCH_3COO (f) KClO

4. Arrange the following salts in order of increasing acidity.

   (a) NH_4Cl (b) $(CH_3)_3NHCl$ (c) $C_6H_5NH_3Cl$ (d) $(CH_3)_2NH_2Cl$

5. Calculate the pH of each of the solutions below.

   (a) 0.15 M NaBrO
   (b) 0.020 M NH_4Cl
   (c) 0.50 M RbF
   (d) 0.063 M anilinium hydrobromide, $C_6H_5NH_3Br$
   (e) 1.3×10^{-4} M NaCN
   (f) 8.2×10^{-3} M $Ca(NO_2)_2$

6. What is the hydrolysis constant, K_b, for the hypothetical anion X^- if 0.10 M aqueous NaX solution has a pH of 8.20?

7. Write equations, showing coordinated water molecules, for the first step in the hydrolysis of:

   (a) $[Bi(OH_2)_6]^{3+}$ in aqueous $BiCl_3$
   (b) $[Cu(OH_2)_4]^{2+}$ in aqueous $CuCl_2$
   (c) $[Co(OH_2)_6]^{3+}$ in aqueous $Co(NO_3)_3$

8. Calculate the pH and concentrations of all species in the following solutions.

   (a) 0.050 M $MgBr_2$
   (b) 0.025 M $AlCl_3$
   (c) 0.10 M $ZnCl_2$
   (d) 0.10 M $FeCl_2$

9. Given the following indicators and their ionization constants, estimate the pH ranges over which their colors change.

	Acid Range Color	Base Range Color
(a) bromocresol green, $K_a = 3 \times 10^{-5}$	yellow	blue
(b) methyl red, $K_b = 2 \times 10^{-9}$	red	yellow
(c) thymolphthalein, $K_a = 2 \times 10^{-10}$	colorless	blue

10. What will be the color exhibited by each indicator in Problem 9 in solutions of the following pH?

(a) pH = 3.00 (b) pH = 7.00 (c) pH = 11.00

11. Calculate $[H^+]$, $[OH^-]$, pH and pOH at the indicated points for the titration of 50.00 mL of 0.01000 M HCl by 0.0100 M KOH, i.e., KOH is added to HCl. Plot the titration curve, i.e., pH (vertical axis) versus mL of KOH added (horizontal axis).

Total mL KOH solution added	$[H^+]$	$[OH^-]$	pH	pOH
none				
5.00				
15.0				
25.0				
35.0				
45.0				
47.5				
50.0				
52.5				
55.0				
75.0 (50% excess KOH)				
100.0 (100% excess KOH)				

Consult the table of indicators in Table 16-5 (text), and list those that could be used satisfactorily in this titration.

12. Calculate the $[H^+]$, $[OH^-]$, pH, and pOH at the indicated points for the titration of 50.00 mL of 0.0100 M HClO by 0.0100 M NaOH, i.e., NaOH is added to HClO. Draw the titration curve, i.e., pH (vertical axis) versus mL of NaOH added (horizontal axis).

Total mL NaOH solution added	$[H^+]$	$[OH^-]$	pH	pOH
none				
5.00				
15.0				
25.0				
35.0				
45.0				
47.5				
50.0				
52.5				
55.0				
75.0 (50% excess NaOH)				
100.0 (100% excess NaOH)				

Consult the table of indicators in Table 16-5 (text), and list those that could be used satisfactorily in this titration.

ANSWERS FOR EXERCISES

1. (a) no hydrolysis (b) acidic, NH_4^+ hydrolyzes (c) basic, CN^- hydrolyzes (d) basic, NO_2^- hydrolyzes (e) acidic, $(CH_3)_2NH^+$ hydrolyzes (f) no hydrolysis (g) neutral, NH_4^+ and CH_3COO^- hydrolyze equally (h) basic, CH_3COO^- hydrolyzes (i) acidic, NH_4^+ hydrolyzes

2. (b) $NH_4^+ + H_2O \rightleftharpoons NH_3 + H_3O^+$

$$K_a = \frac{[NH_3][H_3O^+]}{[NH_4^+]} = \frac{K_w}{K_b} = \frac{1.0 \times 10^{-14}}{1.8 \times 10^{-5}} = 5.6 \times 10^{-10}$$

(c) $CN^- + H_2O \rightleftharpoons HCN + OH^-$

$$K_b = \frac{[HCN][OH^-]}{[CN^-]} = \frac{K_w}{K_a} = \frac{1.0 \times 10^{-14}}{4.0 \times 10^{-10}} = 2.5 \times 10^{-5}$$

(d) $NO_2^- + H_2O \rightleftharpoons HNO_2 + OH^-$

$$K_b = \frac{[HNO_2][OH^-]}{[NO_2^-]} = \frac{K_w}{K_a} = \frac{1.0 \times 10^{-14}}{4.5 \times 10^{-4}} = 2.2 \times 10^{-11}$$

(e) $(CH_3)_3NH^+ + H_2O \rightleftharpoons (CH_3)_3N + H_3O^+$

$$K_a = \frac{[(CH_3)_3N][H_3O^+]}{[(CH_3)_3NH^+]} = \frac{1.0 \times 10^{-14}}{7.4 \times 10^{-5}} = 1.4 \times 10^{-10}$$

(g) $NH_4^+ + CH_3COO^- \rightleftharpoons NH_3 + CH_3COOH$

$$K = \frac{[NH_3][CH_3COOH]}{[NH_4^+][CH_3COO^-]} = \frac{K_w}{K_aK_b} = \frac{1.0 \times 10^{-14}}{(1.8 \times 10^{-5})(1.8 \times 10^{-5})} = 3.1 \times 10^{-5}$$

(h) $CH_3COO^- + H_2O \rightleftharpoons CH_3COOH + OH^-$

$$K_b = \frac{[CH_3COOH][OH^-]}{[CH_3COO^-]} = \frac{K_w}{K_a} = \frac{1.0 \times 10^{-14}}{1.8 \times 10^{-5}} = 5.6 \times 10^{-10}$$

(i) same as 2(b)

3. $KCl < KNO_2 < KCH_3COO < KClO < KCN < K_2S$ (strongest base)
4. $(CH_3)_2NH_2Cl < (CH_3)_3NHCl < NH_4Cl < C_6H_5NH_3Cl$ (strongest acid)
5. (a) pH = 10.89 (b) pH = 5.48 (c) pH = 8.41 (d) pH = 2.92
(e) pH = 9.68 (f) pH = 7.77 6. $K_b = 2.6 \times 10^{-11}$
7. (a) $[Bi(OH_2)_6]^{3+} + H_2O \rightleftharpoons [Bi(OH_2)_5(OH)]^{2+} + H_3O^+$
(b) $[Cu(OH_2)_4]^{2+} + H_2O \rightleftharpoons [Cu(OH_2)_3(OH)]^+ + H_3O^+$
(c) $[Co(OH_2)_6]^{3+} + H_2O \rightleftharpoons [Co(OH_2)_5(OH)]^{2+} + H_3O^+$
8. (a) pH = 6.41, $[Mg(OH_2)_6^{2+}] = 0.050$ M, $[H_3O^+] = [Mg(OH_2)_5(OH)^+] = 3.9 \times 10^{-7}$ M, $[Br^-] = 0.100$ M, $[OH^-] = 2.6 \times 10^{-8}$ M (b) pH = 3.26, $[Al(OH_2)_6^{3+}] = 0.024$ M, $[H_3O^+] = [Al(OH_2)_5(OH)^{2+}] = 5.5 \times 10^{-4}$ M, $[Cl^-] = 0.075$ M, $[OH^-] = 1.8 \times 10^{-11}$ M (c) pH = 5.30, $[Zn(OH_2)_6^{2+}] = 0.10$ M, $[H_3O^+] = [Zn(OH_2)_5(OH)^+] = 5.0 \times 10^{-6}$ M, $[Cl^-] = 0.20$ M,

$[OH^-] = 2.0 \times 10^{-9}$ M (d) pH = 5.26, $[Fe(OH_2)_6^{2+}] = 0.10$ M,
$[H_3O^+] = [Fe(OH_2)_5(OH)^+] = 5.5 \times 10^{-6}$ M, $[Cl^-] = 0.20$ M,

9. (a) pH of ~3.5-5.5 (b) pH of ~4.3-6.3 (c) pH of ~8.7-10.7

10. (a) bromcresol green: yellow, methyl red:red, thymolphthalein: colorless
(b) bromcresol green: blue, methyl red: yellow, thymolphthalein: colorless
(c) bromcresol green: blue, methyl red: yellow, thymolphthalein: blue

11.

mL KOH added	$[H^+]$	$[OH^-]$	pH	pOH
none	1.0×10^{-2} M	1.0×10^{-12} M	2.00	12.00
5.00	8.18×10^{-3} M	1.22×10^{-12} M	2.09	11.91
15.0	5.38×10^{-3} M	1.86×10^{-12} M	2.27	11.73
25.0	3.33×10^{-3} M	3.00×10^{-12} M	2.48	11.52
35.0	1.76×10^{-3} M	5.68×10^{-12} M	2.75	11.25
45.0	5.26×10^{-4} M	1.90×10^{-11} M	3.28	10.72
47.5	2.56×10^{-4} M	3.91×10^{-11} M	3.59	10.41
50.0	1.0×10^{-7} M	1.0×10^{-7} M	7.00	7.00
52.5	4.10×10^{-11} M	2.44×10^{-4} M	10.39	3.61
55.0	2.10×10^{-11} M	4.76×10^{-4} M	10.68	3.32
75.0	5.00×10^{-12} M	2.00×10^{-3} M	11.30	2.70
100.0	3.00×10^{-12} M	3.33×10^{-3} M	11.52	2.48

Suitable indicators include all those listed in Table 16-5 except methyl orange.

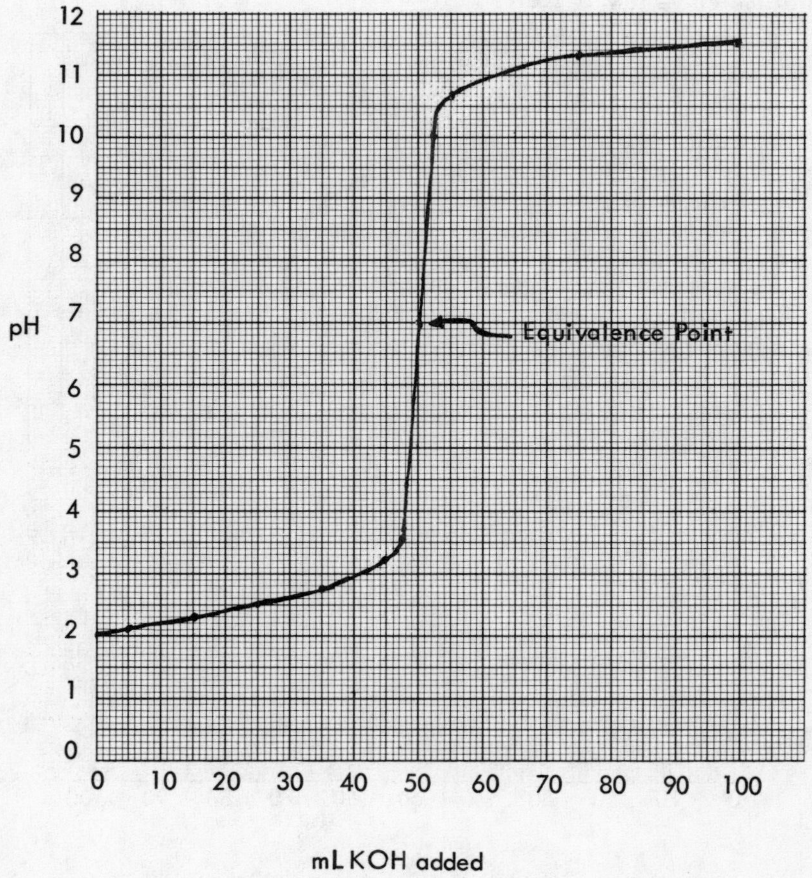

12. mL NaOH added	$[H^+]$	$[OH^-]$	pH	pOH
none	1.87×10^{-5} M	5.35×10^{-10} M	4.73	9.27
5.00	3.15×10^{-7} M	3.17×10^{-8} M	6.50	7.50
15.0	8.15×10^{-8} M	1.22×10^{-7} M	7.09	6.91
25.0	3.50×10^{-8} M	2.86×10^{-7} M	7.46	6.54
35.0	1.50×10^{-8} M	6.67×10^{-7} M	7.82	6.18
45.0	3.88×10^{-9} M	2.58×10^{-6} M	8.41	5.59
47.5	1.84×10^{-9} M	5.43×10^{-6} M	8.74	5.26
50.0	2.63×10^{-10} M	3.80×10^{-5} M	9.58	4.42
52.5	4.10×10^{-11} M	2.44×10^{-4} M	10.39	3.61
55.0	2.10×10^{-11} M	4.76×10^{-4} M	10.68	3.32
75.0	5.00×10^{-12} M	2.00×10^{-3} M	11.30	2.70
100.0	3.00×10^{-12} M	3.33×10^{-3} M	11.52	2.48

As the following titration curve shows, visual indicators are not really suitable for the titration of acids as weak as HClO, $K_a = 3.5 \times 10^{-8}$. The "vertical" portion of the curve is very short.

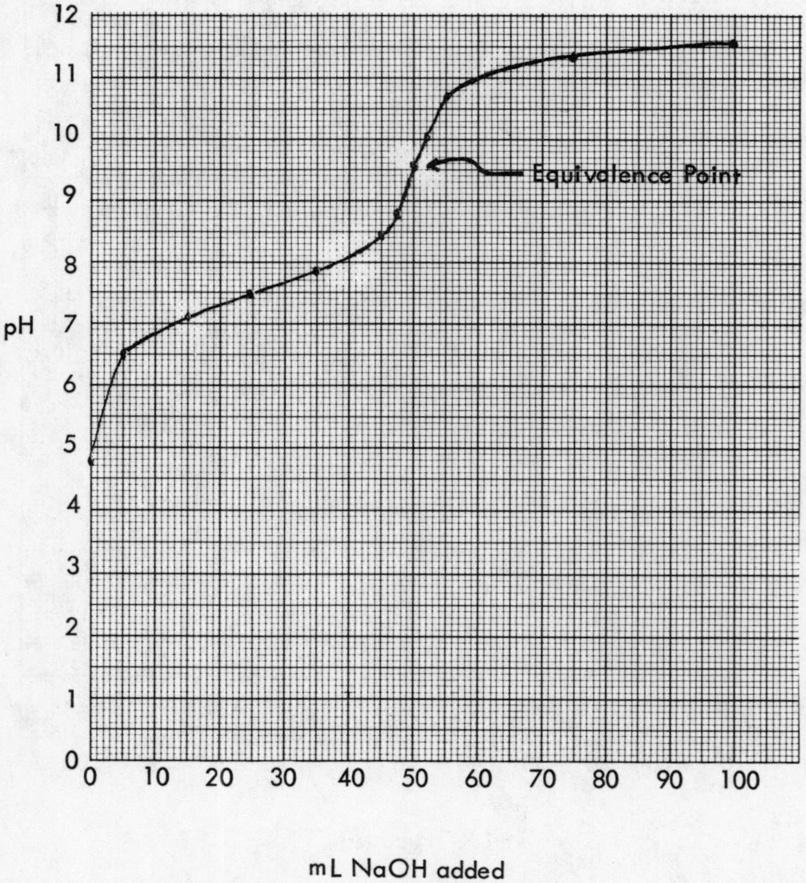

Chapter Eighteen

EQUILIBRIA IN AQUEOUS SOLUTIONS-III

D, G, W: Sec. 18-6 through 18-11

18-1 Introduction

"Insoluble" substances are really very slightly soluble. Ionic compounds dissociate into ions as they dissolve in water. The equilibrium constant for the dissolution process is called a solubility product constant, K_{sp}. The general dissolution reaction may be represented as:

$$M_nX_y\ (s) \rightleftharpoons n\,M^{y+}\ (aq) + y\,X^{n-}\ (aq)$$

The corresponding solubility product constant expression,

$$K_{sp} = [M^{y+}]^n[X^{n-}]^y$$

does not include the concentration of the solid. Solubility product constants apply only to saturated solutions in equilibrium with the solid, M_nX_y. A few specific examples are given below. Appendix H is a table of K_{sp} values at 25°C.

$$BaCO_3\ (s) \rightleftharpoons Ba^{2+}\ (aq) + CO_3^{2-}\ (aq) \qquad K_{sp} = [Ba^{2+}][CO_3^{2-}] = 8.1 \times 10^{-9}$$

$$Al(OH)_3\ (s) \rightleftharpoons Al^{3+}\ (aq) + 3\ OH^-\ (aq) \qquad K_{sp} = [Al^{3+}][OH^-]^3 = 1.9 \times 10^{-33}$$

$$Bi_2S_3\ (s) \rightleftharpoons 2\ Bi^{3+}\ (aq) + 3\ S^{2-}\ (aq) \qquad K_{sp} = [Bi^{3+}]^2[S^{2-}]^3 = 1.6 \times 10^{-72}$$

Once a solution is saturated (in equilibrium with the solid compound) no more solid dissolves. The "concentration of pure solid" is a meaningless term in a saturated solution because excess solid has no effect on the equilibrium.

Like other equilibrium constants, solubility product constants must be determined experimentally. The next two examples illustrate how they may be calculated from experimentally determined data.

Example 18-1: At 25°C the molar solubility of magnesium fluoride, MgF_2, is 1.17×10^{-3} M, i.e., 1.17×10^{-3} mole of MgF_2 dissolves to give one liter of saturated solution. Calculate K_{sp} for MgF_2.

The dissolution of MgF_2 in water can be represented as follows:

$$MgF_2\ (s) \quad \rightleftharpoons \quad Mg^{2+} \quad + \quad 2\ F^-$$

$$\text{molar sol'y} = 1.17 \times 10^{-3}\ \underline{M} \longrightarrow 1.17 \times 10^{-3}\ \underline{M} \qquad 2(1.17 \times 10^{-3}\ \underline{M})$$

In saturated MgF_2 solution: $[Mg^{2+}] = 1.17 \times 10^{-3}\ \underline{M}$, $[F^-] = 2(1.17 \times 10^{-3}\ \underline{M})$, and no undissociated MgF_2 is present.

$$K_{sp} = [Mg^{2+}][F^-]^2 = (1.17 \times 10^{-3})(2 \times 1.17 \times 10^{-3})^2 = 6.4 \times 10^{-9}$$

$$\underline{\underline{[Mg^{2+}][F^-]^2 = 6.4 \times 10^{-9}}}$$

Example 18-2: The solubility of barium phosphate, $Ba_3(PO_4)_2$, is 3.94×10^{-5} g/100 mL of water at 25°C. Evaluate K_{sp}.

We first convert solubility in g/100 mL to molar solubility.

$$\frac{?\ \text{mol}}{\text{L}} = \frac{3.94 \times 10^{-5}\ \text{g}}{100\ \text{mL}} \times \frac{1000\ \text{mL}}{\text{L}} \times \frac{1\ \text{mol}\ Ba_3(PO_4)_2}{601.9\ \text{g}} = 6.55 \times 10^{-7}\ \underline{M}\ Ba_3(PO_4)_2$$

Now that the molar solubility of $Ba_3(PO_4)_2$ is known, the equilibrium concentrations of the ions can be calculated.

$$Ba_3(PO_4)_2\ (s) \quad \rightleftharpoons \quad 3\ Ba^{2+} \quad + \quad 2\ PO_4^{3-}$$

$$6.55 \times 10^{-7}\underline{M} \longrightarrow 3(6.55 \times 10^{-7}\ \underline{M}) \qquad 2(6.55 \times 10^{-7}\ \underline{M})$$

$$K_{sp} = [Ba^{2+}]^3[PO_4^{3-}]^2 = (3 \times 6.55 \times 10^{-7})^3\ (2 \times 6.55 \times 10^{-7})^2 = \underline{\underline{1.3 \times 10^{-29}}}$$

Molar solubilities of solids and concentrations of constituent ions in saturated solutions of slightly soluble compounds may be calculated from the tabulated K_{sp} values.

Example 18-3: The solubility product constant for CuI is 5.1×10^{-12} at 25°C. Calculate the molar solubility of CuI, $[Cu^+]$, and $[I^-]$ in saturated aqueous solution. How many grams of CuI dissolve to give 500 mL of saturated solution? CuI is copper(I) iodide or cuprous iodide.

The dissolution reaction follows. We let x represent the molar solubility of CuI.

$$CuI\ (s) \quad \rightleftharpoons \quad Cu^+ \quad + \quad I^-$$

$$\text{molar sol'y} = x\ \underline{M} \longrightarrow x\ \underline{M} \qquad x\ \underline{M}$$

$K_{sp} = [Cu^+][I^-] = (x)(x) = 5.1 \times 10^{-12}$ $\qquad x^2 = 5.1 \times 10^{-12}$

$x = \underline{\underline{2.3 \times 10^{-6}\ M = \text{molar solubility of CuI in water}}}$.

Also, $x = \underline{\underline{[Cu^+] = [I^-] = 2.3 \times 10^{-6}\ \underline{M}}}$

Now that we know the molar solubility of CuI we can determine how many grams dissolve to give 500 mL of saturated solution.

$$\underline{?}\ g\ CuI = 500\ mL \times \frac{2.3 \times 10^{-6}\ mol\ CuI}{1000\ mL} \times \frac{190.4\ g\ CuI}{1\ mol\ CuI} =$$

$$\underline{\underline{2.2 \times 10^{-4}\ g\ CuI\ \text{per}\ 500\ mL\ \text{sat'd sol'n}}}$$

18-2 Extraction of Higher Order Roots

Frequently it is necessary to extract higher order roots, as Example 18-4 illustrates. A general method for extracting a higher order root is (1) take the log of the number of interest, (2) divide the log by the number corresponding to the root, (3) take the antilog of the result. Or symbolically,

$$x^n = a \qquad x = \sqrt[n]{a} = \text{antilog}\left(\frac{\log a}{n}\right)$$

Example 18-4: Solve the following equation for x: $218\ x^5 = 1.64 \times 10^{-29}$

Dividing both sides of the equation by 218 gives:

$$x^5 = 7.52 \times 10^{-32}$$

Shifting the decimal so that the exponent is divisible by 5 gives:

$$x^5 = 7520 \times 10^{-35} \qquad x = (\sqrt[5]{7520}) \times 10^{-7}$$

We find the fifth root of 7520 by (1) taking its logarithm,

$$\log 7520 = 3.8762$$

(2) dividing the result by 5, $\frac{3.8762}{5} = 0.7752$ and (3) taking the antilog of that number. Antilog $0.7752 = 5.96$

Thus, $x = \sqrt[5]{7520 \times 10^{-35}} = \underline{\underline{5.96 \times 10^{-7}}}$

Example 18-5: Calculate the concentrations of Tl^{3+} and OH^- ions in saturated thallium(III) hydroxide, $Tl(OH)_3$, solution. K_{sp} for $Tl(OH)_3 = 1.5 \times 10^{-44}$.

Let x = molar solubility of $Tl(OH)_3$

$$Tl(OH)_3\ (s) \rightleftharpoons Tl^{3+} + 3\ OH^-$$

molar sol'y = $x\ \underline{M}$ $x\ \underline{M}$ $3 \times \underline{M}$

$$K_{sp} = [Tl^{3+}][OH^-]^3 = 1.5 \times 10^{-44}$$
$$(x)(3x)^3 = 1.5 \times 10^{-44}$$
$$27\ x^4 = 15000 \times 10^{-48}$$
$$x^4 = 556 \times 10^{-48}$$
$$x = [\text{antilog}\ (\frac{\log 556}{4})] \times 10^{-12}$$
$$x = [\text{antilog}\ (0.6863) \times 10^{-12}$$
$$x = 4.9 \times 10^{-12}\ \underline{M} = \text{molar sol'y}.$$
$$[Tl^{3+}] = x\ \underline{M} = 4.9 \times 10^{-12}\ \underline{M}$$
$$[OH^-] = 3x\ \underline{M} = 1.5 \times 10^{-11}\ \underline{M}\ \text{(from } Tl(OH)_3 \text{ alone)}$$

Note that the concentration of OH^- furnished by $Tl(OH)_3$ is much smaller than the concentration of OH^- furnished by the ionization of water, 1.0×10^{-7} $\underline{M}$. Thus, $Tl(OH)_3$ is so slightly soluble that its saturated solutions are very nearly neutral.

18-3 Common Ion Effect

The solubility of a slightly soluble compound in pure water (or pure solvent, in general), is always greater than in a solution containing a common ion, i.e., an ion produced by the slightly soluble compound as well as by another compound in the solution. Application of LeChatelier's Principle to solubility equilibria is thus illustrated. The presence of a common ion represses dissolution of the slightly soluble compound as Example 18-6 illustrates.

Example 18-6: Answer the questions of Example 18-3 using 0.10 $\underline{M}$ NaI as solvent instead of pure H_2O.

The 0.10 $\underline{M}$ solution NaI is completely ionized and a small amount of CuI dissolves to saturate the solution with CuI.

$$NaI \xrightarrow{100\%} Na^+ + I^- \leftarrow \text{from NaI only}$$
$$0.10\ \underline{M} \qquad 0.10\ \underline{M} \qquad 0.10\ \underline{M}$$

$$CuI\ (s) \rightleftharpoons Cu^+ + I^- \leftarrow \text{from CuI only}$$
Let molar sol'y = $x\ \underline{M}$ $\qquad x\ \underline{M} \qquad x\ \underline{M}$

The total concentration of I^- in the saturated solution is $(0.10 + x)\ \underline{M}$.

$$K_{sp} = [Cu^+][I^-] = (x)(0.10 + x) = 5.1 \times 10^{-12}.$$

Since the molar solubility of CuI in water is small, we may assume: $x \ll 0.10$ and therefore $(0.10 + x) \approx 0.10$, since the solubility of CuI is less in NaI solution than in water.

$$0.10x = 5.1 \times 10^{-12}$$
$$x = 5.1 \times 10^{-11}\ \underline{M} = [Cu^+] = \text{molar sol'y of CuI in } 0.10\ \underline{M}\ \text{NaI}$$
$$[I^-] = (0.10 + x)\ \underline{M} \approx 0.10\ \underline{M}$$

Note that the molar solubility of CuI in water (2.3×10^{-6} mol/L from Ex. 18-3) is approximately 4.5×10^4 times greater than in 0.10 $\underline{M}$ NaI. The mass of CuI that dissolves in 500 mL of 0.10 $\underline{M}$ NaI can be calculated.

$$?\ g\ CuI = 500\ mL \times \frac{5.1 \times 10^{-11}\ mol\ CuI}{1000\ mL} \times \frac{190.4\ g\ CuI}{1\ mol\ CuI} = \underline{\underline{4.9 \times 10^{-9}\ g\ CuI}}$$

The results of Examples 18-3 and 18-6 are summarized below. Recall K_{sp} for $CuI = 5.1 \times 10^{-12}$.

Solvent	Molar Sol'y of CuI	$[Cu^+]$	$[I^-]$	Mass of CuI in 500 mL
H_2O	$2.3 \times 10^{-6}\ \underline{M}$	$2.3 \times 10^{-6}\ \underline{M}$	$2.3 \times 10^{-6}\ \underline{M}$	2.2×10^{-4} g CuI
0.10 $\underline{M}$ NaI	$5.1 \times 10^{-11}\ \underline{M}$	$5.1 \times 10^{-11}\ \underline{M}$	0.10 $\underline{M}$	4.9×10^{-9} g CuI

The following examples illustrate how K_{sp} values can be used to determine whether precipitation will occur in a given solution.

Example 18-7: Will $Ni(OH)_2$ precipitate if 50.0 mL of 0.0010 $\underline{M}$ $NiCl_2$ and 50.0 mL of 0.0030 $\underline{M}$ KOH solutions are mixed? K_{sp} for nickel hydroxide, $Ni(OH)_2$, is 2.8×10^{-16}.

After mixing, but before any reaction occurs, the concentrations are:

$$\underline{M}_{NiCl_2} = 1/2\ (0.0010\ \underline{M}) = 5.0 \times 10^{-4}\ \underline{M} \text{ and}$$

$$\underline{M}_{KOH} = 1/2\ (0.0030\ \underline{M}) = 1.5 \times 10^{-3}\ \underline{M}$$

Both nickel chloride and potassium hydroxide are soluble and completely ionized in dilute solutions.

$NiCl_2$	$\longrightarrow$	Ni^{2+}	+	$2\ Cl^-$
$5.0 \times 10^{-4}\ \underline{M}$		$5.0 \times 10^{-4}\ \underline{M}$		$2(5.0 \times 10^{-4}\ \underline{M})$
KOH	$\longrightarrow$	K^+	+	OH^-
$1.5 \times 10^{-3}\ \underline{M}$		$1.5 \times 10^{-3}\ \underline{M}$		$1.5 \times 10^{-3}\ \underline{M}$

Precipitation of $Ni(OH)_2$ will occur if the solution is "more than saturated" with Ni^{2+} and OH^- ions, i.e., if the ion product, Q_{sp}, exceeds K_{sp}.

$$Q_{sp} = [Ni^{2+}]_{initial}\ [OH^-]^2_{initial}$$

$$Q_{sp} = (5.0 \times 10^{-4})(1.5 \times 10^{-3})^2 = 1.1 \times 10^{-9}$$

Since the ion product, 1.1×10^{-9}, exceeds K_{sp}, 2.8×10^{-16}, then $Ni(OH)_2$ precipitates until K_{sp} is just satisfied by the concentrations of ions remaining in solution.

Example 18-8: How many grams of solid $CoCl_2 \cdot 6\,H_2O$, cobalt(II) chloride hexahydrate, must be added to 100 mL of $1.0 \times 10^{-5}\ \underline{M}$ NaOH solution to just begin precipitation of $Co(OH)_2$? K_{sp} for $Co(OH)_2$ is 2.5×10^{-16}.

Sodium hydroxide is soluble and completely ionized.

$NaOH$	$\xrightarrow{H_2O}$	Na^+	+	OH^-
$1.0 \times 10^{-5}\ \underline{M}$		$1.0 \times 10^{-5}\ \underline{M}$		$1.0 \times 10^{-5}\ \underline{M}$

The solubility product expression for $Co(OH)_2$ is:

$$K_{sp} = [Co^{2+}][OH^-]^2 = 2.5 \times 10^{-16}$$

$$[Co^{2+}](1.0 \times 10^{-5})^2 = 2.5 \times 10^{-16}$$

$$[Co^{2+}] = \frac{2.5 \times 10^{-16}}{(1.0 \times 10^{-5})^2} = 2.5 \times 10^{-6}\ \underline{M}$$ ← minimum $[Co^{2+}]$ necessary to satisfy K_{sp}

Therefore, to initiate precipitation, $[Co^{2+}] > 2.5 \times 10^{-6}$ M. We now calculate the weight of $CoCl_2 \cdot 6\,H_2O$ required to prepare 100 mL of solution with $[Co^{2+}] = 2.5 \times 10^{-6}$ M.

$$\underline{?}\ g\ CoCl_2 \cdot 6\,H_2O = 100\ mL \times \frac{2.5 \times 10^{-6}\ mol\ Co^{2+}}{1000\ mL} \times \frac{238\ g\ CoCl_2 \cdot 6\,H_2O}{1\ mol\ Co^{2+}}$$

$$= 6.0 \times 10^{-5}\ g = \underline{\underline{0.000060\ g\ CoCl_2 \cdot 6\,H_2O}}$$

Ever-so-slightly more than 6.0×10^{-5} g of $CoCl_2 \cdot 6\,H_2O$ is required to exceed K_{sp} for $Co(OH)_2$.

Example 18-9: It is desired to recover quantitatively the nickel from 50.0 mL of 0.1000 M nickel chloride, $NiCl_2$, solution by adding solid sodium hydroxide, NaOH. How much pure NaOH must be added to precipitate 99.999% of the Ni^{2+} from the solution as $Ni(OH)_2$? Assume no volume change due to addition of solid NaOH. K_{sp} for $Ni(OH)_2 = 2.8 \times 10^{-16}$.

If 99.999% of the Ni^{2+} is to be removed, then 0.0010% (1.0×10^{-5} as a fraction) must remain in solution. The concentration of Ni^{2+} that remains in solution will be:

$$[Ni^{2+}] = 1.0 \times 10^{-5} \times 0.1000\ M = 1.0 \times 10^{-6}\ M$$

The solution will be saturated with $Ni(OH)_2$, and we may substitute into the K_{sp} expression for $Ni(OH)_2$ and find the equilibrium concentration of OH^-.

$$[Ni^{2+}][OH^-]^2 = 2.8 \times 10^{-16}$$

$$[OH^-]^2 = \frac{2.8 \times 10^{-16}}{[Ni^{2+}]} = \frac{2.8 \times 10^{-16}}{1.0 \times 10^{-6}} = 2.8 \times 10^{-10}$$

$$\underline{\underline{[OH^-] = 1.7 \times 10^{-5}\ M}} \qquad \text{(equilibrium conc'n)}$$

Since 99.999% of the Ni^{2+} ions will be precipitated, we assume that the reaction is complete. The calculation of the equilibrium concentration of OH^- ions shows that very little excess NaOH (1.7×10^{-5} mol/L) is required. The equation for the precipitation of $Ni(OH)_2$,

$$Ni^{2+} + 2\,OH^- \longrightarrow Ni(OH)_2\ (s)$$

shows that two moles of OH^- ions (and therefore two moles of NaOH) are required for each mole of Ni^{2+} ions.

$$? \text{ mol NaOH} = 50.0 \text{ mL} \times \frac{0.100 \text{ mol } Ni^{2+}}{1000 \text{ mL}} \times \frac{2 \text{ mol NaOH}}{1 \text{ mol } Ni^{2+}} \times \frac{40.0 \text{ g NaOH}}{1 \text{ mol NaOH}}$$

$$= \underline{\underline{0.400 \text{ g NaOH}}} \quad \text{required for reaction}$$

The excess amount of NaOH 1.7×10^{-5} mol/L, (3.4×10^{-5} g NaOH/50 mL) is so small that we may say that ever-so-slightly more than 0.400 g NaOH is required.

18-4 Fractional Precipitation

The process by which one ion is separated from a solution of chemically similar ions by precipitation is called fractional precipitation.

Example 18-10: A solution of 1.0×10^{-3} M magnesium nitrate, $Mg(NO_3)_2$, 1.0×10^{-3} M cadmium nitrate, $Cd(NO_3)_2$, and 1.0×10^{-3} M, $Co(NO_3)_2$, cobalt(II) nitrate, is treated with solid NaOH. Determine the order in which the metal hydroxides precipitate and the concentration of OH^- necessary just to begin precipitation of each. K_{sp} for $Mg(OH)_2 = 1.5 \times 10^{-11}$, K_{sp} for $Cd(OH)_2 = 1.2 \times 10^{-14}$ and K_{sp} for $Co(OH)_2 = 2.5 \times 10^{-16}$.

The solution contains three soluble, ionic salts. The initial concentrations of metal ions are: $[Mg^{2+}] = [Cd^{2+}] = [Co^{2+}] = 1.0 \times 10^{-3}$ M. Each metal hydroxide begins to precipitate when its ion product exceeds its K_{sp}. We will calculate the concentration of OH^- necessary for $Mg(OH)_2$ to precipitate.

$$[Mg^{2+}][OH^-]^2 = 1.5 \times 10^{-11}$$

$$[OH^-]^2 = \frac{1.5 \times 10^{-11}}{[Mg^{2+}]} = \frac{1.5 \times 10^{-11}}{1.0 \times 10^{-3}} = 1.5 \times 10^{-8}$$

$$[OH^-] = 1.2 \times 10^{-4} \text{ M} \quad \text{required to satisfy } K_{sp} \text{ for } Mg(OH)_2$$

In order to initiate precipitation of $Mg(OH)_2$, $\underline{\underline{[OH^-] > 1.2 \times 10^{-4} \text{ M}}}$.

Similarly, for $Cd(OH)_2$ precipitation,

$$[Cd^{2+}][OH^-]^2 = 1.2 \times 10^{-14}$$

$$[OH^-]^2 = \frac{1.2 \times 10^{-14}}{[Cd^{2+}]} = \frac{1.2 \times 10^{-14}}{1.0 \times 10^{-3}} = 1.2 \times 10^{-11}$$

$$\underline{\underline{[OH^-] = 3.5 \times 10^{-6} \text{ M}}} \text{ required to satisfy } K_{sp} \text{ for } Cd(OH)_2$$

And for $Co(OH)_2$ precipitation,

$$[Co^{2+}][OH^-]^2 = 2.5 \times 10^{-16}$$

$$[OH^-]^2 = \frac{2.5 \times 10^{-16}}{1.0 \times 10^{-3}} = 2.5 \times 10^{-13}$$

$$[OH^-] = 5.0 \times 10^{-7}\ \underline{M} \quad \text{required to satisfy } K_{sp} \text{ for } Co(OH)_2$$

Thus, the order of precipitation is $Co(OH)_2$, then $Cd(OH)_2$, and then $Mg(OH)_2$.

Example 18-11: Refer to Example 18-10 and determine the percentage of Co^{2+} still in solution at the point at which just enough NaOH has been added to remove as much Co^{2+} as possible without precipitating any $Cd(OH)_2$. Also, determine the percentages of Cd^{2+} and Co^{2+} still in solution at the point at which the maximum amount of Cd^{2+} has been removed without precipitation of any $Mg(OH)_2$.

Precipitation of $Co(OH)_2$ will commence when $[OH^-]$ just exceeds 5.0×10^{-7} $\underline{M}$. Only $Co(OH)_2$ will precipitate until $[OH^-]$ just exceeds 3.5×10^{-6} $\underline{M}$, the concentration of OH^- required to satisfy K_{sp} for $Cd(OH)_2$. The concentration of Co^{2+} remaining in solution at this point is calculated below.

$$[Co^{2+}][OH^-]^2 = 2.5 \times 10^{-16}$$

$$[Co^{2+}] = \frac{2.5 \times 10^{-16}}{[OH^-]^2} = \frac{2.5 \times 10^{-16}}{(3.5 \times 10^{-6})^2} = 2.0 \times 10^{-5}\ \underline{M}$$

$$[Co^{2+}] = 2.0 \times 10^{-5}\ \underline{M} \leftarrow \text{remaining in sol'n just before } Cd(OH)_2 \text{ ppt's.}$$

$$\%Co^{2+} \text{ still in sol'n} = \frac{[Co^{2+}]\text{ sol'n}}{[Co^{2+}]\text{ orig.}} \times 100\% = \frac{2.0 \times 10^{-5}\ M}{1.0 \times 10^{-3}\ M} \times 100\%$$

$$= 2.0\% \text{ of } Co^{2+} \leftarrow \text{remains in sol'n as sol'n becomes sat'd with respect to } Cd(OH)_2$$

Similarly, in order for precipitation of $Mg(OH)_2$ to begin, $[OH^-]$ must just exceed 1.2×10^{-4} $\underline{M}$. We calculate $[Cd^{2+}]$ and $[Co^{2+}]$ still in solution at this point.

$$[Cd^{2+}][OH^-]^2 = 1.2 \times 10^{-14}$$

$$[Cd^{2+}] = \frac{1.2 \times 10^{-14}}{(1.2 \times 10^{-4})^2} = 8.3 \times 10^{-7}\ \underline{M} \leftarrow \text{remaining in sol'n just before } Mg(OH)_2 \text{ ppt's}$$

$$\% \; Cd^{2+} \text{ still in sol'n} = \frac{[Cd^{2+}] \text{ sol'n}}{[Cd^{2+}] \text{ orig}} \times 100\% = \frac{8.3 \times 10^{-7} \underline{M}}{1.0 \times 10^{-3} \underline{M}} \times 100\%$$

$$= \underline{\underline{0.083\% \text{ of } Cd^{2+}}}$$ ← remains in sol'n as sol'n becomes sat'd. with respect to $Mg(OH)_2$

$$[Co^{2+}][OH^-]^2 = 2.5 \times 10^{-16}$$

$$[Co^{2+}] = \frac{2.5 \times 10^{-16}}{(1.2 \times 10^{-4})^2} = 1.7 \times 10^{-8} \underline{M}$$ ← remaining in sol'n just before $Mg(OH)_2$ ppt's

$$\% Co^{2+} \text{ still in sol'n} = \frac{[Co^{2+}] \text{ sol'n}}{[Co^{2+}] \text{ orig}} \times 100\% = \frac{1.7 \times 10^{-8} \underline{M}}{1.0 \times 10^{-3} \underline{M}} \times 100\%$$

$$= \underline{\underline{0.0017\% \text{ of } Co^{2+}}}$$ ← remains in sol'n as sol'n becomes sat'd with respect to $Mg(OH)_2$

Thus we see that:

98% of Co^{2+} precipitates before $Cd(OH)_2$ begins to precipitate, but 99.917% of Cd^{2+} and 99.9983% of Co^{2+} precipitate before $Mg(OH)_2$ begins to precipitate.

Example 18-12: If a solution is to be made 0.15 M in $Mg(NO_3)_2$ and 0.10 M in aqueous NH_3, what concentration of NH_4NO_3 is necessary to prevent precipitation of $Mg(OH)_2$? K_{sp} for $Mg(OH)_2 = 1.5 \times 10^{-11}$ and K_b for $NH_3 = 1.8 \times 10^{-5}$.

Since we know K_{sp} for $Mg(OH)_2$ and the concentration of Mg^{2+} in solution [from the complete ionization of $Mg(NO_3)_2$] we can calculate the minimum $[OH^-]$ that would cause precipitation of $Mg(OH)_2$.

$$[Mg^{2+}][OH^-]^2 = 1.5 \times 10^{-11} \qquad [Mg^{2+}] = 0.15\ M$$

$$[OH^-]^2 = \frac{1.5 \times 10^{-11}}{[Mg^{2+}]} = \frac{1.5 \times 10^{-11}}{0.15} = 1.0 \times 10^{-10}$$

$$[OH^-] = 1.0 \times 10^{-5}\ M$$ ← max. $[OH^-]$ possible without exceeding K_{sp} for $Mg(OH)_2$

We now calculate the concentration of NH_4NO_3 required to buffer 0.10 M aqueous NH_3 solution to give $[OH^-] = 1.0 \times 10^{-5}$ M. Ammonium nitrate ionizes completely and represses the ionization of ammonia by the common ion effect. Let x M be the concentration of NH_4NO_3 required.

$$\underset{x\ M}{NH_4NO_3} \xrightarrow{100\%} \underset{x\ M}{NH_4^+} + \underset{x\ M}{NO_3^-}$$

The ionization of NH_3 must produce no more than 1.0×10^{-5} M OH^-.

$$\underset{(0.10 - 1.0 \times 10^{-5}\ M)}{NH_3} + H_2O \rightleftharpoons \underset{1.0 \times 10^{-5}\ M}{NH_4^+} + \underset{1.0 \times 10^{-5}\ M}{OH^-}$$

Note that NH_4^+ is produced by both NH_4NO_3 and aqueous NH_3. We can now substitute into the ionization constant expression for aqueous NH_3.

$$K_b = \frac{[NH_4^+][OH^-]}{[NH_3]} = 1.8 \times 10^{-5}$$

$$\frac{(x + 1.0 \times 10^{-5})(1.0 \times 10^{-5})}{(0.10 - 1.0 \times 10^{-5})} = 1.8 \times 10^{-5}$$

Assume that $x \gg (1.0 \times 10^{-5})$ and that $(0.10 - 1.0 \times 10^{-5}) \approx 0.10$.

$$1.8 \times 10^{-5} = \frac{x(1.0 \times 10^{-5})}{0.10}$$

$x = \underline{\underline{0.18\ M}} = [NH_4^+] =$ concentration of NH_4NO_3 necessary to prevent precipitation of $Mg(OH)_2$.

Clearly $0.18 \gg 1.0 \times 10^{-5}$ and our assumptions are valid.

18-6 Dissolution of Precipitates

Precipitates can be dissolved by three main methods: (1) converting the ions in a saturated solution of a precipitate to a weak (predominantly un-ionized) electrolyte, which causes the ion product, Q_{sp}, to become less than K_{sp}, (2) converting one of the ions to another species by oxidation or reduction, thus decreasing the ion product, and (3) formation of a soluble complex compound from one ion that effectively removes it from solution.

18-7 Complex Ion Formation

Complex ion formation occurs when certain electron donating species replace water molecules coordinated to aqueous metal ions. For example, Fe^{3+} (aq) ions, or more accurately $[Fe(OH_2)_6]^{3+}$ ions, react with cyanide ions to form the hexacyanoferrate(III) complex ion.

$$[Fe(OH_2)_6]^{3+} + 6\ CN^- \rightleftharpoons [Fe(CN)_6]^{3-} + 6\ H_2O$$

Many complex ions are very stable, i.e., they dissociate only very slightly. Their stabilities are indicated by their dissociation constants, K_d, which are equilibrium constants for dissociation reactions. For example, the overall reaction for the dissociation of $[Fe(CN)_6]^{3-}$ ions can be represented as:

$$[Fe(CN)_6]^{3-} + 6\ H_2O \rightleftharpoons [Fe(OH_2)_6]^{3+} + 6\ CN^-$$

or, simply,

$$[Fe(CN)_6]^{3-} \rightleftharpoons Fe^{3+} + 6\ CN^-$$

$$K_d = \frac{[Fe^{3+}][CN^-]^6}{[[Fe(CN)_6]^{3-}]} = 1.3 \times 10^{-44}$$

Dissociation constants for some complex ions are tabulated in Appendix I in the text.

18-8 Calculations Involving Complex Ions

Example 18-13: Calculate the concentrations of the various species in 0.15 M $[Co(NH_3)_6](OH)_3$ solution. K_d for $[Co(NH_3)_6]^{3+}$ is 2.2×10^{-34}.

The soluble complex compound is essentially completely dissociated into its constituent ions,

$$[Co(NH_3)_6](OH)_3 \xrightarrow{100\%} [Co(NH_3)_6]^{3+} + 3\ OH^-$$
$$0.15\ \underline{M} \qquad\qquad 0.15\ \underline{M} \qquad\qquad 0.45\ \underline{M}$$

but the complex ion dissociates only slightly. Let x be the concentration of free Co^{3+} (aq) produced by the dissociation of the hexamminecobalt(III) ion.

$$[Co(NH_3)_6]^{3+} \rightleftharpoons Co^{3+} + 6\ NH_3$$
$$(0.15-x)\ \underline{M} \qquad x\ \underline{M} \qquad 6\ x\ \underline{M}$$

$$K_d = \frac{[Co^{3+}][NH_3]^6}{[[Co(NH_3)_6]^{3+}]} = 2.2 \times 10^{-34}$$

$$\frac{(x)(6x)^6}{0.15-x} = 2.2 \times 10^{-34}$$

Assume $(0.15-x) \approx 0.15$ since K_d is very small.

$$\frac{46656\ x^7}{0.15} = 2.2 \times 10^{-34}$$

$$x^7 = 7.07 \times 10^{-40} = 707 \times 10^{-42}$$

Taking the seventh root of both sides of this equation gives:

$$[Co^{3+}] = x = 2.6 \times 10^{-6}\ \underline{M}$$

$$[NH_3] = 6x = 1.6 \times 10^{-5}\ \underline{M}$$

(Assumption is valid, i.e., $[Co(NH_3)_6]^{3+}$ dissociates only very slightly.)

$$[Co(NH_3)_6{}^{3+}] = 0.15\ \underline{M} - 2.6 \times 10^{-6}\ \underline{M} = 0.15\ \underline{M}$$

$$[OH^-] = 0.45\ \underline{M}$$

Note: The presence of OH^- (0.45 M) in solution from dissociation of the complex compound into its constituent ions prevents any significant ionization of NH_3 as a base.

Example 18-14: Calculate the amount of gaseous NH_3 that must be added to one liter of pure water to dissolve 0.0050 mole of $Cu(OH)_2$. Assume no volume change due to addition of $Cu(OH)_2$ and NH_3. K_{sp} for $Cu(OH)_2$ is 1.6×10^{-19} and K_d for $[Cu(NH_3)_4]^{2+}$ is 8.5×10^{-13}.

The overall reaction may be represented as:

$$Cu(OH)_2\ (s) + 4\ NH_3\ (aq) \rightleftharpoons [Cu(NH_3)_4]^{2+}\ (aq) + 2\ OH^-\ (aq)$$

rxn. ratio: $0.0050\ \underline{M}$ $\quad 4(0.0050\ \underline{M}) = 0.0200\ \underline{M}$ $\quad 0.0050\ \underline{M}$ $\quad 2(0.0050\ \underline{M}) = 0.0100\ \underline{M}$

The reaction ratio indicates that $4(0.0050\ \underline{M}) = 0.0200\ \underline{M}\ NH_3$ actually combines with $0.0050\ \underline{M}\ Cu(OH)_2$ to dissolve all the $Cu(OH)_2$. Additional NH_3 must also be in solution in equilibrium with the complex ion. The overall reaction really consists of two equilibria. The first is the equilibrium for the dissolution of $Cu(OH)_2$.

$$Cu(OH)_2\ (s) \rightleftharpoons Cu^{2+}\ (aq) + 2\ OH^-\ (aq)$$

$$K_{sp} = [Cu^{2+}][OH^-]^2 = 1.6 \times 10^{-19}$$

We know from the overall reaction that $[OH^-]$ must be $0.0100\ \underline{M}$ if all the $Cu(OH)_2$ dissolves. (The $[OH^-]$ produced from ionization of H_2O is negligible in comparison.) Thus,

$$[Cu^{2+}] = \frac{1.6 \times 10^{-19}}{[OH^-]^2} = \frac{1.6 \times 10^{-19}}{(0.0100)^2} = 1.6 \times 10^{-15}\ \underline{M}$$

We may now use $[Cu^{2+}] = 1.6 \times 10^{-15}\ \underline{M}$ in the dissociation equilibrium for $[Cu(NH_3)_4]^{2+}$ to calculate the $[NH_3]$ in solution, i.e., the equilibrium concentration of NH_3.

$$[Cu(NH_3)_4]^{2+} \rightleftharpoons Cu^{2+} + 4\ NH_3$$

$(0.0050 - 1.6 \times 10^{-15})\ \underline{M}$ $\qquad 1.6 \times 10^{-15}\ \underline{M}$

Since $(0.0050 - 1.6 \times 10^{-15})\ \underline{M} \approx 0.0050\ \underline{M}$, we have:

$$\frac{[Cu^{2+}][NH_3]^4}{[[Cu(NH_3)_5]^{2+}]} = K_d = 8.5 \times 10^{-13} = \frac{(1.6 \times 10^{-15})[NH_3]^4}{0.0050}$$

Rearranging gives:

$$[NH_3]^4 = \frac{(8.5 \times 10^{-13})(0.0050)}{1.6 \times 10^{-15}} = 2.6\bar{6}$$

Taking the fourth root of both sides of the equation gives:

$[NH_3]_{equil} = 1.3\ \underline{M}$

And, $[NH_3]_{Total} = [NH_3]_{complexed} + [NH_3]_{equil}$

$$= 0.0200\ \underline{M} + 1.3\ \underline{M} = 1.32\ \underline{M} \approx 1.3\ \underline{M}$$

Or, the total amount of ammonia necessary to dissolve 0.0050 mole of $Cu(OH)_2$ in one liter of water is slightly more than 1.3 mole.

Example 18-15: It is desired to dissolve 0.010 mole of CuI in one liter of water which is buffered at pH = 6.50. How many moles of KCN must be added to the solution? Assume no volume change. K_{sp} for CuI = 5.1×10^{-12}, K_d for $[Cu(CN)_2^-]$ = 1.0×10^{-16}, K_a for HCN = 4.0×10^{-10}.

Since CN^- is the anion of a very weak acid, the H^+ and the Cu^+ ions will compete for the CN^- ion. In order to dissolve 0.010 mole CuI in a liter of solution, 0.020 mole of CN^- must react with the CuI.

	CuI	+	2 CN^-	$\longrightarrow$	$[Cu(CN)_2]^-$	+	I^-
rxn ratio:	0.010 $\underline{M}$		0.020 $\underline{M}$		0.010 $\underline{M}$		0.010 $\underline{M}$

The I^- will not combine with H^+ because HI is a strong acid. We can calculate the maximum $[Cu^+]$ that can exist in the solution, which is 0.010 $\underline{M}$ in I^-, using the solubility product constant expression for CuI.

$$CuI\ (s) \rightleftharpoons Cu^+ + I^-$$

$$K_{sp} = [Cu^+][I^-] = 5.1 \times 10^{-12}$$

$$[Cu^+] = \frac{5.1 \times 10^{-12}}{[I^-]} = \frac{5.1 \times 10^{-12}}{0.010} = 5.1 \times 10^{-10}\ \underline{M} \leftarrow \text{max. } [Cu^+]$$

This $[Cu^+]$ may be substituted into the dissociation equilibrium constant expression for $[Cu(CN)_2]^-$ to calculate $[CN^-]$. The concentration of $[Cu(CN)_2]^-$ that dissociates is small since K_d is so small, so the equilibrium concentration of $[Cu(CN)_2]^-$ is essentially 0.010 $\underline{M}$.

$$[Cu(CN)_2]^- \rightleftharpoons Cu^+ + 2\,CN^-$$

$$K_d = \frac{[Cu^+][CN^-]^2}{[[Cu(CN)_2]^-]} = 1.0 \times 10^{-16} = \frac{(5.1 \times 10^{-10})[CN^-]^2}{0.010}$$

$$[CN^-]^2 = \frac{(1.0 \times 10^{-16})(0.010)}{5.1 \times 10^{-10}} = 2.0 \times 10^{-9}\ \underline{M}$$

$$[CN^-] = 4.5 \times 10^{-5}\ \underline{M}$$

From the given pH = 6.50, we can substitute $[H^+]$ into the HCN equilibrium constant expression to determine how much cyanide ion is tied up as HCN. Since pH = 6.50, $[H^+] = 10^{-6.50} = 3.16 \times 10^{-7}\ \underline{M}$,

$$HCN \rightleftharpoons H^+ + CN^-$$

$$K_a = \frac{[H^+][CN^-]}{[HCN]} = 4.0 \times 10^{-10} = \frac{(3.16 \times 10^{-7})(4.5 \times 10^{-5})}{[HCN]}$$

$$[HCN] = \frac{(3.16 \times 10^{-7})(4.5 \times 10^{-5})}{4.0 \times 10^{-10}} = 3.6 \times 10^{-2}\ \underline{M}$$ ← which tells us that 0.036 mole of CN^- will be tied up in un-ionized HCN

The total CN^- concentration is:

$$[CN^-]_{Total} = [CN^-]_{complexed} + [CN^-]_{free} + [CN^-]_{in\ HCN}$$

$$[CN^-]_{Total} = 0.020\ \underline{M} + 4.5 \times 10^{-5}\ \underline{M} + 0.036\ \underline{M} = \underline{\underline{0.056\ M}}$$

Thus, slightly more than <u>0.056 mole of KCN</u> must be added to one liter of solution buffered to pH = 6.50 to dissolve 0.010 mole of copper(I) iodide, CuI, <u>if</u> the pH is to be maintained at 6.50.

EXERCISES

You may consult the Appendices in the text for constants, as necessary.

1. Write solubility product expressions for the following slightly soluble compounds.

   (a) $BaCO_3$ (b) $Pb(OH)_2$ (c) Sb_2S_3 (d) $Ca_3(PO_4)_2$
   (e) $CaC_2O_4 \cdot H_2O$ (f) $Be_3Al_2(SiO_3)_6$ (g) $CsAl(SO_4)_2 \cdot 12\,H_2O$

2. From the water solubilities given at 25°C, calculate solubility product constants for the following compounds.

   (a) $Cu(N_3)_2$, 0.00075 g/100 mL
   (b) $CaC_2O_4 \cdot 1/2\ H_2O$, 0.00253 g/100 mL
   (c) $PbCrO_4$, 5.8×10^{-6} g/100 mL
   (d) $In(IO_3)_3$, 0.067 g/100 mL
   (e) $FeCO_3$, 0.0067 g/100 mL

3. Calculate the molar solubilities of the following solid compounds in pure water from their K_{sp} values which follow their formulas.

   (a) $Sn(OH)_4$, 1×10^{-57} (b) Ag_2CrO_4, 9.0×10^{-12}
   (c) $Tl(OH)_3$, 1.5×10^{-44} (d) $Ca_3(PO_4)_2$, 1.0×10^{-25}
   (e) $Ba_3(AsO_4)_2$, 1.1×10^{-13} (f) $Ag_4[Fe(CN)_6]$, 1.6×10^{-41}

4. Calculate the masses of the compounds in problem 3 in grams per 100 mL of saturated solution.

5. Calculate the molar solubilities of each of the solid compounds below in the solutions indicated.

(a) $Ni(OH)_2$	$K_{sp} = 2.8 \times 10^{-16}$	in 0.10 $\underline{M}$ NaOH
(b) Ag_2SO_4	$K_{sp} = 1.7 \times 10^{-5}$	in 0.20 $\underline{M}$ K_2SO_4
(c) AgBr	$K_{sp} = 3.3 \times 10^{-13}$	in 0.15 $\underline{M}$ $CaBr_2$
(d) MgF_2	$K_{sp} = 6.4 \times 10^{-9}$	in 0.23 $\underline{M}$ NaF
(e) MgF_2	$K_{sp} = 6.4 \times 10^{-9}$	in 0.10 $\underline{M}$ $MgCl_2$
(f) $Pb_3(PO_4)_2$	$K_{sp} = 3.0 \times 10^{-44}$	in 0.10 $\underline{M}$ $Pb(NO_3)_2$

6. Calculate the number of grams of solute that will dissolve to give one liter of saturated solution of the substances listed in question 5.

7. Calculate the maximum concentration of thallium(I) ions that can exist in a solution containing 4.0×10^{-16} $\underline{M}$ of sulfide ions. Neglect hydrolysis. K_{sp} for $Tl_2S = 7.6 \times 10^{-23}$.

8. Calculate the concentration of strontium ions necessary to cause precipitation of strontium chromate, $SrCrO_4$, in 1.00×10^{-4} $\underline{M}$ Na_2CrO_4 solution. K_{sp} for $SrCrO_4$ is 3.6×10^{-5}.

9. How many moles and how many grams of solid KCl must be dissolved in 200 mL of 4.0×10^{-5} $\underline{M}$ Na_2PtCl_6 solution (contains Na^+ and $PtCl_6^{2-}$ ions) to begin precipitation of the relatively insoluble potassium compound, K_2PtCl_6? Assume no volume change. K_{sp} for $K_2PtCl_6 = 1.1 \times 10^{-5}$.

10. Will precipitation occur if the following pairs of aqueous solutions are mixed? Consult Appendix H for solubility products. Ignore hydrolysis.

(a) 500 mL of 3.1×10^{-6} $\underline{M}$ $NiCl_2$ and 500 mL of 6.4×10^{-4} $\underline{M}$ Na_2CO_3

(b) 200 mL of 1.0×10^{-4} $\underline{M}$ NaF and 100 mL of 2.0×10^{-3} $\underline{M}$ $MgCl_2$

(c) 250 mL of 0.0010 $\underline{M}$ $AgNO_3$ and 1.00 L of 3.4×10^{-5} $\underline{M}$ Na_3PO_4

(d) 650 mL of 0.00020 $\underline{M}$ $Ca(OH)_2$ and 400 mL of 4.1×10^{-5} $\underline{M}$ $CrCl_3$

11. How many moles of solid NaI must be added to remove 99.9999% of the Hg_2^{2+} as Hg_2I_2 from 650 mL of 0.616 $\underline{M}$ mercurous nitrate, $Hg_2(NO_3)_2$ solution? K_{sp} is 4.5×10^{-29} for Hg_2I_2.

12. Refer to question 11. How many moles of solid Na_2SO_4 must be added to remove 99.9999% of the Hg_2^{2+} as Hg_2SO_4, whose K_{sp} is 6.8×10^{-7}?

13. A solution is 1.0×10^{-4} $\underline{M}$ in $AgNO_3$. If 2.0×10^{-5} mole of solid $BaCl_2$ is added to 500 mL of the solution how many moles and how many grams of precipitate form? What percentage of the Ag^+ precipitates?

14. A solution is 0.10 $\underline{M}$ in each of the salts, NaCl, NaBr, and NaI. It is desired to separate the halide ions by fractional precipitation by adding solid $Pb(NO_3)_2$. K_{sp} for $PbCl_2 = 1.7 \times 10^{-5}$, for $PbBr_2 = 6.3 \times 10^{-6}$, and for $PbI_2 = 8.7 \times 10^{-9}$.

(a) What will be the order of precipitation?

(b) What is the $[I^-]$ and the percentage of iodide ion precipitated just before any $PbBr_2$ precipitates?

(c) What is the $[I^-]$ and the percentage of iodide ion precipitated just before $PbCl_2$ begins to precipitate?

(d) What is the $[Br^-]$ and the percentage of bromide ion precipitated just before $PbCl_2$ begins to precipitate?

(e) What do you conclude about the feasibility of precipitating these halide ions as lead(II) salts?

15. It is desired to dissolve 8.62 grams of suspended $MgCO_3$ in 500 mL of solution already saturated with $MgCO_3$. If this is to be accomplished by bubbling gaseous HCl through the suspension, how many moles of HCl must react?

16. Calculate the pH of saturated $Mg(OH)_2$ solution. K_{sp} for $Mg(OH)_2$ = 1.5×10^{-11}.

17. (a) What is the concentration of Fe^{2+} in 6.06×10^{-3} $\underline{M}$ $K_4[Fe(CN)_6]$? K_d for $[Fe(CN)_6]^{4-} = 1.3 \times 10^{-37}$. Ignore any hydrolysis.

(b) What is the concentration of CN^- in the solution of (a)?

18. A 5.00 gram sample of $SrSO_4$ is washed with 100 mL of water. Assuming the water becomes saturated, what weight of $SrSO_4$ dissolves? K_{sp} for $SrSO_4 = 2.8 \times 10^{-7}$.

19. How many moles of NaCl would be necessary to dissolve 1.0×10^{-4} mole of HgS to give 1.0 liter of aqueous solution by formation of the complex ion $HgCl_4^{2-}$? Is this possible? K_{sp} for HgS = 3.0×10^{-53} and K_d for $HgCl_4^{2-} = 8.3 \times 10^{-16}$.

20. Will $Cd(OH)_2$ precipitate if 150 mL of 1.0×10^{-3} $\underline{M}$ $Cd(NO_3)_2$ is mixed with 200 mL of 4.0×10^{-3} $\underline{M}$ NaOH? K_{sp} for $Cd(OH)_2$ is 1.2×10^{-14}.

21. Will $Cd(OH)_2$ precipitate if 150 mL of 1.0×10^{-3} $\underline{M}$ $Cd(NO_3)_2$ is mixed with 200 mL of 4.0×10^{-3} $\underline{M}$ aqueous ammonia? K_{sp} for $Cd(OH)_2$ is 1.2×10^{-14} and K_b for NH_3 is 1.8×10^{-5}.

22. Refer to problem 21. How many moles of NH_4Cl would be necessary to prevent precipitation of $Cd(OH)_2$?

ANSWERS FOR EXERCISES

1. (a) $K_{sp} = [Ba^{2+}][CO_3^{2-}]$ (b) $K_{sp} = [Pb^{2+}][OH^-]^2$
(c) $K_{sp} = [Sb^{3+}]^2[S^{2-}]^3$ (d) $K_{sp} = [Ca^{2+}]^3[PO_4^{3-}]^2$
(e) $K_{sp} = [Ca^{2+}][C_2O_4^{2-}]$ (f) $K_{sp} = [Be^{2+}]^3[Al^{3+}]^2[SiO_3^{2-}]^6$
(g) $K_{sp} = [Cs^+][Al^{3+}][SO_4^{2-}]^2$

2. (a) $K_{sp} = 5.3 \times 10^{-13}$ (b) $K_{sp} = 3.4 \times 10^{-8}$ (c) $K_{sp} = 3.2 \times 10^{-14}$
(d) $K_{sp} = 2.7 \times 10^{-11}$ (e) $K_{sp} = 3.4 \times 10^{-7}$

3. (a) 1×10^{-12} mol/L (b) 1.3×10^{-4} mol/L (c) 4.9×10^{-12} mol/L
(d) 3.9×10^{-6} mol/L (e) 1.0×10^{-3} mol/L (f) 2.3×10^{-9} mol/L

4. (a) 2×10^{-11} g (b) 0.0043 g (c) 1.2×10^{-10} g (d) 1.2×10^{-4} g
(e) 0.069 g (f) 1.5×10^{-7} g

5. (a) 2.8×10^{-14} mol/L
(b) 4.6×10^{-3} mol/L (c) 1.1×10^{-12} mol/L (d) 1.2×10^{-7} mol/L
(e) 1.3×10^{-4} mol/L (f) 2.7×10^{-21} mol/L

6. (a) 2.6×10^{-12} g
(b) 1.4 g (c) 2.1×10^{-10} g (d) 7.5×10^{-6} g (e) 8.1×10^{-3} g
(f) 2.2×10^{-18} g

7. $[Tl^+] = 4.4 \times 10^{-4}$ M

8. $[Sr^{2+}] = 0.36$ M

9. 0.10 mol KCl; 7.5 g KCl

10. (a) no (b) no (c) yes, Ag_3PO_4
(d) yes, $Cr(OH)_3$ ppts.

11. the stoichiometric amount (0.800 mol) plus an additional 5.5×10^{-12} mol (insignificant) that remains in solution.

12. the stoichiometric amount (0.400 mol) plus an additional 0.72 mol

13. 5.1×10^{-3} g AgCl precipitate; 72% of Ag^+ precipitates

14. (a) PbI_2 then $PbBr_2$ then $PbCl_2$ (b) $[I^-] = 3.7 \times 10^{-3}$ M; 96.3% of I^- is precipitated (c) $[I^-] = 2.3 \times 10^{-3}$ M; 97.7% of I^- is precipitated (d) $[Br^-] = 6.1 \times 10^{-2}$ M; 39% of Br^- is precipitated
(e) The separations are very poor, particularly the separation of Br^- and Cl^-.

15. 0.205 mol HCl must react with $MgCO_3$.

16. pH = 10.51

17. (a) 5.6×10^{-7} M (b) 3.4×10^{-6} M

18. 0.0097 g dissolves

19. 5.8×10^6 mol NaCl; impossible

20. yes

21. yes

22. 2.7×10^{-3} mol NH_4Cl

Chapter Nineteen

ELECTROCHEMISTRY

D, G, W: Chap. 19

19-1 Introduction

There are two kinds of electrochemical cells: (1) electrolytic and (2) voltaic or galvanic cells. In electrolytic cells nonspontaneous redox reactions are forced to occur by passage of electrical current from an external source. The process is called electrolysis. In voltaic cells electrical current (energy) is produced by spontaneous redox reactions. The cathode is the electrode at which reduction occurs, and the anode is the electrode at which oxidation occurs in both kinds of cells.

19-2 Electrolytic Cells

We can deduce the essential features of electrolytic cells from experimental observations. These include identification of oxidation and reduction half-reactions, the overall cell reaction, the anode and cathode, the positive and negative electrodes, and the direction of flow of electrons.

Example 19-1: Make a simplified but complete diagram of the cell used for the electrolysis of molten (melted) potassium bromide, KBr (melting point = 730°C), from the following experimental observations. Inert (nonreactive) electrodes are used.

(1) Molten potassium metal forms at one electrode and floats on top of the molten KBr.

(2) Reddish-brown gaseous bromine, Br_2 (B. P. = 59°C), is produced at the other electrode.

The half-reactions involve the reduction of K^+ to K at the cathode and the oxidation of Br^- to Br_2 at the anode.

$$2\,(K^+ + e^- \rightarrow K) \qquad \text{(reduction at cathode)}$$
$$2\,Br^- \rightarrow Br_2 + 2e^- \qquad \text{(oxidation at anode)}$$
$$2\,K^+ + 2\,Br^- \rightarrow 2\,K(\ell) + Br_2\,(g) \qquad \text{(cell reaction)}$$

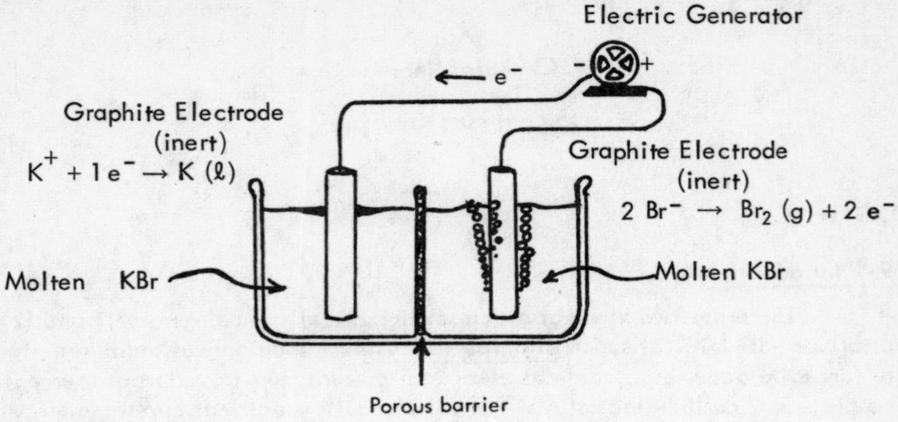

Since electrons are forced to flow by application of an external source of direct current, they move away from the positive electrode (toward which they would flow spontaneously) toward the negative electrode through the wire. Thus the anode is the positive electrode and the cathode is the negative electrode. Such is the case in all electrolytic cells.

Example 19-2: The following observations have been made during the electrolysis of aqueous sulfuric acid, H_2SO_4, using inert electrodes. Diagram and label the cell.

(1) Gaseous hydrogen, H_2, is produced at one electrode and the solution becomes less acidic around this electrode.

(2) Gaseous oxygen, O_2, is produced at the other electrode and the solution becomes more acidic around this electrode.

The cathode half-reaction must be the reduction of H^+ from the acid to H_2 and the anode half-reaction must be the oxidation of H_2O to O_2 and H^+.

$$2\,(2\,H^+ + 2\,e^- \longrightarrow H_2) \qquad \text{(reduction at cathode)}$$
$$2\,H_2O \longrightarrow O_2 + 4\,H^+ + 4\,e^- \qquad \text{(oxidation at anode)}$$
$$2\,H_2O\,(\ell) \longrightarrow 2\,H_2\,(g) + O_2\,(g) \qquad \text{(cell reaction)}$$

Note that H^+ is consumed at the cathode and produced at the anode. The net reaction is the same as the electrolysis of water, and as the electrolysis proceeds the concentration of sulfuric acid increases. The fact that water is oxidized at the anode in preference to sulfate ion, SO_4^{2-}, or hydrogen sulfate ion, HSO_4^-, while H^+ ion is reduced at the cathode illustrates one of the basic principles of electrochemistry: the most easily oxidized species present is oxidized and the most easily reduced species is reduced.

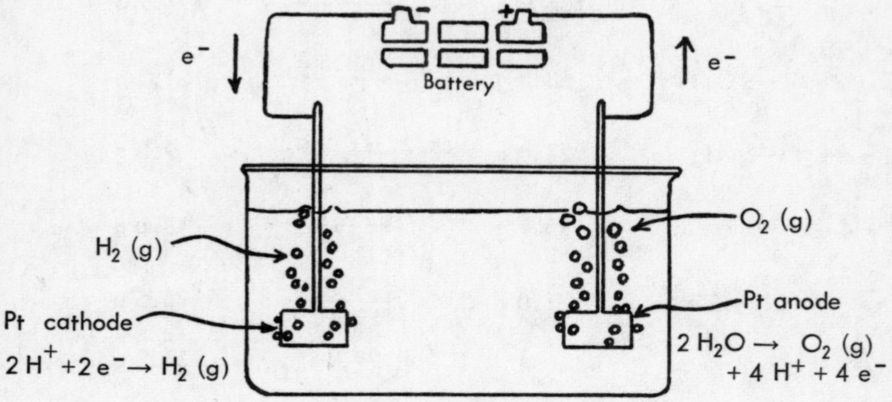

19-3 Faraday's Law

Faraday's Law may be stated as follows: one equivalent weight of oxidizing agent is reduced and one equivalent weight of reducing agent is oxidized by the passage of 96,487 coulombs of electrical charge (one faraday) through an electrochemical cell. This corresponds to ($\ominus$) the charge on 6.02×10^{23} electrons and is ordinarily rounded off to 96,500 coulombs in calculations involving only three significant figures. Table 19-1 illustrates this point.

$$\text{one faraday} \ominus 96,487 \text{ coulombs} \ominus 6.02 \times 10^{23}\ e^- \ominus \text{one eq}$$

Amounts of electrical current are usually expressed in units of amperes. One ampere corresponds to the passage of one coulomb of charge per second.

$$\text{one ampere} = \frac{\text{one coulomb}}{\text{second}}$$

Table 19-1. Amounts of Substances Produced at Electrodes By Passage of One Faraday of Electricity

Half-Reaction	Atomic or Formula Weight of Product	1 Eq Product	Mass of Products per 96,487 coulombs
$Ag^{+} + 1\ e^{-} \longrightarrow Ag$	107.9 g	$\frac{107.9\ g}{1}$	107.9 g
$Cu^{2+} + 2\ e^{-} \longrightarrow Cu$	63.5 g	$\frac{63.5\ g}{2}$	31.8 g
$Al^{3+} + 3\ e^{-} \longrightarrow Al$	27.0 g	$\frac{27.0\ g}{3}$	9.0 g
$Pt^{4+} + 4\ e^{-} \longrightarrow Pt$	195.1 g	$\frac{195.1\ g}{4}$	48.8 g
$2\ Cl^{-} \longrightarrow Cl_2 + 2\ e^{-}$	71.0 g	$\frac{71.0\ g}{2}$	35.5 g
		$\frac{22.4\ L_{STP}}{2}$	$11.2\ L_{STP}$

Example 19-3: Metallic aluminum can be obtained by electrolysis of aluminum fluoride, AlF_3, dissolved in molten cryolite, K_3AlF_6. How much aluminum is obtained at the cathode by electrolysis under a 4.00 ampere current for 2.50 hours?

The half-reaction and appropriate relationships are:

Al^{3+}	+	$3\ e^{-}$	$\longrightarrow$	Al
1 mol		$3(6.02 \times 10^{23})$		1 mol
		3(96,500 coul)		27.0 g = 3 eq

First we calculate the total charge passing through the cell:

$$\underline{?}\ \text{coul} = 2.50\ \text{hr} \times \frac{60\ \text{min}}{1\ \text{hr}} \times \frac{60\ \text{s}}{1\ \text{min}} \times \frac{4.00\ \text{coul}}{\text{s}} = 3.60 \times 10^{4}\ \text{coul}$$

The passage of 3.60×10^{4} coulombs of charge deposits 3.36 g Al at the cathode.

$$\underline{?}\ \text{g Al} = 3.60 \times 10^{4}\ \text{coul} \times \frac{27.0\ \text{g Al}}{3(96{,}500\ \text{coul})} = \underline{\underline{3.36\ \text{g Al}}}$$

Example 19-4: Refer to Example 19-3. How many grams and what volume of fluorine, F_2, (measured at STP) could be liberated at the anode during the same length of time?

$$2\,F^- \longrightarrow F_2 + 2\,e^-$$

2 mol	1 mol	$2 \times (6.02 \times 10^{23})$
	38.0 g	2 (96,500 coul)
	22.4 L_{STP}	

From Example 19-3 we know that 3.60×10^4 coulombs pass through the cell in 2.50 hours.

$$\underline{?}\ g\ F_2 = 3.60 \times 10^4\ coul \times \frac{38.0\ g\ F_2}{2(96,500\ coul)} = \underline{\underline{7.09\ g\ F_2}}$$

$$\underline{?}\ L_{STP}\ F_2 = 3.60 \times 10^4\ coul \times \frac{(22.4\ L_{STP})F_2}{2(96,500\ coul)} = \underline{\underline{4.18\ L_{STP}\ F_2}}$$

Example 19-5: Refer to Example 19-3. How long would the electrolysis have to continue to produce 100 L_{STP} of F_2 with a current of 4.00 amperes?

$$\underline{?}\ hr = 100\ L_{STP}\ F_2 \times \frac{2(96,500\ coul)}{22.4\ L_{STP}} \times \frac{1 s}{4.00\ coul} \times \frac{1\ hr}{3600\ s} = \underline{\underline{59.8\ hr}}$$

Example 19-6: A 2.00 ampere current is passed through an electrolytic cell containing an aqueous divalent (2+) metal salt for 4.75 hours. During that time 10.40 grams of the metal are deposited at the cathode. Identify the metal.

The number of coulombs of charge passing through the cell is:

$$\underline{?}\ coul = 4.75\ hr \times \frac{3600\ s}{hr} \times \frac{2.00\ coul}{s} = 3.42 \times 10^4\ coul$$

To identify the metal we must determine its atomic weight (g/mol). The metal ion has a 2+ charge so the appropriate relationships are:

$$M^{2+} + 2\,e^- \longrightarrow M$$

1 mol	$2\,(6.02 \times 10^{23})$	1 mol
	2(96,500 coul)	? g

Since 10.40 g of the metal are produced by 3.42×10^4 coul, we calculate the weight of metal produced by 2(96,500 coul), i.e., 1 mole of the metal.

$$\frac{?\text{ g M}}{\text{mol}} = \frac{2(96,500\text{ coul})}{\text{mol}} \times \frac{10.40\text{ g M}}{3.42 \times 10^4\text{ coul}} = \underline{\underline{58.7\text{ g/mol}}}$$

Therefore, the metal is nickel, atomic weight = 58.7 amu.

Example 19-7: Determine the charge on an ion of rhodium, Rh, if 1.919 grams of metallic rhodium are deposited during the electrolysis of an aqueous solution of a rhodium salt under a 1.500 ampere current for 1.00 hour.

We represent the electrolysis reaction as follows where n is the charge on the rhodium ion.

Rh^{n+}	+	$n e^-$	$\longrightarrow$	Rh
1 mol		$n(6.02 \times 10^{23})$		1 mol
		n(96,500 coul)		102.9 g

$$1.919\text{ g Rh} = 1.00\text{ hr} \times \frac{3600\text{ s}}{\text{hr}} \times \frac{1.500\text{ coul}}{\text{s}} \times \frac{102.9\text{ g Rh}}{n(96,500\text{ coul})}$$

We must solve the equation for n, the charge on the ion.

$$n = \frac{(1.500)(1.00)(3600)(102.9)}{(96,500)(1.919)} = \underline{\underline{3.00}} \quad \therefore \text{ The ion is } \underline{\underline{Rh^{3+}}}.$$

19-4 Voltaic Cells (Galvanic Cells)

In voltaic cells two half-reactions that are physically separated occur spontaneously to produce electricity when they are connected by a wire or other conducting surface. A salt bridge usually completes the circuit. No external source of electrical energy is necessary. Voltaic cells can also be diagrammed on the basis of experimental observations.

Example 19-8: One electrode consists of a strip of magnesium immersed in 1.00 M aqueous magnesium nitrate, $Mg(NO_3)_2$, solution. The other electrode consists of a strip of silver metal immersed in 1.00 M aqueous silver nitrate, $AgNO_3$. The cell is completed by a wire and a 5% agar- KNO_3 salt bridge. A voltmeter is inserted between the electrodes in the external circuit. Construct a diagram of the cell from the following experimental observations.

(1) The initial cell potential read from the voltmeter is 3.16 volts.

(2) The magnesium strip loses mass and the Mg^{2+} concentration increases in the solution surrounding the magnesium strip.

(3) The silver strip gains mass and the concentration of Ag^{+} decreases in the solution surrounding the silver strip.

The half-reactions and cell reaction must be as follows.

$Mg \longrightarrow Mg^{2+} + 2\,e^{-})$	(oxidation at anode)
$2\,(Ag^{+} + e^{-} \longrightarrow Ag)$	(reduction at cathode)
$Mg + 2\,Ag^{+} \longrightarrow Mg^{2+} + 2\,Ag$	(cell reaction)

The cell is diagrammed below.

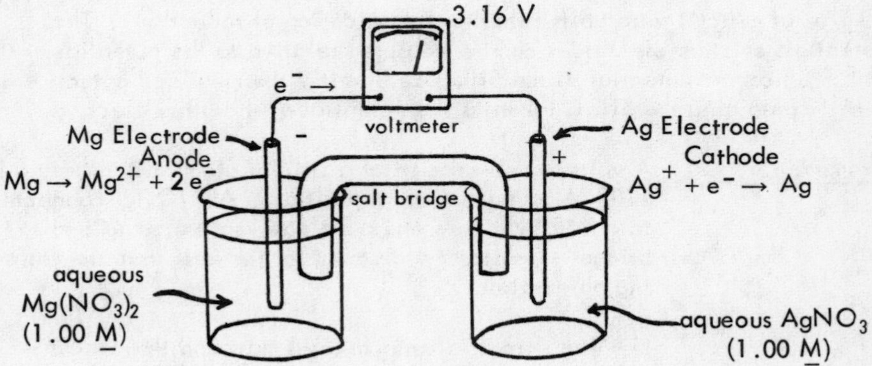

The cell reaction is spontaneous and the flow of electrons is from the anode to the cathode. Therefore the anode must be the negative electrode and the cathode must be the positive electrode. This is true in all voltaic cells. The opposite is true in electrolytic cells. Remember, however, that in both cases oxidation occurs at the anode and reduction occurs at the cathode.

19-5 Standard Electrodes

Standard electrodes consist of oxidized and reduced forms of a substance in contact with each other at unit activities. For our purposes, unit activities are one molar concentrations of all dissolved species, one atmosphere partial pressures for gases, and pure solids and liquids.

19-6 The Standard Hydrogen Electrode (SHE)

The standard hydrogen electrode is shown in Figure 19-11 in the text. A strip of (inert) platinum is immersed in 1.00 M H^+ solution. Hydrogen gas, H_2, is bubbled through a glass envelope and over the platinum at a pressure of one atmosphere.

Either of two half-reactions can occur at the standard hydrogen electrode, depending upon whether it functions as an anode or cathode. This, in turn, depends upon the electrode to which it is connected.

		E°
(as cathode)	$2\,H^+ + 2\,e^- \longrightarrow H_2$	0.0000 volt
(as anode)	$H_2 \longrightarrow 2\,H^+ + 2\,e^-$	0.0000 volt

The standard electrode potential, E°, of the SHE is arbitrarily assigned a value of exactly zero volts (whether for oxidation or reduction). The potentials of other electrodes can be measured relative to the potential of the SHE. Since the potential of the SHE is zero volts, the measured potential of a cell containing the SHE is taken as the potential of the other electrode.

Example 19-9: A voltaic cell consists of a strip of aluminum immersed in 1.00 M aqueous aluminum nitrate, $Al(NO_3)_3$, connected to a SHE by a wire and a 5% agar-saturated KCl salt bridge. Construct a diagram for the cell from the following observations.

(1) The strip of aluminum loses mass and the concentration of Al^{3+} increases in solution around the aluminum.

(2) The partial pressure of H_2 increases and the concentration of H^+ decreases at the SHE.

(3) The initial voltage (potential) of the cell is 1.66 volts.

Since the cell potential is 1.66 volts, this is also the potential of the standard aluminum electrode (for oxidation). The half-reactions and net cell reaction must be the following:

		E°
(oxidation at anode)	$2\,(Al \longrightarrow Al^{3+} + 3\,e^-)$	1.66 volts
(reduction at cathode)	$3\,(2\,H^+ + 2\,e^- \longrightarrow H_2)$	0.00 volt
(net cell reaction)	$2\,Al + 6\,H^+ \longrightarrow 2\,Al^{3+} + 3\,H_2$	E°_{cell} = 1.66 volts

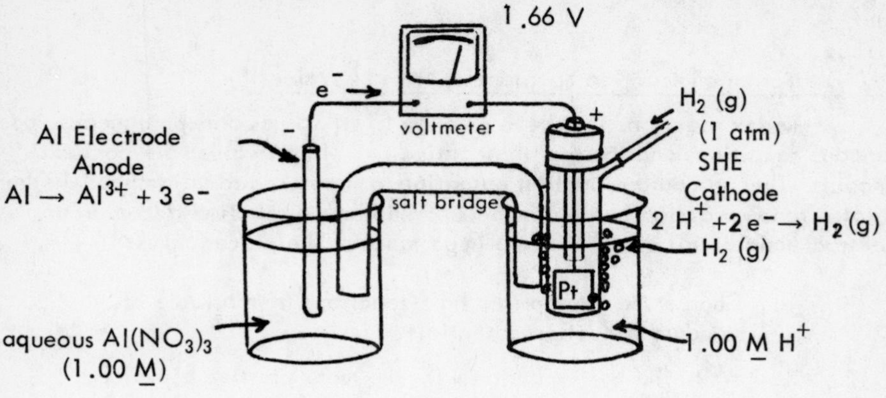

19-7 Standard Reduction Potentials

By measuring the potentials of other standard electrodes versus the SHE (or versus other electrodes whose potentials have already been determined versus the SHE) a table of standard electrode potentials can be generated. By international convention, all standard electrode potentials, E^o's, are tabulated as reduction potentials. That is, they indicate the relative tendencies of electrodes to behave as cathodes versus the SHE. Electrodes that act as cathodes versus the SHE are assigned positive E^o values and those that act as anodes versus the SHE are assigned negative E^o values. The more positive the E^o value for a half-reaction is, the greater its tendency to occur in that direction is. It is also important to realize that reversing a half-reaction changes the sign but not the magnitude of E^o. The superscript "o" signifies standard electrochemical conditions, or unit activities.

		E^o
(reduction)	$Al^{3+} + 3e^- \longrightarrow Al$	-1.66 volt
(oxidation)	$Al \longrightarrow Al^{3+} + 3e^-$	+1.66 volt

Table 19-2 in the text gives the standard reduction potentials for several half-reactions, while Appendix J contains many more.

The species listed on the left side of such tabulations can act as oxidizing agents and those on the right can act as reducing agents. The strongest oxidizing agents are at the lower left. They have the most positive E^o values for reduction. The weakest oxidizing agents are at the upper left. The strongest reducing agents are at the upper right; they have the most negative standard reduction potentials and, therefore, the most positive standard oxidation potentials. The weakest reducing agents are at the lower right.

19-8 Prediction of Reaction Spontaneity from E^o Values

Redox reactions that have positive E^o_{cell} values are spontaneous under standard conditions and those with negative E^o_{cell} values are nonspontaneous. Thus standard reduction potentials can be used to determine whether or not a redox reaction is spontaneous. Whether the reaction occurs in an electrochemical cell or not is of no importance. The procedure is:

1. Choose the appropriate half-reactions from a table of standard reduction potentials.

2. Write the half-reaction with the more positive E^o value, along with its E^o value, first.

3. Write the other half-reaction as an oxidation, i.e., reverse the tabulated reduction half-reaction and change the sign of E^o.

4. Balance the electron transfer.

5. Add the half-reactions and E^o values and eliminate common terms. When this procedure is followed E^o_{cell} will be positive, indicating spontaneity of the forward reaction.

Example 19-10: Can metallic silver reduce Mg^{2+} to metallic magnesium and be oxidized to Ag^+ ions, or will the reverse reaction be spontaneous when all species are present at unit activities?

Either the forward or reverse reaction will be spontaneous; the other will be nonspontaneous. We can determine the spontaneous reaction by following the procedure outlined above.

$$2(Ag^+ + 1\,e^- \longrightarrow Ag) \qquad +0.799 \text{ volt}$$
$$Mg \longrightarrow Mg^{2+} + 2\,e^- \qquad -(-2.37 \text{ volts})$$
$$2\,Ag^+ + Mg \longrightarrow 2\,Ag + Mg^{2+} \qquad E^o_{cell} = +3.17 \text{ volts}$$

Notice that when the reduction of Mg^{2+} to Mg is reversed it becomes an oxdiation and the sign of its potential is also changed. Since E^o_{cell} is positive, the forward reaction as written is spontaneous; magnesium metal reduces Ag^+ to silver metal and is oxidized to Mg^{2+}.

Example 19-11: Can permanganate ions, MnO_4^-, oxidize iron to iron(II) ions, Fe^{2+}, and be reduced to Mn^{2+} in acidic solution when all species are at unit activity?

Again we follow the procedure outlined above.

	E°
$2\,(MnO_4^- + 8\,H^+ + 5\,e^- \longrightarrow Mn^{2+} + 4\,H_2O)$	+1.51 volts
$5(Fe \longrightarrow Fe^{2+} + 2\,e^-)$	-(-0.44 volt)
$2\,MnO_4^- + 16\,H^+ + 5\,Fe \longrightarrow 2\,Mn^{2+} + 8\,H_2O + 5\,Fe^{2+}$	E°_{cell} = +1.95 volts

The answer is yes, because E°_{cell} is positive for the oxidation of Fe to Fe^{2+} by MnO_4^- in acidic solution.

Example 19-12: Can the sulfate ion, SO_4^{2-}, oxidize metallic zinc to zinc hydroxide, $Zn(OH)_2$, and be reduced to the sulfite ion, SO_3^{2-}, in basic solution with all species at unit activity?

	E°
$SO_4^{2-} + H_2O + 2\,e^- \longrightarrow SO_3^{2-} + 2\,OH^-$	-0.93 volt
$Zn + 2\,OH^- \longrightarrow Zn(OH)_2 + 2\,e^-$	-(-1.24 volts)
$SO_4^{2-} + H_2O + Zn \longrightarrow SO_3^{2-} + Zn(OH)_2$	E°_{cell} = +0.31 volt

The answer is yes, because E°_{cell} for the oxidation of Zn by SO_4^{2-} (reduction of SO_4^{2-} by Zn) is positive, and the reaction is spontaneous. Note that E°_{cell} is not very positive.

Example 19-13: Can Cr^{3+} ions reduce Pb^{2+} to metallic lead and form dichromate ions, $Cr_2O_7^{2-}$, in acidic solution when all species are present at unit activity?

	E°
$Cr_2O_7^{2-} + 14\,H^+ + 6\,e^- \longrightarrow 2\,Cr^{3+} + 7\,H_2O$	+1.33 volts
$3\,(Pb \longrightarrow Pb^{2+} + 2\,e^-)$	-(-0.126 volt)
$Cr_2O_7^{2-} + 14\,H^+ + 3\,Pb \longrightarrow 2\,Cr^{3+} + 7\,H_2O + 3\,Pb^{2+}$	E°_{cell} = +1.46 volts

The answer is no. The spontaneous reaction (E°_{cell} = +1.46 volts) is just the reverse of the one in question which, therefore, must be nonspontaneous (E°_{cell} = -1.46 volts).

19-9 The Nernst Equation

The tabulated standard reduction potentials, E°, apply only when all species are present at unit activity, i.e., 1.00 M concentrations of dissolved species, 1.00 atmosphere partial pressures of gases, and pure solids and liquids. Under other conditions at 25°C the electrode potential, E, can be calculated from the Nernst equation.

$$E = E^{\circ} - \underbrace{\frac{0.0592}{n} \log Q}_{\text{correction factor for nonstandard conditions}}$$

E = the electrode potential under nonstandard conditions

E° = the standard (tabulated) electrode potential

n = number of electrons gained or lost

Q = reaction quotient (same form as thermodynamic equilibrium constant, K, but not necessarily equilibrium concentrations)

Example 19-14: Calculate the potential for the Cu^{+}/Cu electrode as a cathode when $[Cu^{+}] = 1.00 \times 10^{-3}$ M.

The half-reaction is $Cu^{+} + e^{-} \longrightarrow Cu$

$$E = E^{\circ} - \frac{0.0592}{n} \log Q, \qquad n = 1,\ E^{\circ} = +0.521 \text{ volt}$$

$$Q = \frac{1}{[Cu^{+}]} = \frac{1}{1.00 \times 10^{-3}} = 1.00 \times 10^{3}$$

$$E = +0.521 - \frac{0.0592}{1} \log (1.00 \times 10^{3})$$

$$= +0.521 - \frac{0.0592}{1} (3) = +0.521 - 0.178 \text{ (correction factor)}$$

$$= +0.343 \text{ volt}$$

Example 19-15: Calculate the potential for the hydrogen electrode as a cathode when the $[H^{+}]$ is 5.00 M and the partial pressure of H_2 is 0.250 atmosphere.

The appropriate half-reaction and standard electrode potential are:

$$2\,H^{+} + 2\,e^{-} \longrightarrow H_2, \qquad n = 2,\ E^{\circ} = 0.000 \text{ volt}$$

$$Q = \frac{P_{H_2}}{[H^{+}]^2}$$

$$E = E^{\circ} - \frac{0.0592}{n} \log Q$$

$$= 0.000 - \frac{0.0592}{2} \log \frac{0.250}{(5.00)^2}$$

$$= 0.000 - \frac{0.0592}{2} \log(0.01) = 0.000 - \frac{0.0592}{2}(-2)$$

$$= \underline{\underline{-0.059 \text{ volt}}}$$

Example 19-16: Refer to Examples 19-14 and 19-15. If these two electrodes are connected in a voltaic cell under the stated conditions what will be the cell potential?

Since the Cu^+/Cu electrode has a positive potential to function as a cathode and the hydrogen electrode has a negative potential to function as a cathode, the former must be the cathode and the latter the anode. The spontaneous reaction and cell potential are determined in the same way as if E^o values were used under standard conditions.

	E ← not E^o
$2\,(Cu^+ + e^- \longrightarrow Cu)$	+0.343 volt
$H_2 \longrightarrow 2\,H^+ + 2\,e^-$	−(−0.059 volt)
$2\,Cu^+ + H_2 \longrightarrow 2\,Cu + 2\,H^+$	E_{cell} = +0.402 volt

The reduction of cuprous ions, Cu^+, to metallic copper and the oxidation of H_2 to H^+ is spontaneous at 25°C with $E_{cell} = 0.402$ volt when the $[Cu^+] = 1.00 \times 10^{-3}$ M, $[H^+] = 5.00$ M, and $P_{H_2} = 0.250$ atm.

19-10 Relationships Among E^o_{cell}, ΔG^o, and K

In our study of thermodynamics we learned that the standard Gibbs free energy change, ΔG^o, and the equilibrium constant, K, for a reaction are related by the equation below.

$$\Delta G^o = -RT\ell nK = -2.303\ RT \log K$$

$$R = 8.314 \text{ J/mol K or } 1.987 \text{ cal/mol K}$$

$$T = 298 \text{ K (for standard state conditions)}$$

The standard cell potential, E^o_{cell}, for a redox reaction is related to ΔG^o by the following equation.

$$\Delta G^o = -nFE^o_{cell}$$

n = number of electrons transferred

$$F = \text{the Faraday, } \frac{96,487 \text{ coul}}{\text{mol } e^-} \times \frac{1 \text{ J}}{\text{volt-coul}} = 96,487 \frac{\text{J}}{\text{volt-mol } e^-}$$

We can summarize these two relationships as follows:

$$\Delta G^{o} = -nFE^{o}_{cell} = -2.303\ RT \log K$$

The relationships can also be summarized as shown in the following tabulation.

Forward Reaction	ΔG^{o}	E^{o}_{cell}	K
spontaneous	−	+	>1
equilibrium	0	0	=1
nonspontaneous	+	−	<1

Under nonstandard state conditions ΔG (not ΔG^{o}) is related to E_{cell} (not E^{o}_{cell})

$$\Delta G = -nFE_{cell}$$

The value of an equilibrium constant changes only with temperature, and does not change with concentrations or partial pressures. Therefore it is related only to ΔG^{o} and E^{o}_{cell}.

Example 19-17: Calculate the equilibrium constant, K_c, at 25°C from standard reduction potentials for the reaction below.

$$Sn\ (s) + 2\ H^{+} + S\ (s) \rightleftharpoons Sn^{2+} + H_2S\ (g)$$

The appropriate half-reactions and standard reduction potentials are:

$$S + 2\ H^{+} + 2\ e^{-} \longrightarrow H_2S \qquad E^{o} = +0.14 \text{ volt}$$

$$Sn^{2+} + 2\ e^{-} \longrightarrow Sn \qquad E^{o} = -0.14 \text{ volt}$$

In order to generate the equation of interest the Sn^{2+}/Sn half-reaction must be reversed, i.e., changed to an oxidation, and the sign of its E^{o} value is changed.

	E^{o}
$S + 2\ H^{+} + 2\ e^{-} \longrightarrow H_2S$	+0.14 volt
$Sn \longrightarrow Sn^{2+} + 2\ e^{-}$	−(−0.14 volt)
$Sn + 2\ H^{+} + S \longrightarrow Sn^{2+} + H_2S$	E^{o}_{cell} = +0.28 volt

$$-nFE^{o}_{cell} = -2.303\ RT \log K$$

$$\log K = \frac{nFE^{o}_{cell}}{2.303\ RT}$$

$$\log K = \frac{(2)(96,487 \frac{J}{\text{volt-mol}})(+0.28 \text{ volt})}{(2.303)(8.314 \frac{J}{\text{mol K}})(298 \text{ K})}$$

$$\log K = 9.5$$

$$\underline{\underline{K = 3 \times 10^9}} \quad \text{(thermodynamic equilibrium constant)}$$

$$\underline{\text{At equilibrium}} \quad K = \frac{[Sn^{2+}]\, P_{H_2S}}{[H^+]^2} = 3 \times 10^9$$

Since the standard state of H_2S is gaseous, H_2S appears in the equilibrium constant expression in terms of its partial pressure. This value of K (the so-called "thermodynamic equilibrium constant") is really a hybrid of K_c and K_p. It could be converted to K_c by multiplying both sides of the equation by 1/RT which converts the partial pressure of H_2S to $[H_2S]$.

$$P_{H_2S} = (\frac{n}{V})RT = [H_2S]RT, \quad [H_2S] = \frac{P_{H_2S}}{RT}$$

$$K_c = \frac{[Sn^{2+}](\frac{P_{H_2S}}{RT})}{[H^+]^2} = 3 \times 10^9 (\frac{1}{RT})$$

$$K_c = \frac{[Sn^{2+}][H_2S]}{[H^+]^2} = 3 \times 10^9 \left(\frac{1}{(0.0821)(298)}\right) = \underline{\underline{1 \times 10^8}}$$

Example 19-18: Calculate ΔG^o for the reaction of Example 19-17:

$$\Delta G^o = -nFE^o_{cell}$$

$$= -(2 \text{ mol})(96,487 \frac{J}{\text{volt-mol}})(+0.28 \text{ volt})$$

$$= \underline{\underline{-5.4 \times 10^4 \text{ J} = -54 \text{ kJ}}}$$

The negative sign for ΔG^o indicates that the reaction is spontaneous under standard conditions at 25°C. This is consistent with a large value of K and a positive E^o_{cell} value.

Example 19-19: Refer to Example 19-16. Calculate ΔG^o and ΔG values and K_c at 25°C from the information given. The reaction is

$$2\, Cu^+ \,(1.00 \times 10^{-3}\, \underline{M}) + H_2\, (0.250 \text{ atm}) \longrightarrow 2\, Cu + 2\, H^+ \,(5.00\, \underline{M})$$

The half-reactions and E^o values are:

	E^o
$2(Cu^+ + e^- \longrightarrow Cu)$	+0.52 volt
$H_2 \longrightarrow 2H^+ + 2e^-$	0.00 volt
$2Cu^+ + H_2(g) \longrightarrow Cu(s) + 2H^+$	E^o_{cell} = +0.52 volt (spontaneous)

ΔG^o can be calculated directly from E^o_{cell}.

$$\Delta G^o = -nFE^o_{cell}$$

$$= -(2\ \text{mol})(96{,}487\ \frac{J}{\text{volt-mol}})(+0.52\ \text{volt}) = -1.0 \times 10^5\ J$$

$$\underline{\underline{\Delta G^o = -1.0 \times 10^2\ kJ}}\ \text{(reaction spontaneous)}$$

The thermodynamic equilibrium constant can also be calculated.

$$\Delta G^o = -2.303\ RT \log K$$

$$\log K = \frac{\Delta G^o}{-2.303\ RT}$$

$$= \frac{-1.0 \times 10^5\ J/mol}{(-2.303)(8.314\ \frac{J}{mol\ K})(298\ K)} = 18$$

$$K = 1 \times 10^{18}$$

$$K = \frac{[H^+]^2}{[Cu^+]^2 P_{H_2}} = 1 \times 10^{18}$$

The standard state of H_2 is gaseous, so it appears in terms of its partial pressure in the thermodynamic equilibrium constant, which can be converted to K_c by multiplying both sides by RT, i.e., 1/1/RT . This converts P_{H_2} to $[H_2]$. (See Example 19-17).

$$K_c = \frac{[H^+]^2}{[Cu^+]^2[H_2]} = 1 \times 10^{18}(RT) = 1 \times 10^{18}[(0.0821)(298)]$$

$$K_c = \frac{[H^+]^2}{[Cu^+]^2[H_2]} = 2 \times 10^{19}$$

Example 19-20: Given the following information, estimate K_{sp} for zinc hydroxide, $Zn(OH)_2$, at 25°C.

$$Zn(OH)_2\ (s) + 2\ e^- \longrightarrow Zn\ (s) + 2\ OH^- \qquad E^\circ = -1.245 \text{ volts}$$

$$Zn^{2+} + 2\ e^- \longrightarrow Zn\ (s) \qquad E^\circ = -0.763 \text{ volt}$$

The solubility equilibrium and solubility product constant expression for $Zn(OH)_2$ are:

$$Zn(OH)_2\ (s) \rightleftharpoons Zn^{2+} + 2\ OH^-, \qquad K_{sp} = [Zn^{2+}][OH^-]^2$$

The equation for the dissolution equilibrium can be obtained by subtracting the Zn^{2+}/Zn half-cell reaction (making it an oxidation) from the first half-cell reaction given above.

		E°
$Zn(OH)_2\ (s) + 2\ e^- \longrightarrow$	$Zn\ (s) + 2\ OH^-$	-1.245 volts
$Zn\ (s) \longrightarrow$	$Zn^{2+} + 2\ e^-$	$-(-0.763$ volt$)$
$Zn(OH)_2\ (s) \rightleftharpoons$	$Zn^{2+} + 2\ OH^-$	$E^\circ_{cell} = -0.482$ volt

Now that we know $E^\circ_{cell} = -0.482$ volt, which tells us that the process is non-spontaneous, we can calculate K_{sp}.

$$-nFE^\circ_{cell} = -2.303\ RT \log K_{sp}$$

$$\log K_{sp} = \frac{nFE^\circ_{cell}}{2.303\ RT}$$

$$= \frac{(2)(96,487 \frac{J}{\text{volt-mol}})(-0.482 \text{ volt})}{(2.303)(8.314 \frac{J}{\text{mol K}})(298\ K)} = -16.3$$

$$K_{sp} = 5 \times 10^{-17} = [Zn^{2+}][OH^-]^2$$

Note that this value compares favorably with the value listed for K_{sp} for $Zn(OH)_2$ in Appendix H, 4.5×10^{-17}. The very negative value for the equilibrium constant (K_{sp}) tells us that the reverse reaction is favored at standard electrochemical conditions.

EXERCISES

1. A sample of molten lithium hydride, LiH, (melting point, 680°C) was electrolyzed using inert electrodes. Diagram the cell completely as in the illustrative examples and write balanced equations for half-reactions and the overall reaction from the observations below. When the cell is in operation: (a) molten silvery white metallic lithium forms at one electrode, and (b) gaseous hydrogen, H_2, bubbles off at the other electrode.

2. Do the same as Problem 1 for the electrolysis of aqueous calcium bromide at 98°C using inert electrodes. The observations are: (a) reddish-brown gaseous bromine, Br_2, (a liquid at room temperature) bubbles off at one electrode, and (b) gaseous hydrogen is produced at the other electrode and the solution becomes basic around that electrode.

3. How many grams of the following elements could be plated out at the cathode by the passage of a 2.25 ampere current (through a liquid that contains the indicated species) for 3.00 hours?

   (a) H_2 from H_2O
   (b) Na from Na^+
   (c) Ni from Ni^{2+}
   (d) Al from Al^{3+}
   (e) Pd from Pd^{4+}

4. What volume (at STP) of the following gases could be produced at an electrode of an electrolytic cell by the passage of a 1.75 ampere current for 10.0 hours? Assume no other half-reaction occurs at the electrode of interest.

   (a) Cl_2 from Cl^- (b) O_2 from H_2O (c) NO from NO_3^-

5. How many hours would a 4.15 ampere current have to flow to produce the following?

   (a) 10.0 grams of Ca from molten $CaCl_2$
   (b) 5.65 grams of Ag from aqueous $AgNO_3$
   (c) 4.16 grams O_2 from H_2O
   (d) 33.6 L_{STP} of Cl_2 from molten NaCl
   (e) 4.86 grams of Ga from molten $GaCl_3$

6. What are the equivalent weights of each of the reactants of problem 5?

7. A 3.18 ampere current is passed for 0.431 hours through a solution containing a gold salt. If 3.36 g of gold plates out at the cathode, what is the charge on the gold ions in solution?

8. A solution of an acetate salt having a cation with a 1+ charge is electrolyzed with a 4.50 ampere current for 1.24 hours. As a result 13.2 grams of metal plate out at the cathode. What is the metal?

In problems 9-11 voltaic cells are described and observations after circuit completion are given. In each case diagram the cell completely from the observations, as in illustrative examples, and write balanced equations for half-reactions and the overall cell reaction.

9. One electrode consists of a strip of lead metal immersed in 1.0 M $Pb(NO_3)_2$ solution. The other consists of a strip of copper metal immersed in 1.0 M $Cu(NO_3)_2$ solution. The cell is completed by a wire and a salt bridge. Observations:

   (a) The lead electrode loses mass and the $[Pb^{2+}]$ increases in the solution around this electrode.
   (b) The copper electrode gains mass (due only to copper) and the surrounding solution becomes lighter blue as $[Cu^{2+}]$ decreases.

10. One electrode is the standard hydrogen electrode and the other consists of an inert platinum electrode in contact with an oxygen-free solution that is 1.0 M in both $FeCl_2$ and $FeCl_3$. A wire and a salt bridge complete the circuit. Observations:

    (a) The pH decreases at the SHE.
    (b) The concentration of ferric ions decreases at the platinum electrode.

11. One electrode consists a strip of metallic silver immersed in 1.0 M NaCl and also in contact with solid AgCl. The other electrode is a strip of platinum immersed into a solution which is 1.0 M in Cr^{3+}, 1.0 M in $Cr_2O_7^{2-}$, and 1.0 M in H^+. A wire and a salt bridge complete the circuit. Observations:

    (a) The $[Cr^{3+}]$ increases, the $[Cr_2O_7^{2-}]$ decreases, and the pH increases at the platinum electrode.
    (b) The silver electrode loses mass and more silver chloride is produced.

12. Calculate the standard cell potentials for the following reactions. Which ones are spontaneous under standard conditions?

    (a) $I_2\ (s) + 2\ Cl^-\ (aq) \longrightarrow Cl_2\ (g) + 2\ I^-\ (aq)$
    (b) $Zn\ (s) + H_2S\ (g) \longrightarrow ZnS\ (s) + H_2\ (g)$
    (c) $3\ Cu^{2+}\ (aq) + 2\ NO\ (g) + 4\ H_2O\ (\ell) \longrightarrow 3\ Cu\ (s) + 8\ H^+\ (aq) + 2\ NO_3^-\ (aq)$
    (d) $Zn\ (s) + HgO\ (s) + H_2O\ (\ell) \longrightarrow Hg\ (\ell) + Zn(OH)_2\ (s)$
    (e) $Cr_2O_7^{2-}\ (aq) + 14\ H^+\ (aq) + 6\ Fe^{2+}\ (aq) \longrightarrow 2\ Cr^{3+}\ (aq) + 6\ Fe^{3+}\ (aq) + 7\ H_2O\ (\ell)$

13. Answer the following questions. Assume 1.0 M concentrations of all ions and one atmosphere partial pressures of all gases at 25°C.

    (a) Will palladium(II) ions oxidize mercury to mercurous ions, Hg_2^{2+}, in aqueous solution or will Hg_2^{2+} ions oxidize palladium to Pd^{2+}?
    (b) Will hydrogen reduce nickel(II) ions to metallic nickel and be oxidized to hydrogen ions in aqueous solution?

13. (c) Will hexaamminecobalt(III) ion, $Co(NH_3)_6^{3+}$, oxidize zinc in basic solution to the tetrahydroxyzincate(II) ion, $[Zn(OH)_4]^{2-}$ and be reduced to the hexaamminecobalt(II) ion, $Co(NH_3)_6^{2+}$?

14. The standard cell potential, E° cell for the reaction

$$2\ Cr\ (s) + 6\ OH^-\ (aq) + 3\ Cl_2\ (g) \longrightarrow 2\ Cr(OH)_3\ (s) + 6\ Cl^-\ (aq)$$

is +2.66 volts. E° for the reduction of Cl_2 to Cl^- is +1.36 volt. What is the standard reduction potential for the half-reaction below?

$$Cr(OH)_3\ (s) + 3\ e^- \longrightarrow Cr\ (s) + 3\ OH^-\ (aq)$$

15. Calculate the reduction potentials for the following electrodes under the conditions stated at 25°C. (You may have to balance the appropriate half-reactions.)

(a) Fe^{3+} (0.070 M)/Fe^{2+} (0.00010 M)
(b) H^+ (1.00 M)/H_2 (5.00×10^3 atm)
(c) Sn^{4+} (2.40×10^{-1} M)/Sn^{2+} (1.00×10^{-6} M)
(d) $[RhCl_6]^{3-}$ (1.65×10^{-1} M)/Rh,Cl^- (1.00×10^{-4} M)
(e) ZnS/Zn, S^{2-} (1.50×10^{-8} M)

16. Calculate the initial cell potentials, E_{cell}, for voltaic cells constructed from the following combinations of electrodes listed in Problem 15. Write the balanced equation for the cell reaction.

(a) 15(a) and 15(b)
(b) 15(b) and 15(c)
(c) 15(d) and 15(e)
(d) 15(c) and 15(e)

17. Calculate the thermodynamic equilibrium constant at 25°C for each of the reactions below from standard electrode potentials.

(a) $2\ MnO_4^- + 6\ H^+ + 5\ H_2S \rightleftharpoons 2\ Mn^{2+} + 8\ H_2O + 5\ S$
(b) $2\ MnO_2 + 3\ NO_3^- + 2\ OH^- \rightleftharpoons 2\ MnO_4^- + 3\ NO_2^- + H_2O$
(c) $3\ NiO_2 + 2\ Cr(OH)_3 + 4\ OH^- \rightleftharpoons 3\ Ni(OH)_2 + 2\ CrO_4^{2-} + 2\ H_2O$

18. Calculate ΔG° at 25°C for each of the reactions of Problem 17 from E°_{cell} values.

19. Calculate ΔG (nonstandard conditions) at 25°C for the reactions below under the conditions stated. Also calculate the thermodynamic equilibrium constant for each at 25°C.

(a) $3\ Fe^{2+}\ (2.5 \times 10^{-4}\ M) + 3\ S + 2\ NO_3^-\ (6.00\ M) + 8\ H^+\ (1.00 \times 10^{-3} M)$
$\rightleftharpoons 3\ FeS + 2\ NO\ (1\ atm) + 4\ H_2O$

(b) $5\ BrO_3^-\ (1.00\ M) + 3\ H_2O + 3\ I_2 \rightleftharpoons 5\ Br^-\ (1.00 \times 10^{-5}\ M) + 6\ IO_3^-$
$(1.00 \times 10^{-3}\ M) + 6\ H^+\ (1.00\ M)$

20. Calculate the thermodynamic equilibrium constant, K, at 25°C for the reaction below from E° values. Write the mass action expression for K.

$$2\,MnO_4^{2-}\,(aq) + Cl_2\,(g) \rightleftharpoons 2\,MnO_4^-\,(aq) + 2\,Cl^-\,(aq)$$

21. Use the following information to calculate K_{sp} for FeS.

				E°
FeS (s) +	2 e⁻	$\longrightarrow$	Fe (s) + S^{2-}	−1.01 V
Fe^{2+} +	2 e⁻	$\longrightarrow$	Fe (s)	−0.44 V

22. Given the following information,

			E°
$[Zn(CN)_4]^{2-}$ (aq) + 2 e⁻	$\longrightarrow$	Zn (s) + 4 CN^- (aq)	−1.26 V
Zn^{2+} (aq) + 2 e⁻	$\longrightarrow$	Zn (s)	−0.763 V

calculate K_d for $[Zn(CN)_4]^{2-}$.

$[Zn(CN)_4]^2\,(aq) \rightleftharpoons Zn^{2+}\,(aq) + 4\,CN^-\,(aq)$ $\qquad K_d = ?$

ANSWERS FOR EXERCISES

1. Cathode (−): $2(Li^+ + e^- \longrightarrow Li)$
Anode (+) : $2\,H^- \longrightarrow H_2 + 2\,e^-$
Overall : $2\,Li^+ + 2\,H^- \longrightarrow 2\,Li + H_2$

Electrons flow through wire (external circuit) from anode to cathode.

2. Cathode (−): $2\,e^- + 2\,H_2O \longrightarrow H_2 + 2\,OH^-$
Anode (+) : $2\,Br^- \longrightarrow Br_2 + 2\,e^-$
Overall : $2\,H_2O + 2\,Br^- \longrightarrow Br_2 + H_2 + 2\,OH^-$

3. (a) 0.254 g H_2 (b) 5.79 g Na (c) 7.39 g Ni (d) 2.27 g Al (e) 6.67 g Pd 4. (a) 7.31 L Cl_2 (b) 3.66 L O_2 (c) 4.87 L NO

5. (a) 3.22 hr (b) 0.338 hr (c) 3.36 hr (d) 19.4 hr (e) 1.35 hr 6. (a) 55.6 g $CaCl_2$ (b) 170 g $AgNO_3$ (c) 9.0 g H_2O (d) 58.5 g NaCl (e) 58.7 g $GaCl_3$ 7. 3+ 8. Cu

9. Anode (−) : $Pb \longrightarrow Pb^{2+} + 2\,e^-$
Cathode (+) : $Cu^{2+} + 2\,e^- \longrightarrow Cu$
Overall : $Pb + Cu^{2+} \longrightarrow Pb^{2+} + Cu$

Electrons flow from anode to cathode through the external circuit.

10. Anode (−) : $H_2 \longrightarrow 2\,H^+ + 2\,e^-$
Cathode (+): $2(Fe^{3+} + e^- \longrightarrow Fe^{2+})$
Overall : $H_2 + 2\,Fe^{3+} \longrightarrow 2\,H^+ + 2\,Fe^{2+}$

11. Anode (−) : $6(Ag + Cl^- \longrightarrow AgCl + e^-)$
Cathode (+): $Cr_2O_7^{2-} + 14\,H^+ + 6\,e^- \longrightarrow 2\,Cr^{3+} + 7\,H_2O)$
Overall : $6\,Ag + 6\,Cl^- + Cr_2O_7^{2-} + 14\,H^+ \longrightarrow AgCl + 2\,Cr^{3+} + 7\,H_2O$

12. (a) E^o_{cell} = −0.825V; nonspontaneous
(b) E^o_{cell} = +1.21 V; spontaneous
(c) E^o_{cell} = −0.62 V; nonspontaneous
(d) E^o_{cell} = +1.343 V; spontaneous
(e) E^o_{cell} = +0.56 V; spontaneous

13. (a) Palladium(II) ions oxidize mercury to mercury(I) ions, Hg_2^{2+}, in aqueous solution; E^o_{cell} = +0.198 V (b) No, hydrogen will not reduce nickel(II) ions to metallic nickel, E^o_{cell} = −0.25 V (c) Yes, E^o_{cell} = +1.1V

14. E^o = −1.30V 15. (a) E = +0.939 V (b) E = −0.109 V (c) E = +0.31 V (d) E = +0.90 V (e) E = −1.21 V

16. (a) E_{cell} = +1.048 V (b) E_{cell} = +0.42 V (c) E_{cell} = +2.11 V (d) E_{cell} = +1.52 V

17. (a) $$K = \frac{[Mn^{2+}]^2}{[MnO_4^-]^2[H^+]^6\,P_{H_2S}^{\;5}} = 5 \times 10^{231}$$

(b) $$K = \frac{[MnO_4^-]^2[NO_2^-]^3}{[NO_3^-]^3[OH^-]^2} = 1 \times 10^{-59}$$

(c) $$K = \frac{[CrO_4^{2-}]^2}{[OH^-]^4} = 8 \times 10^{61}$$

18. $\Delta G^o = -1.32 \times 10^3$ kJ for 17(a)
$\Delta G^o = +337$ kJ for 17(b)
$\Delta G^o = -353$ kJ for 17(c)

19. (a) $\Delta G = -1.54 \times 10^3$ kJ; $K = 4 \times 10^{201}$
(b) $\Delta G = -9.6 \times 10^2$ kJ; $K = 10^{122}$

20. $$K = \frac{[Cl^-]^2[MnO_4^-]^2}{P_{Cl_2}[MnO_4^{2-}]^2} = 1.6 \times 10^{-54}$$

21. $K_{sp} = 10^{-19}$

22. $K_d = 10^{-17}$

FOUR-PLACE TABLE OF LOGARITHMS

	0	1	2	3	4	5	6	7	8	9
1.0	.0000	.0043	.0086	.0128	.0170	.0212	.0253	.0294	.0334	.0374
1.1	.0414	.0453	.0492	.0531	.0569	.0607	.0645	.0682	.0719	.0755
1.2	.0792	.0828	.0864	.0899	.0934	.0969	.1004	.1038	.1072	.1106
1.3	.1139	.1173	.1206	.1239	.1271	.1303	.1335	.1367	.1399	.1430
1.4	.1461	.1492	.1523	.1553	.1584	.1614	.1644	.1673	.1703	.1732
1.5	.1761	.1790	.1818	.1847	.1875	.1903	.1931	.1959	.1987	.2014
1.6	.2041	.2068	.2095	.2122	.2148	.2175	.2201	.2227	.2253	.2279
1.7	.2304	.2330	.2355	.2380	.2405	.2430	.2455	.2480	.2504	.2529
1.8	.2553	.2577	.2601	.2625	.2648	.2672	.2695	.2718	.2742	.2765
1.9	.2788	.2810	.2833	.2856	.2878	.2900	.2923	.2945	.2967	.2989
2.0	.3010	.3032	.3054	.3075	.3096	.3118	.3139	.3160	.3181	.3201
2.1	.3222	.3243	.3263	.3284	.3304	.3324	.3345	.3365	.3385	.3404
2.2	.3424	.3444	.3464	.3483	.3502	.3522	.3541	.3560	.3579	.3598
2.3	.3617	.3636	.3655	.3674	.3692	.3711	.3729	.3747	.3766	.3784
2.4	.3802	.3820	.3838	.3856	.3874	.3892	.3909	.3927	.3945	.3962
2.5	.3979	.3997	.4014	.4031	.4048	.4065	.4082	.4099	.4116	.4133
2.6	.4150	.4166	.4183	.4200	.4216	.4232	.4249	.4265	.4281	.4298
2.7	.4314	.4330	.4346	.4362	.4378	.4393	.4409	.4425	.4440	.4456
2.8	.4472	.4487	.4502	.4518	.4533	.4548	.4564	.4579	.4594	.4609
2.9	.4624	.4639	.4654	.4669	.4683	.4698	.4713	.4728	.4742	.4757
3.0	.4771	.4786	.4800	.4814	.4829	.4843	.4857	.4871	.4886	.4900
3.1	.4914	.4928	.4942	.4955	.4969	.4983	.4997	.5011	.5024	.5038
3.2	.5051	.5065	.5079	.5092	.5105	.5119	.5132	.5145	.5159	.5172
3.3	.5185	.5198	.5211	.5224	.5237	.5250	.5263	.5276	.5289	.5302
3.4	.5315	.5328	.5340	.5353	.5366	.5378	.5391	.5403	.5416	.5428
3.5	.5441	.5453	.5465	.5478	.5490	.5502	.5514	.5527	.5539	.5551
3.6	.5563	.5575	.5587	.5599	.5611	.5623	.5635	.5647	.5658	.5670
3.7	.5682	.5694	.5705	.5717	.5729	.5740	.5752	.5763	.5775	.5786
3.8	.5798	.5809	.5821	.5832	.5843	.5855	.5866	.5877	.5888	.5899
3.9	.5911	.5922	.5933	.5944	.5955	.5966	.5977	.5988	.5999	.6010
4.0	.6021	.6031	.6042	.6053	.6064	.6075	.6085	.6096	.6107	.6117
4.1	.6128	.6138	.6149	.6160	.6170	.6180	.6191	.6201	.6212	.6222
4.2	.6232	.6243	.6253	.6263	.6274	.6284	.6294	.6304	.6314	.6325
4.3	.6335	.6345	.6355	.6365	.6375	.6385	.6395	.6405	.6415	.6425
4.4	.6435	.6444	.6454	.6464	.6474	.6484	.6493	.6503	.6513	.6522
4.5	.6532	.6542	.6551	.6561	.6571	.6580	.6590	.6599	.6609	.6618
4.6	.6628	.6637	.6646	.6656	.6665	.6675	.6684	.6693	.6702	.6712
4.7	.6721	.6730	.6739	.6749	.6758	.6767	.6776	.6785	.6794	.6803
4.8	.6812	.6821	.6830	.6839	.6848	.6857	.6866	.6875	.6884	.6893
4.9	.6902	.6911	.6920	.6928	.6937	.6946	.6955	.6964	.6972	.6981
5.0	.6990	.6998	.7007	.7016	.7024	.7033	.7042	.7050	.7059	.7067
5.1	.7076	.7084	.7093	.7101	.7110	.7118	.7126	.7135	.7143	.7152
5.2	.7160	.7168	.7177	.7185	.7193	.7202	.7210	.7218	.7226	.7235
5.3	.7243	.7251	.7259	.7267	.7275	.7284	.7292	.7300	.7308	.7316
5.4	.7324	.7332	.7340	.7348	.7356	.7364	.7372	.7380	.7388	.7396
5.5	.7404	.7412	.7419	.7427	.7435	.7443	.7451	.7459	.7466	.7474
5.6	.7482	.7490	.7497	.7505	.7513	.7520	.7528	.7536	.7543	.7551
5.7	.7559	.7566	.7574	.7582	.7589	.7597	.7604	.7612	.7619	.7627
5.8	.7634	.7642	.7649	.7657	.7664	.7672	.7679	.7686	.7694	.7701
5.9	.7709	.7716	.7723	.7731	.7738	.7745	.7752	.7760	.7767	.7774